Adaptive Strategies for Water Heritage

Carola Hein
Editor

Adaptive Strategies for Water Heritage

Past, Present and Future

Springer Open

Editor
Carola Hein
Faculty of Architecture and the Built Environment
Technical University Delft
Delft, Zuid-Holland, The Netherlands

ISBN 978-3-030-00267-1 ISBN 978-3-030-00268-8 (eBook)
https://doi.org/10.1007/978-3-030-00268-8

Library of Congress Control Number: 2019934522

This Springer imprint is published by the registered company Springer Nature Switzerland AG
The registered company address is: Gewerbestrasse 11, 6330 Cham, Switzerland

Foreword by Giulio Boccaletti

That the world may be facing a water crisis is an idea now firmly entrenched in global discourse. The World Economic Forum has ranked the risks associated with water as among the highest to global prosperity. The UN has declared 2018–2028 as the Decade for Action on Water for Sustainable Development. Indeed, the symptoms of the current moment point to a society that has not come to terms with its own water insecurity: chronic scarcity and over-extraction are the norms in about one-third of the world's basins. Some twenty million people per year are displaced by natural catastrophes caused by water, an amount comparable to that of war. Billions do not enjoy safe, reliable access to water in their homes. As the human population tripled over the last forty years, the number of animals in freshwater systems—such as fish, amphibians, and birds—has dropped by more than three-quarters. It is predicted that the risks to both people and nature will worsen as climate change modifies the hydrology of the planet.

What makes any discussion about water complicated is that it carries multiple economic, legal, political, and cultural values. Water is a public good; at times, it is a private good; it is often a resource held in common. Access to water and sanitation is a human right. In some cases, water is subject to public trust, in others to private ownership. Its most complicated attribute is its delivery, which has very little to do with the substance itself. And protection from excessive quantities of water is likewise essential. A society's water security is a product of its landscape, infrastructure, and institutions. Because the impact of choices about these key issues may last over long periods of time, often outliving generations, cultural values, and even economic systems, the historical record is not simply instrumental to our understanding of how water issues have evolved over time: it is an essential component of the architecture societies used to manage water, whether they realize it or not. Framing water as heritage defines it as an object of study and positions it for preservation.

Adaptive Strategies to Water Heritage is a welcome addition to the growing literature on the world's water heritage. The broad scope of the papers in this volume reflects the pervasiveness of water-related issues across societies, as well as the universality of solutions. The methodological heterogeneity it embraces, which lies at the boundary of conservation practice, historical analysis, anthropology, and sociology, accurately reflects the multi-disciplinary nature of the issue.

The relationship between society and its water landscape is dialectical and deeply contextual. Wrestling with it, examining it to understand our water past, and recognizing its role in defining our present are essential to preparing for what is to come.

Giulio Boccaletti
Chief Strategy Officer and Global Managing Director
for Water, The Nature Conservancy
London, United Kingdom

Giulio Boccaletti is the Chief Strategy Officer and Global Managing Director for Water at The Nature Conservancy. He has been an academic and an executive in the private sector, and has spent the last fifteen years working on water issues at the intersection of public policy, economic strategy, and the environment. He is a member of the World Economic Forum's Global Futures Council on Environment and Natural Resource Security.

Foreword by Henk Ovink

The future is rapidly changing, the present is in high-speed transition, and complexity is increasing every day. Complexity is in the challenges we face, in their interdependence across political boundaries, in the systems—environmental, social, and economic—we use to organize ourselves, and our personal interests. Challenges are at all levels exacerbated by climate change, increasingly and always worse every day, every year. If we add up the numbers, the future looks bleak: every year a new record in rains, droughts, floods, migrants, economic, environmental, and humanitarian destruction and despair. More deaths, conflicts, extreme events, and dollars lost. These extremes become more and more extreme and impact on the world's vulnerable places ever more harshly. The future is here, grounded in the past. Can we learn from the past to help us tackle our future?

Learning from the past—so easily said, so hard to do. It is tempting to look back to the past and simplify the world, to imagine that things were once simple and focused. This is where populism looms, in its nostalgic longing to control, to surveil, to quash disturbing surprises. But looking back and simplifying do not give us a clear picture of history, nor an honest perspective on our future. Learning from the past does not mean we only look back, but that we also look ahead. Using history and our capacity to understand, we learn to value the past. History is the broker between us and the past, our aid as we try to explore and exploit that past, to use it to help us leapfrog into the future. We need an equal, just, and sustainable society that takes care of the planet, of everything and everyone, and leaves behind no one. The UN Agenda 2030 sketches out this path forward, littered with challenges and barriers—none of which is easy to overcome. We will have to reinvent ourselves a multitude of times. Yet, this change can only come when we, collectively, embrace the past as a perspective on the future.

Learning from the history of water is one of the most amazing journeys one can take: To see, to know, the course of the river, a drop of water, humankind's inventions and interventions for managing water in nature and in our cities. Amazing deltas dotted the planet, mitigating water extremes long before we had to learn the words *climate change*. For centuries before the ecological crisis, water-wise and water-rich cities proliferated, where water was an equal partner, where it was celebrated and valued. It is this capacity of water to unite, to bring together the multiple values of society—environmental, economic, social, and cultural—that stands out as an inspiration to us to understand the past in order to learn for the future. Valuing water means bringing together all interests that have to do with water. It means embracing all partners and their perspectives, protecting all our sources, building trust and capacity, learning and empowering, innovating, testing, and investing.

The interdependencies and complexities of climate change demand a comprehensive approach, cutting through silos, vested interests, and political positions. Our planet, our cities, our built systems have all the core values and principles of complexity we need. But, gradually, we have lost track of these. Our growing demand led us to abandon our growing capacity to learn, to look back while stepping ahead. Rethinking the future through the past can help us reinvent ourselves and our systems, restore our core values, and build a just, equal, and sustainable society. Values drawn from water, from nature, and from culture intertwine with our capacity to understand this complexity, allow us to strengthen the relationships between our environmental, societal, and economic systems, and build upon them. Leaving behind our stubborn convictions, we reach to an adaptive, flexible, sustainable, and ever-changing way forward. Un-certainty is our certainty, new extremes are the new normal, and changing interventions for the future, not failed repetitions of the past, are the road ahead.

We have no time to waste. The future is here. We can change, collectively, if only we learn to connect the past with the future, embrace and exploit complexity, live and work together, and act now.

Henk Ovink
Rotterdam, The Netherlands

Henk Ovink is Special Envoy for International Water Affairs for the Netherlands and Sherpa to the UN High-Level Panel on Water. He advocates for water awareness and builds coalitions to initiate transformative interventions, most recently in his new initiative, Water as Leverage. His book *Too Big. Rebuild by Design: A Transformative Approach to Climate Change* reports on his post-Hurricane Sandy recovery work.

Foreword by Diederik Six and and Henk van Schaik

The blue marble photograph, taken in 1972 by the Apollo 17 astronauts, shows Earth—our four-billion-year-old planet—floating in the void of space, its most striking feature, the omnipresent blue of water. On our planet, water is *Life*. Water is a friend and foe of life. From this Life, human beings emerged in Africa about 200,000 years ago, initiating agriculture and water management activities, thereby increasing the production of food, providing water services, and reducing the vulnerability of settlements to the perils of drought and flood. Water management innovations spanned the gamut of activity: structural, tangible measures such as reservoirs and dams; organizational arrangements for developing and operating structures; and intangible cultural–spiritual–ethical–ritual meanings and practices. These material, conceptual, and spiritual connections in water management made it possible for cities to develop in Mesopotamia and the Indus valley as well as along the banks of the Nile and China's rivers. Today, the remnants of ancient water cultures are found on every continent. Archeological and anthropological research tells us about these ancient water cultures, these origins of our present cultural identities.

The Industrial Revolution, largely made possible by abundant natural resources, including water, brought to human life an unprecedented growth of population, life expectancy, and economies. Despite our resources and our vast experience with water management, however, since the 1970s, we have come to realize the limits to this heralded growth. The over-exploitation of natural resources, irreversible pollution, and climate change all threaten biodiversity, fossil fuels, the water cycle, and the planet itself.

Since 2012, ICOMOS Netherlands has been exploring what can be learned from water-related heritage rooted in culture and nature. What insights can we derive from ancient water structures such as the dams of the Middle East or the *qanats* of arid regions? governance arrangements such as the water boards of the Netherlands? or, the ethico-spiritual frameworks of those of the Incas? How can these varied forms of water-related heritage teach and inspire future planners, architects, politicians, design engineers, and others as they address present and future water-related challenges? It is relatively easy to point to water-related heritage as a source of inspiration for innovation and creativity. In practice, however, this claim needs and must be held to scientific validation and specification through practical and illustrative cases—as well as to methodological guidance for policy makers, planners, designers, and training programs.

This publication addresses that need. It presents twenty-one chapters on water-related heritage that encompass a broad spectrum of theoretical and methodological approaches and studies of water-related heritage from a range of the world's regions. It contributes to the ongoing dialogue on water and heritage between policy makers, scientists, civil society, and spiritual leaders. The case studies it presents bridge the divides between the scientific disciplines of engineering, architecture, history, archeology, anthropology, and theology.

It is our hope that this publication will also make decision makers and citizens more aware of the importance of water-related heritage for the future. We also hope that it will contribute to the work of the International Scientific Committee at ICOMOS and to the recognition of Water and Cultural Heritage in both the Water Agenda and the Water Goals of the Sustainable Development Agenda.

In this spirit, we thank the editor and authors for their commitment to this project as well as for their perseverance in highlighting the importance of humanity's water heritage to the future. We look forward to continuing to work with them to establish a concerted, global effort for water-related cultural heritage that is locally and globally acknowledged as a source of inspiration and information for the future.

Jhr. Ir. Diederik Six
President, ICOMOS Netherlands
Doorn, The Netherlands

Ir. Henk van Schaik
Ambassador Water and Heritage
ICOMOS Netherlands
Wageningen, The Netherlands

Acknowledgements

A book of this size would not be possible without the engagement and dedication of a large number of people. Particular thanks go to the board and the two leaders of ICOMOS (International Council on Monuments and Sites) Netherlands, Henk van Schaik, who has tirelessly pushed the water and heritage agenda for many years, and Diederik Six, who both initiated the project with Erik Luijendijk and has consistently supported and inspired the program at every step. They have been helped in this endeavor, notably by Jan (J. C. A.) Kolen and Mara de Groot from the Center for Global Heritage and Development (CGHD) of Leiden University, Erasmus University Rotterdam, and Delft University of Technology. Linde Egberts of University of Amsterdam and Delft University of Technology helped organize the workshop on behalf of the Heritage and Environment Group of the CGHD she then leads with Carola Hein.

The book proposal particularly benefited from the input of Maurits Ertsen, Steffen Nijuis, and Gerdy Verschuure-Stuip of Delft University of Technology. Other researchers provided feedback on chapters: Particular thanks go to Tino Mager of Delft University of Technology and Maaike van Berkel of Radboud University Nijmegen. Irene Curulli of University of Eindhoven and Rutgerd Boelens of University of Wageningen and Amsterdam provided additional insights into the establishment of a research agenda on water and heritage as well on the creation of the International Scientific Group on Water and Heritage within ICOMOS.

Several additional people participated in the workshop and scholarly exchange, providing valuable insights, but for various reasons did not participate in the final book. I am particularly grateful to Hanna Pennock, Ian Lilley, Eric Luiten, Michiel Korthals, Reinout Rutte, Heleni Porfyriou, Stefan Uhlenbrook, and Paul van de Laar.

Such a project needs extensive organizational and financial support. The CGHD has greatly contributed to the development of the water and heritage theme, by both building networks and financially supporting a workshop, Water and Heritage for the Future, held at Delft University of Technology and Fort Vechten in November

2016. The Chair of History of Architecture and Urban Planning at Delft University of Technology provided additional administrative and financial support for the conference and the editing of the book. Thanks go to Kaiyi Zhu from Delft University of Technology for help with permissions and other works to finalize the book. Particular thanks go to Laura Helper-Ferris, who did excellent work to help harmonize, polish, and finalize the various chapters.

Once again, the project would not have been possible without the patience and support of my family. With love to Patrick, Caya, Aliya, Jolan, Joris, and my parents Wuppi and Walter.

Contents

Editor and Contributors

About the Editor

Carola Hein is Professor and Head of the History of Architecture and Urban Planning at Delft University of Technology. Her books include *The Capital of Europe*, *Rebuilding Urban Japan after 1945*, *Port Cities*, and *The Routledge Handbook of Planning History*. She currently works on the transmission of planning ideas among port cities and within landscapes of oil.

Contributors

Nahuel Beccan Dávila Beccan Davila Urbanismo, Leiden, The Netherlands

Csaba Centeri Institute of Nature Conservation and Landscape Management, Szent István University, Gödöllö, Hungary

Yu-Chuan Chang HsingWu University, Taiwan, Republic of China

Xi Chen Brunel University, London, England, UK

Tom A. Daamen Delft University of Technology, Delft, The Netherlands

Hildebrand P. G. de Boer Dutch Foundation for Industrial Culture, Amsterdam, The Netherlands

Arie den Boer Erasmus University Rotterdam, Rotterdam, The Netherlands

Martin Delgado Rotterdam, The Netherlands

Linde Egberts Vrije Universteit Amsterdam, Amsterdam, The Netherlands

Sebastian Eiter Department of Landscape Monitoring, NIBIO, Aas, Norway

Maurits Ertsen Delft University of Technology, Delft, The Netherlands

Bénédicte Gaillard Brandenburg University of Technology, Senftenberg, Germany

Carola Hein Delft University of Technology, Delft, The Netherlands

Wenyao Hsu Taiwan International Institute for Water Education, Taiwan, Republic of China

Azadeh Arjomand Kermani Delft University of Technology, Delft, The Netherlands

Yasunori Kitao Kyoto Women's University, Kyoto, Japan

Jan (J. C. A.) Kolen Leiden University, Leiden, The Netherlands

Alexandra Kruse insitu World Heritage Consulting, Buc, France

Zdeněk Kučera Charles University, Prague, Czech Republic

Izumi Kuroishi Aoyama Gakuin University, Tokyo, Japan

Andrew M. Law Newcastle University, Newcastle upon Tyne, England, UK

Chung-Hsi Lin National Yunlin University of Science and Technology, Taiwan, Republic of China

Suzanne Loen Delft University of Technology, Delft, The Netherlands

Felix Madrazo Rotterdam, The Netherlands

Tino Mager Delft University of Technology, Delft, The Netherlands

Meisha Hunter Burkett Li/Saltzman Architects, New York, USA

Han Meyer Delft University of Technology, Delft, The Netherlands

Steffen Nijhuis Delft University of Technology, Delft, The Netherlands

José M. Pagés Sánchez Hamburg Hafencity University, Hamburg, Germany

Bernd Paulowitz Hamburg, Germany

Oskar Puschmann Department of Landscape Monitoring, NIBIO, Aas, Norway

Hans Renes Faculty of Geosciences, Utrecht University, Utrecht, The Netherlands

Araceli Rojas University of Warsaw, Warsaw, Poland

Michael Roth Nürtingen-Geislingen Polytechnic, Nürtingen, Germany

Sofia Saavedra Bruno Rotterdam, The Netherlands

Arash Salek Rotterdam Municipality, Rotterdam, The Netherlands

Negar Sanaan Bensi Delft University of Technology, Delft, The Netherlands

Diederik Six ICOMOS, Amsterdam, The Netherlands

Martina Slámová Technical University in Zvolen, Zvolen, Slovakia

Sander van Alphen Vrije Universiteit Amsterdam, Amsterdam, The Netherlands

Henk van Schaik ICOMOS, Amsterdam, The Netherlands

Wout van der Toorn Vrijthoff Delft University of Technology, Delft, The Netherlands

Gerdy Verschuure-Stuip Delft University of Technology, Delft, The Netherlands

Hsiaoen Wu Taiwan International Institute for Water Education, Taiwan, Republic of China

Sinite Yu Taiwan International Institute for Water Education, Taiwan, Republic of China

The flood gate Oostertscheldekering, a Dutch Delta work, Wikimedia, released under a creative commons attribution 3.0 Unported license

Chapter 1
Introduction: Connecting Water and Heritage for the Future

Carola Hein, Henk van Schaik, Diederik Six, Tino Mager, Jan (J. C. A.) Kolen, Maurits Ertsen, Steffen Nijhuis and Gerdy Verschuure-Stuip

Abstract Water has served and sustained societies throughout the history of humankind. People have actively shaped its course, form, and function for human settlement and the development of civilizations. Around water, they have created socioeconomic structures, policies, and cultures; a rich world of narratives, laws, and practices; and an extensive tangible network of infrastructure, buildings, and urban form. Today, the complex and diverse systems of the past are necessarily the framework for preservation and reuse as well as for new systems. Through twenty-one chapters in five thematic sections, this book links the practices of the past to a present in which heritage and water are largely two separate disciplinary and professional fields. It describes an alternative emerging present in which policymaking and design work together to recognize and build on traditional knowledge and skills while imagining how such efforts will help us develop sustainable futures for cities, landscapes, and bodies of water.

Keywords Water heritage · Adaptive reuse · Policymaking · ICOMOS · Center for Global Heritage and Development · Dutch water and heritage practice · Water and heritage agenda

---

C. Hein (✉) · T. Mager · M. Ertsen · S. Nijhuis · G. Verschuure-Stuip
Delft University of Technology, Delft, The Netherlands
e-mail: C.M.Hein@tudelft.nl

H. van Schaik · D. Six
ICOMOS, Amsterdam, The Netherlands

J. (J. C. A.) Kolen
Leiden University, Leiden, The Netherlands

C. Hein (ed.), *Adaptive Strategies for Water Heritage*,
https://doi.org/10.1007/978-3-030-00268-8_1

Introduction

Water has always been a central human concern. The earliest prehistoric hunters and gatherers, although nomadic, also settled down along rivers, lakes, and coastlines to ensure access to crucial resources–including water. Later, prehistoric farming societies manipulated water systems and redirected water to meet their essential needs. Early states managed water resources with large-scale facilities like aqueducts, irrigation systems, and polders. They used water to cultivate the earth for drinking, food, and agriculture. No less importantly, water became a key element in their complex social organizations and political ideologies. People around the world have both used water as a means of defense and have learned to defend themselves against water, in the form of floods from river and sea. At the same time, the water in rivers, lakes, and oceans facilitated trade networks and the exchange of goods, people, and ideas. This led, among others, to the development of extensive port cities that connected local communities with the rest of the world.

Over millennia, people have created immensely rich and varied, often interconnected, systems to manage water: in lowlands and mountains, wetlands and deserts, agricultural landscapes, urban networks, and on urban waterfronts. Today, those systems are heritage—a term used here broadly to encompass both recognized World Heritage Sites as well as the historic built environment that people have chosen to preserve. This heritage is often also still vital and functional. It can range in size from ingenious small-scale water harvesting systems and aqueducts to larger water pumping facilities and irrigation and drainage networks, dike systems, and defense systems (Steenhuis 2015; Labanca Correa de Araujo 2015). It includes maritime cultural landscapes, canals, harbors, and waterfronts (Daly 2015; Hein 2011; Meyer 1999) as well as local knowledge and skills and regional traditions in water engineering (Sugiura et al. 2015; Scarborough 2003).

Water heritage is found in spaces that are closely linked to traditions, rituals, and narratives. The hydraulic network at Angkor served both the physical infrastructure and the ritual network of sacred places and temples (Hang 2015). In the Dutch delta, the elements of the infrastructure for water management–dikes, river forelands, polders, locks, and drawbridges–are icons of Dutch historical identity (Steenhuis 2015). Maritime heritage is spiritually important to local coastal communities in South Africa (Sharfman 2017). Indeed, port cities have a distinctive shared culture (Hein 2016). This heritage is a crucial source of information both for understanding how water systems worked in the past and discerning their impact on the present. It is also a source of knowledge for water managers and environmental engineers; an integral part of architectural and urban design; as well as a site of cultural identification, historical experience, public engagement, leisure, and tourism.

A few scholars have examined select aspects of the management of water and heritage. Maritime archeologist Christer Westerdahl introduced the notion of *maritime cultural landscape* to name and better explore, study, and preserve the networks present between communities based on travel and trade over water, be they oceans, seas, inland lakes, rivers, or artificial waterways. These networks included social and political relationships as well as their associated ancient routes, harbors, shipyards, settlements, and other physical structures (Westerdahl 1992). The historians Jerry Bentley, Renate Bridenthal, and Kären Wigen have coined the term *seascape* to capture the history of maritime regions around the world (2007). The journal *Water History* explores the historical relationship between people and water resource use, but does not specifically engage heritage or the seas and oceans. Planning historians have published extensively on waterfront redevelopment and the role of port-related heritage structures. Architectural and planning historian Carola Hein has proposed the concept of *port cityscapes*, arguing that the reach of the port into its neighboring city and region merits comprehensive investigation (2011, 2016). Planner Han Meyer and landscape architect Steffen Nijhuis have pointed to the need to study urbanized deltas and the dual challenges of river and sea water (2014). Other planners and landscape architects have explored issues of design, water, and heritage. The work on hydro-biographies stands as an example (Land-id, Beek and Kooiman 2014; Bosch and Soree 2016). At the institutional level, the Ramsar Convention is an intergovernmental treaty for the preservation and wise use of wetlands; its work encompasses their natural and cultural heritage (Ramsar 1994). In addition, UNESCO's World Heritage Center has published a special issue, *Living with Water* (2011). And the International World Water System Heritage Program, launched in 2016 by the World Water Council in collaboration with the International Commission on Irrigation and Drainage, has initiated a registrar for the intangible values of water-related heritage.

But these initiatives and studies in historical, urban, and geographical research have had only tangential influence, if any, on the practice and policy of water heritage management of diverse typologies—from buildings to landscapes, from engineered structures to nature conservation. Overall, academics, policymakers, designers, and the public alike largely perceive heritage and water as separate worlds, represented by different sectors and organizations; informed by different philosophies, scientific disciplines, policy frameworks, and design concepts.

Although water and cultural heritage are linked through complex interrelationships, each is approached from siloed perspectives. Water is examined along the disciplinary lines of science, engineering, governance, and management, whereas cultural heritage is often looked at as comprising isolated structures rather than as consisting of elements of a larger system. And researchers often miss water itself altogether. Overall, water's potential to connect sites of living heritage with each other; water-related heritage's capacity to connect past, present, and future; and water's role as heritage in spatial developments, landscape design, and urban planning remain underestimated and underexplored. Moreover, water-related policymaking is highly segregated within itself, with different specialists dedicated to investigating drinking water, tourism and recreation, nature and biodiversity, transport and mobility, safety and security, and so on. Top-down approaches dominate all of these fields. Moreover, many are primarily land-based, that is, connected to national agendas and focused upon water that is on or related to land.

Today, global climate change, pollution, and changing political and societal patterns affect both water and heritage on multiple scales; these include systems for drinking water, irrigation, and drainage as well as the heritage of coastal areas, deltas, and port cities (Lieske et al. 2015; Okamura 2015; Comer 2015; also see **Statement of Amsterdam** (1999; Willems and Van Schaik 2015). Rising seas challenge Pacific archipelagos (Peterson 2015) and the coastal plains and major port cities of the southern and eastern parts of the USA, while flooding rivers threaten cities and towns in the Low Countries, Cambodia, and Bangladesh. Conversely, severe droughts and desertification, resulting in land degradation in other parts of the world, challenge the livelihood of millions of people. Other climate-change-driven challenges, including expected food shortages and mass migration, underscore the need to rethink our longstanding relationship with water, culture, and our built heritage. The future of water and heritage structures also depends on political, economic, environmental, cultural, and spatial frameworks, including globalization and the privatization of water and heritage structures. The growing and changing pollution of canals, rivers, and seas—notably, the threat of plastic waste to nature, people, and structures—also warrants new forms of inquiry and design. The energy transition, and the design steps needed to achieve it, will also create new kinds of heritage in the future. Vast areas of ports and petroleum installations are just one example of potential future water-related heritage sites.

History and heritage matter when we design new relationships with water. Water-related heritage preserves and transmits forgotten best practices and catastrophic events. It harbors the long histories of water systems and safeguards our cultural memory for generations to come. New investigations of water history and heritage can serve as a source of information, inspiration, and identity-building in water management, wetland recreation, and marine engineering; they are relevant to the redevelopment, redesign, and reuse of existing and ancient water systems as well as to the design of new systems. The reuse, adaptation, or redesign of old systems can contribute to the quality of life of communities and other groups, and to their sense of place and self-identification. Finally, understanding and analyzing the relationship between water and heritage can also help us refine our understanding of tangible and intangible heritage more broadly.

This volume brings new voices to this important and urgent multilateral project at the interface of water and cultural heritage and shows how we might address its concerns in both scientific research and research-based policymaking. It is one of many undertakings carried out by ICOMOS Netherlands, the Dutch branch of the International Council of Monuments and Sites, which has sponsored a range of initiatives: conferences, events, books, and even this volume at all geographical and governmental scales, from the local and regional to the international and global, in order to stimulate thinking about the interrelationships between cultural heritage and water management. The rich and agenda-setting contributions of ICOMOS Netherlands to this challenge are discussed more extensively below. It is, nevertheless, worth noting that its work, along with that of the Centre for Global Heritage and Development (CGHD) of Leiden University, Erasmus University Rotterdam, and Delft University of Technology, is rooted in the long Dutch history of water-related heritage.

The Dutch Connection: Water and Heritage in the Netherlands

The Netherlands is an exemplary delta region in which a dynamic society with a growing population and an increasingly urbanized landscape—rich in heritage—has long faced the challenges of large-scale water management. The country has a centuries-long history of water management which encompasses drinking water supply systems, irrigation and drainage for agriculture, coastal protection, river management, canals, ports, and the use of water as a means of military defense. What was once pioneering infrastructure is now cultural heritage: The Dutch delta illustrates the peculiar position water and heritage occupy well. On the one hand, water management heritage in the Netherlands is extremely rich, historically layered, large-scale, omnipresent, and strongly tied to national and regional issues of cultural identification (Van Gorp and Renes 2003). Dutch contributions to UNESCO's World Heritage List primarily consist of monuments and structures from the country's longstanding engagement with water and the sea. Amsterdam's seventeenth-century canal ring, the Beemster Polder, reclaimed from a lake with the help of windmills by 1612; Kinderdijk, a concentration of windmills for draining polderland from around 1740; the Defense Line of Amsterdam, a military defense line built from 1883 to 1920; the Ir. D. F. Woudagemaal, a steam-driven water pumping station from 1920; and the former island of Schokland, a strip of peatland that lost its island character with the construction of the Noordoostpolder in the 1940s animate the list. On the other hand, this water-related heritage is, at best, only introduced into discussions about the delta's future and the quality of life of its inhabitants and visitors in a meager way. Few consider the area's water-related cultural heritage as a source of information or inspiration for the necessary development and innovation of water governance, engineering, or design—a topic which is also explored by the Organization for Economic Co-operation and Development (OECD 2014).

The Dutch have long debated the very definition of water management heritage. Since the early 1990s, when the Dutch delta was threatened by floods, two different perspectives have developed on the history and heritage of water management, as well as on how to deal with the rapidly changing hydrological situation (Landschap als Geheugen 1993; Van de Cammen and De Klerk 2003; Ruimte voor de Rivier 2007; Van Toorn 2011). One group sees heritage within the continuity of cutting-edge engineering and strategies of water management, while the other aims to preserve the heritage of the physical water structures of the past. Engineers, planners, and nature conservationists propose that the water-related heritage of the Dutch delta is, in fact, largely, an intangible tradition of innovative engineering and skillful water management, that is, of finding new solutions for changing environmental and hydrological conditions. From this perspective, it is necessary to reshape the river landscape and its heritage at a profound level. It is also essential to fundamentally make the delta "climate-change proof" for the future (Kolen et al. 2014, 178). The typical Dutch river landscape and management system with its closed dikes, artificial river forelands (the *uiterwaarden*), and sunken polders, now the pride of Dutch identity and water management, is outdated and could even have catastrophic effects. In keeping with the emphasis on innovation as against the physicality of water structures, the Rijkswaterstaat—literally, the agency in charge of the national water level and effectively in charge of water management and infrastructure—set up a program called Room for the River to provide rivers experiencing high water levels more space free from development. The IJssel River, for example, is being deepened for navigation, and a high water channel between two dikes has been constructed to run parallel to it. Space is created to enable the river to move more freely through the landscape and to facilitate nature's own restoration of biodiversity within newly created wetlands. What continues and what counts as water management heritage to these actors is the successful tradition of innovation: of finding smart technical solutions for complex water issues (Nienhuis 2008; Hoeksema 2006; Huisman 2004). In contrast, some heritage managers have adopted a somewhat different perspective on water-related heritage. They stress the importance of *existing* water management structures as valuable cultural heritage and icons of Dutch national identity. For them, it is the physical structures of water management, such as the closed dike systems which Dutch engineers and planners started in the Middle Ages and which gained their current form in the second half of the nineteenth century, that have to be preserved. Occasional floods are considered to be part of a system which has never been and will never be perfectly safe (Kolen et al. 2014, p. 179).

This binary dialogue between water and heritage sectarians is unproductive. In recent years, a new approach has emerged in the Netherlands that combines both kinds of heritage work: technological creativity and historic preservation. The Belvedere Memorandum in heritage management and its incentive program (1999–2009) laid out a decade of national policy to integrate heritage management with new spatial developments through historically informed design. It has produced promising experiments in connecting heritage and water issues on a local scale (Belvedere Memorandum 1999; Janssen et al. 2014). With this dynamic approach, new solutions can be found through reusing old hydraulic systems or creating new ones by applying historic approaches. The approach even facilitates interventions in line with sustainable practices, as can be seen in recent research on large-scale water systems—some of which has been published in the *Polder Atlas* (Steenbergen, et al. 2009). The Dutch Heritage Department pursues a similar strategy, exploring sustainable practices for classified heritage structures (Rijksdienst 2018). Heritage groups and planners around the world can benefit from these experiments and insights.

The Water and Heritage Agenda at ICOMOS

Given the obvious ties between the Netherlands and water-related heritage, ICOMOS Netherlands first sought to solidify the few existing relationships between the worlds of water and heritage management. Between 2013 and 2018, it convened a number of expert meetings and workshops to develop its water and heritage agenda. The group presented many lectures on the topic at international conferences outside the Netherlands. Its goals aligned with the UN agenda, as identified in the Sustainable Development Goals (SDG) (UN 2015). In September 2013, ICOMOS Netherlands organized a five-day international conference entitled "Protecting Deltas, Heritage Helps!" which brought together many experts and representatives of governmental, nongovernmental, and intergovernmental organizations from all over the world to share experiences. All invited representatives, partners, and experts expressed urgency regarding the integration of heritage and water management at local, regional, and international scales, while recognizing the equally urgent need for a global exchange of experiences and best practices. To celebrate efforts and achievements in connecting water and heritage for the future, ICOMOS Netherlands also developed the "water and heritage monument shield", an award it first presented in 2013 to the city of Amsterdam (Six and Luijendijk 2015, p. 12). The conference resulted in the Statement of Amsterdam, which called on water and heritage stakeholders, institutions, and specialists to collaborate on active research, education, and communication in order to advocate for the recognition of water and heritage as one connected theme rather than as two independent fields (Willems and Van Schaik 2015).

In 2015, ICOMOS Netherlands published an edited and peer-reviewed volume, *Water & Heritage: Material, Conceptual, and Spiritual connections* (Willems and Van Schaik 2015), based partly on papers presented at the 2013 conference. It opens with this statement by Mrs. Irina Bokova, then, the Director-General of UNESCO:

> From the beginning of time, humanity has sought out sources of water to sustain life, health and the ecosystems on which they depend. This is especially true today, in this turning point year for the international community, as States shape a new global sustainability agenda. Limiting the impacts of floods, landslides, and droughts, water security and cooperation are basic requirements to improving lives and to empowering people to overcome hunger and disease. The stakes are high. Peace and democracy thrive when people and cultures cooperate for water. Literacy, gender equality, economic development, respect for human rights, freedoms, and diversity–all of these depend on water security (Bokova 2015, p. 9).

Following from this theme of crisis, the authors argued that:

> Saving the deltas of the world will be one of the most critical challenges for a sustainable future of humankind. Exposure to water-related hazards, especially due to climate change resulting in higher frequency and intensity of disasters, together with an increasing population density and richness in cultural and natural heritage puts communities, particularly in the world's deltas, at high risk. Rapid urbanization of delta areas without respecting the historic water structures accumulated over the centuries make these areas, in which economic, social, and cultural values are concentrated, even more vulnerable (Six and Luijendijk 2015, 11).

The current volume heeds this call and builds upon these events and publications to investigate deeply a range of heritage sites and to explore the implications and opportunities they offer to future design. It is also partly based on a two-day conference entitled, "Water and Heritage for the Future," held in November, 2016. The conference was organized through close collaboration by ICOMOS Netherlands and the Center for Global Heritage and Development.

The first day of the conference, held at Delft University of Technology's Faculty of Architecture and the Built Environment, brought together a large and varied group of scientific researchers in spatial planning, urban design, landscape architecture, civil engineering, water management, history, anthropology, and archeology. The second day was held at Fort Vechten, located where the New Dutch Waterline—a water-based defense line—intersects with the Limes, the former Roman frontier near the city of Utrecht. The conference gathered experts in policymaking, legislation, and applied research. Following these events, a team of Dutch researchers collaborated to develop a research agenda; a foundation for collaboration and dialogue with researchers from other disciplines, areas of work, and nationalities; a charter for an international scientific group; and the initiation of an international scientific board. Along with our current book, the draft agenda identifies a number of research topics on water heritage that partly coincide with those identified here and go beyond them. They include: water for services, that is, for systems of drinking water provision and sewage; irrigation and drainage infrastructure; and natural and man-made waterscapes, including reclaimed land areas, defense structures, and the larger matter of water-related transport systems—concerns which this book addresses. The agenda adds water power, that is, water as a means of energy generation and as a destructive force, along with worldviews: encompassing the role of water in the philosophy of life or conception of the world and its built spaces. The draft agenda is designed so that future issues may be added including, for example, environmental pollution and climate change.

In December 2017, at the nineteenth ICOMOS General Assembly in Delhi, ICOMOS Netherlands presented an informal proposal to a meeting of the larger body's scientific board meeting to initiate **the International Scientific Group on Water and Heritage.** The scientific board expressed appreciation for the initiative and encouraged ICOMOS Netherlands to prepare a formal proposal. During the General Assembly in Delhi, the Taiwan International Institute for Water Education (TIIWE) offered to host an international conference on water and heritage as part of the preparatory process for the International Scientific Group.

This Book and Its Structure

This book is a stepping stone in the process of developing international scientific interest and an international scientific agenda on water and heritage. It brings into discussion water and heritage issues through the lens of international cases, while providing deeper insight into the Dutch case. It explores five thematic areas related to water heritage: infrastructure designed for drinking water; agricultural sites engineered for irrigation and drainage; areas gained by poldering and other land reclamation in agriculture, settlement, and defense systems; river and coastline planning; and urban and engineered structures in ports and on waterfronts. Each chapter first addresses larger themes of water heritage, ranging from policymaking to narratives and from sociocultural meaning to subjectivity, before exploring case studies and concluding with future-oriented solutions for heritage practice.

The book opens with an exploration of freshwater services through time and space, the preservation of infrastructure, its redesign, and potential for inspiring future design. As Meisha Hunter's examination shows, intricately engineered systems have served large populations. Partly due to their service role in traditional water systems, these systems, distinguished for their utility, have received less attention than World Heritage Sites listed for their aesthetics, although infrastructural sites can teach us much that will help us respond to future crises. Areas in which the provisioning of water suffers from ongoing or accelerated desertification are equally threatened. Systems such as the *qanat* of the Middle East and northern Africa are often extremely vulnerable to relatively small changes in climate, precipitation, political and social organization, and the exchange and transmission of local specialized knowledge (as noted in Pangare and Pangare 2015). Negar Sanaan Bensi explains that the qanat system is not just heritage to be preserved as reminder of a past, but an ongoing element of the culture and civilization on the Iranian Plateau. Relatedly, what seem essentially to be water management interventions, such as the drilling of deep wells after World War II, have larger governmental and cultural implications, which must be acknowledged so as to benefit future interventions.

Araceli Rojas and Nahuel Beccan Dávila demonstrate the relevance for the future of design proposals that build on and derive from historic water systems. In Monte Albán, a site which originates in ancient Oaxaca and has been a UNESCO World Heritage Site since 1993, the supply system of water—mainly consisting of natural rivers and tributaries—has defined the infrastructure of settlements while serving as vessels of ritual meaning. Looking ahead, they suggest that design solutions based on the historic water system can inspire designers to formulate new strategies for preserving the natural environment and archeological heritage, while improving living conditions for local people. Suzanne Loen deepens understanding of the Dutch heritage in freshwater management, a field whose traditional decentralized practices of public and private rainwater harvesting largely disappeared at the advent of centralized water supply systems. Her goal is to show the potential this heritage contains for creating an integrated approach to water supply, landscape conservation, and water-secure livable cities.

Other historic engineered water infrastructure systems include those that improved agricultural land like meadows and rice paddies, structures that are intimately related to modes of societal organization and narrative construction. Hans Renes, Csaba Centeri, Sebastian Eiter, Bénédicte Gaillard, Alexandra Kruse, Zdeněk Kučera, Oskar Puschmann, Michael Roth, and Martina Slamova explore the ways in which the restoration of derelict water meadows in northwestern and central Europe, Slovakia, and Norway can help create and advance regional identity on a European scale and, at the same time, restore biodiversity, improve water retention capacity, and promote tourism and local *understanding* of historical cultural values. Alexandra Kruse and Bernd Paulowitz complement this investigation with insight into the ways in which Dutch land reclamation technology expanded throughout Europe in the form of the Holler colonies, tangible evidence of a common European economic and social history. Izumi Kuroishi rounds out this investigation into agricultural irrigation by exploring the history of irrigation in Japan's Sanbonkihara rice paddy region in relation to cultural practices, narratives, and festivals that have shaped the community around agricultural heritage. Many historical water structures both addressed the water-related needs of a location and created social communities. Modern technological interventions often ignored this intricate balance. Recent climate shifts have emphasized the shortcomings of these systems, as the case of the Taiyuan Tableland illustrates—where a pond and canal system originally built under the influence of generations of foreign colonists, immigrants, and experts has deteriorated. Locals, using what authors Sinite Yu, Chung-His Lin, Hsiaoen Wu, Wenyao Hsu, and Yu-Chuan Chang call *participatory narrative weaving*, have successfully challenged further development plans for the area.

Water management on land can take on various forms: creating land for agriculture or urbanization and defending that land against attacks. In coastal and alluvial lowlands all over the world (Nijhuis et al. 2019), historic water management projects blocked water from some areas of land and controlled water levels artificially so people could live and work on the reclaimed land. This often centuries-old interaction between human beings and water has produced a rich variety of polder landscapes. Increasing flood risk due to sea level rise and increased climate turbulence, ongoing subsidence due to intense drainage, and rapid urbanization all call for protective action. Three chapters explore the spatial and social construction and meaning of polder landscapes. Yasunori Kitao sets the stage with a careful analysis of the sociocultural aspects of the construction of the Hachirogata polder in Northern Honshu, the largest and most highly populated island of Japan. The polder is celebrated as an important industrial heritage; however, its narrative rarely acknowledges the traditional fishing practices destroyed by its very construction. Steffen Nijhuis complements the Japanese polder heritage exploration, focusing on the preservation and development of the Dutch Noordoostpolder—built in the twentieth century—and its consequent development as a cultural heritage landscape. The construction of polders, which notably involved Dutch expertise, is a Europe-wide phenomenon and one that may support the creation of a common identity. The Europolder program discussed by Hildebrand de Boer showcases the contemporary benefits of these heritage sites for tourism and regional identity. Other human interventions in water management were designed to protect land against invasion. A unique example of such a large-scale historical water-related site that has been preserved and redesigned is the New Dutch Waterline, an historic defense line. This intervention is examined by Gerdy Verschuure-Stuip. The preservation of this large monument has provided an innovative design connection between water, heritage, and tourism at entirely new scales of intervention.

People around the world have created a broad range of heritage practices along riverbanks and on river waterfronts. Andrew Law examines the Yangtze River as an evolving landscape, what he calls *a heritage of becoming*. His contribution raises the matter of new digital technologies, including augmented reality tools and their potential to shape heritage debates. The necessity of conceiving of heritage as part of a long-lasting creative process in spatial transformation and public and private participation is also at the heart of Arie den Boer's contribution, which argues that cultural heritage in the Netherlands and elsewhere involves construction and reconstruction, use and reuse, public and private stakeholders and civil society over time. A detailed analysis of industry closures, undervalued heritage, along with recent attempts at revitalization shows both the power and opportunities of artistic and cultural projects and of participatory approaches. Sander van Alphen explores how Dutch engineers have responded to the results of centuries of water management by providing new spaces for rivers in the Netherlands, thus addressing questions of both safety and spatial quality. These include the genesis of an attractive living environment and a valuing of the presence of cultural history. Coastal heritage, its natural sources, towns, and buildings are also at the core of Linde Egberts' contribution. She explores the need for coastal regions in Europe to work together to address the common challenges and shared opportunities of coastal tourism. As she appropriately reminds us, old coastal towns were better connected to other port cities over the sea than some of their neighbors on land.

The last group of chapters considers ports and waterfronts. Azadeh Arjomand Kermani, Wout van der Toorn Vrijthoff, and Arash Salek add insights on interactions between ports and their cities, showing how Rotterdam has redeveloped its old harbor heritage. It explores both the history of the former shipbuilding company RDM (Rotterdamsche Droogdok Maatschappij) and the city's renewal of the waterfront to attract cruise ship tourists. The opportunities and challenges posed by cruise ships and their impact on port cities and waterfront heritage become evident in the work of Sofia Saavedra Bruno, Martin Delgado, and Felix Madrazo. Collaborating under the name of Supersudaca, they examine the historical and contemporary logics of the emergence of Caribbean heritage and recent fake heritage buildings in Caribbean cruise destinations. José Manuel Pagés Sánchez and Tom Daamen link waterfront redevelopment to the role of heritage in the sustainable development of the Lisbon historical maritime waterfront, emphasizing the switch from an object-based to a landscape-based approach to heritage. This strategy is based on a governance process that facilities collaboration between port and city authorities. Han Meyer concludes this series of inquiries into port, city, and waterfront relationships by emphasizing that cultural heritage is constructed on our selective understanding of the past. He asserts the need to recognize that buildings, deltas, and nature itself are adaptive and evolutionary, such that we can move from a narrative of human engineering resisting nature, as it has emerged in the twentieth century, to one of dynamic adaptation. Such a reconsideration of cultural heritage is particularly necessary at this time of climate change and the many attendant challenges it holds for urbanized delta regions. Extensive heritage sites on urban waterfronts and working ports and cities are of particular concern. He contends that the cultural and natural heritage of urbanizing deltas itself will help us develop an adaptive approach, not as a complete departure from present ways of doing things, but as a new stage in a centuries-long tradition.

Water heritage systems throughout the world are comprised of physical and functional structures, conceptual and organizational principles, and cultural and spiritual values. This stands despite their many differences in geographical location, climate, cultural and political context, economic and social setting, heritage, and future threat. Scholars and policymakers must closely examine these differences to understand and tune research designs and approaches in politics, policy, and management, as well as future design opportunities (as noted in Katko et al. 2015). Moreover, historical and archeological studies are often able to clarify when and why systems are more or less efficient and what the conditions of exploitation or overexploitation are in the past, writ large, and in the recent past (as in Comer 2015; de Grenade and Varady 2015). When research into former times is closely linked to forward-looking practices in engineering, architectural design, and planning, we are able to make heritage an integral part of the future as well as a means through which design of future sustainable practices can be achieved. The book's aim is to provide further evidence and opportunity for anchoring this claim and ambition. Rather than an end, our effort seeks to catalyze international interest among policymakers, planners, architects, and heritage specialists to integrate planning with the management of water-related heritage. Because substantiation is only in its beginnings, this book does not contain conclusions but rather ends with a promise that the work will be continued.

The book's five themes examine several of the most important purposes of water management: drinking water supply, agriculture, land reclamation, protection, defense, transport, and trade. Many important subjects have not been touched upon: for example, the role of canals and sewage systems in water heritage, among others, merit further examination. Canals have played a pioneering role in human cultural development. Their planning and implementation requires extensive collective effort as well as good hydraulic and topographical knowledge that specifically includes that of the construction of locks, dikes, bridges, and harbors. Sewage systems have also been of great importance in the development of larger settlements. They have facilitated the hygienic conditions that are a prerequisite for the capacity of larger communities to live together in confined spaces. Today, disposal is a fundamental problem in many overpopulated and fast-growing regions of the world. It is a problem which can be seen in the state of many watercourses, rendered as abused flowing dumps that contribute to marine pollution. This development itself often necessitates restructuring watercourses and constructing reservoirs, which, in turn, entail environmental risks.

Other, larger themes, such as water and energy generation, natural, industrial and urbanized waterscapes, water narratives, legal issues, and education also merit additional attention. For example, water has been used for energy generation for thousands of years. And, today, water power can make a significant contribution as a renewable energy source. Further research into the various ways that using water power can help regions solve local energy problems while safeguarding ecological balance would be of great benefit. To date, the discussion on water and heritage has largely neglected issues of the open sea. New scholarship is emerging on the 'urbanization of the oceans' (their increased use for shipping, raw material extraction, energy production, and the siting of pipelines, cables, and other networks are material concerns). The question of whether and how to preserve drilling rigs and other sea-based construction as heritage (Couling 2016; Hein 2018; Couling and Hein 2018) is also now being addressed. These concerns all call for deeper research into historically grounded solutions. The long-term consequences of their consideration can be of help to planners and policymakers in integrating historical knowledge and experience into future-oriented and sustainable solutions that are resilient, balanced, and durable.

Their investigation and that of other objects of water-related heritage requires a broader approach: Water management systems are components of the water cycle and are thus always integrated into other complex natural and cultural systems. When investigating historical objects, as with other heritage sites, scholars must always consider their economic, ecological, organizational, functional, social, and spiritual aspects. Water management systems play a major role in all world religions and spiritual traditions, whether people use water for baptism, ritual purification, or worship it as the source of life. Overall, past, present, and future water cultures merit further investigation. As Rutgerd Boelens has pointed out, politicians and other institutional leaders—and their institutions—have established water cultures and used engineering to then transform historic water cultures and practices (2015). While his particular exploration notes cases that arise from the Andes, where indigenous control over water rights and territories encounters international policymaking, the practice he posits has precedent elsewhere.

Legal issues related to water also deserve further exploration. Innumerable local conflicts surround water control every day. Further research must address the dynamic mix of local, regional, and global norms and conflicts that emerge with the application of traditional and indigenous water control models. Here, very different concepts of efficiency, participation, and authority meet and require closer investigation as to how pluralism and normative diversity relate to the dominant water governance culture.

A crucial point in successfully applying heritage knowledge to current and future issues of water management is education. Academic and professional programs must combine the many strands of isolated expertise to establish a more comprehensive and holistic understanding of how water-related heritage can help build more inclusive and sustainable futures. Water management, civil engineering, hydrology, and urban and spatial planning must therefore be integrated into heritage management education along with social and legal studies.

Heritage is more than historical objects. First and foremost, it encompasses the richest treasure of experience available to humankind. Learning both from best practices and mistakes in history, from the long-term effects of particular decisions and the complex interrelationships between water and society, we can recognize the full potential of water heritage and apply what is gained. Understanding, using, and sharing heritage calls for knowledge gained through research and is best disseminated through technologies of open access. Little more has greater value in sustainably preparing societies for future climatic and demographic changes.

Ultimately, water-related heritage is not only a product of the past, worthy of protection because it allows future generations to pursue their questions through a consideration of historical objects, it is also a product of research and interpretation in the present. Preserving local structures necessitates addressing contemporary challenges, which, in turn, requires strong local cultures and identities. But these take a long period of time to form. Bottom-up processes and participation must be further researched and supported. Local initiatives and targeted support programs may also be helpful in generating interest and thus facilitating identification, the basis of all commitments. In short, artifacts and practices of the past are important as part of the present, to be preserved, to be adapted and reused. They can also become inspirations and foundations for the future; however, we must educate people to see this.

These connections between the material, conceptual, organizational, and cultural–spiritual aspects of water systems, their heritage structures, and future designs form an important point of departure for new initiatives to be executed by ICOMOS, the Centre for Global Heritage and Development of Leiden University, Delft University of Technology, and Erasmus University Rotterdam in the coming years.

References

Belvedere Memorandum (1999) The Belvedere Memorandum—a policy document examining the relationship between cultural history and spatial planning. Ministerie VROM, Nieuwegein

Bentley JH, Bridenthal R, Wigen K (eds) (2007) Seascapes: maritime histories, littoral cultures, and transoceanic exchanges. University of Hawai'i Press, Honolulu

Boelens R (2015) Water, power and identity: the cultural politics of water in the andes (Earthscan studies in water resource management). Routledge, London

Bokova I (2015) Foreword. In: Willems WJH, van Schaik HPJ (eds) Water & heritage. Material, conceptual and spiritual connections. Sidestone Press, Leiden, p 9

Bosch JW, Soree C (2016) Hydrobiografie Schelde-estuarium. Bosch en Slabbers Landschapsarchitecten, Den Haag

Comer DC (2015) Water as an agent of creation and destruction at Petra. In: Willems WJH, van Schaik HPJ (eds) Water & heritage. Material, conceptual and spiritual connections. Sidestone Press, Leiden, pp 231–244

Couling N (2016) Urbanization of the ocean; extractive geometries in the Barents Sea. In: Infrastructure space. LaFargeHolcim forum for sustainable construction. Ruby Press, Berlin

Couling N, Hein C (2018) Blankness: the architectural void of north sea energy logistics. Footprint(23)

Daly K (2015) Preserving New York City's waterfront industrial and maritime heritage through resilient and sustainable development—New York City's coastal development. In: Willems WJH, van Schaik HPJ (eds) Water & heritage. Material, conceptual and spiritual connections. Sidestone Press, Leiden, pp 155–168

de Grenade R, Varady RG (2015) The Santa Cruz river: four millennia of water heritage and security in the US-Mexico border region. In: Willems WJH, van Schaik HPJ (eds) Water & heritage. Material, conceptual and spiritual connections. Sidestone Press, Leiden, pp 371–388

Hang P (2015) Water and heritage in Angkor, Cambodia. The monuments, the ancient hydraulic network and their recent rehabilitation. In: Willems WJH, van Schaik HPJ (eds) Water & heritage. Material, conceptual and spiritual connections. Sidestone Press, Leiden, pp 71–86

Hein C (2011) Port cities: dynamic landscape and global networks. Routledge, New York

Hein C (2016) Port cityscapes: conference and research contributions on port cities. Plann Perspect 31(2):313–326

Hein C (2018) Oil spaces: the global petroleumscape in the Rotterdam/The Hague area. J Urban Hist 44(5):887–929

Hoeksema RJ (2006) Designed for dry feet. Flood protection and land reclamation in the Netherlands. American Society of Civil Engineers, SL

Huisman P (2004) Water in the Netherlands. Managing checks and balances. Netherlands Hydrological Society, Delft

Janssen J, Luiten E, Renes J, Rouwendal J (2014) Heritage planning and spatial development in the Netherlands: changing policies and perspectives. Int J Herit Stud 20(1):1–21

Katko TS, Juuti PS, Pietila PE, Rjala RP (2015) In: Willems WJH, van Schaik HPJ (eds) Water & heritage. Material, conceptual and spiritual connections. Sidestone Press, Leiden, pp 297–312
Kolen JCA, van Manen N, de Kleijn M (2014) History matters: the temporal and social dimension of Geodesign. In: Lee DJ, Dias D, Scholten HJ (eds) Geodesign by integrating design and geospatial sciences. Springer, Cham, pp 173–181
Labanca Correa de Araujo E (2015) Heritage values of water and sea defense in Recife. Challenges for a local governmental approach. In: Willems WJH, van Schaik HPJ (eds) Water & heritage. Material, conceptual and spiritual connections. Sidestone Press, Leiden, pp 169–184
Land-id, Beek and Kooiman (2014) Hydrobiografie van Marken, leven met het water. Arnhem
Landschap als Geheugen (1993) Landschap als geheugen. Opstellen tegen dijkverzwaring. Cadans, Utrecht
Lieske H, Schmidt E, Will T (2015) Flood protection for historic sites. Integrating heritage conservation and flood control concepts. Experiences in Germany. In: Willems WJH, van Schaik HPJ (eds) Water & heritage. Material, conceptual and spiritual connections. Sidestone Press, Leiden, pp 205–216
Meyer H (1999) City and port: urban planning as a cultural venture in London, Barcelona, New York, and Rotterdam: changing relations between public urban space and large-scale infrastructure. International Books, Utrecht
Meyer H, Nijhuis S (2014) Urbanized deltas in transition. Techne Press, Amsterdam
Nienhuis P (2008) Environmental history of the Rhine-Meuse Delta. Springer, Dordrecht
Nijhuis S, Schultz B, Pouderoijen MT (2019) Polder landscapes of the world. Jap Sam Books, Amsterdam (in press)
OECD (2014) Water governance in the Netherlands: fit for the future? OECD studies on water. OECD Publishing, Berlin
Okamura K (2015) Tsunami and heritage after the 2011 Great East Japan Earthquake. In: Willems WJH, van Schaik HPJ (eds) Water & heritage. Material, conceptual and spiritual connections. Sidestone Press, Leiden, pp 245–256
Pangare V, Pangare G (2015) Ancient water wisdom. Traditional water systems in India. In: Willems WJH, van Schaik HPJ (eds) Water & heritage. Material, conceptual and spiritual connections. Sidestone Press, Leiden, pp 57–70
Peterson JA (2015) Pacific islands on the brink of submergence. Rising seas in an age of climate changes. In: Willems WJH, van Schaik HPJ (eds) Water & heritage. Material, conceptual and spiritual connections. Sidestone Press, Leiden, pp 121–140
Ramsar (1994) Convention on wetlands of international importance especially as waterfowl habitat, with the amendments of 28.5.1987 and certified copy of 13.7.1994 by UNESCO. UNESCO, Paris
Rijksdienst voor het Culturel Erfgoed (2018) Erfgoed telt. https://cultureelerfgoed.nl/dossiers/erfgoed-telt. Accessed 25 June 2018
Ruimte voor de Rivier (2007) Planologische Kernbeslissing Ruimte voor de Rivier. Ministerie van V&W, Ministerie van VROM en Ministerie van LNV, The Hague
Scarborough VL (2003) The flow of power. Ancient water systems and landscapes. School of American Research, Santa Fe
Sharfman J (2017) Troubled waters. Developing a new approach to maritime and underwater cultural heritage management in Sub-Saharan Africa, vol 41. Archaeological Studies Leiden University, Leiden
Six DL, Luijendijk E (2015) Preface. In: Willems WJH, van Schaik HPJ (eds) Water & heritage. Material, conceptual and spiritual connections. Sidestone Press, Leiden, pp 11–13
Steenbergen C, Reh W, Nijhuis S, Pouderoijen MT (2009) The Polder Atlas of the Netherlands. Pantheon of the low lands. Thoth, Bussum
Steenhuis M (2015) The Deltaworks: heritage and new space for a changing world. In: Willems WJH, van Schaik HPJ (eds) Water & heritage. Material, conceptual and spiritual connections. Sidestone Press, Leiden, pp 331–350
Sugiura M, Sato Y, Ota S (2015) The framework of skills and knowledge shared in long-enduring organizations in the improvement of irrigation efficiency in Japan. In: Willems WJH, van Schaik HPJ (eds) Water & heritage. Material, conceptual and spiritual connections. Sidestone Press, Leiden, pp 313–330

UN (2015) Sustainable development goals. 17 goals to transform our world. http://www.un.org/sustainabledevelopment/sustainable-development-goals
van de Cammen H, De Klerk L (2003) Ruimtelijke ordening. Van Grachtengordel tot Vinex-wijk. Het Spectrum, Houten
van Gorp B, Renes J (2003) The Dutch landscape: W way of seeing. In: van Gorp B, Hoff M, Renes J (eds) Dutch windows. Cultural geographical essays on the Netherlands. FRW, Utrecht, pp 55–75
van Toorn W (2011) Het grote landschapsboek. Querido, Amsterdam
Westerdahl Chr (1992) The maritime cultural landscape. Int J Naut Archaeol 21(1):5–14
Willems WJH, van Schaik HPJ (eds) (2015) Water & heritage. Material, conceptual and spiritual connections. Sidestone Press, Leiden
World Heritage Center (2011) Living with water. In: World heritage, vol 59, Mar 2011

Carola Hein is Professor and Head of the History of Architecture and Urban Planning at Delft University of Technology. Her books include *The Capital of Europe*, *Rebuilding Urban Japan after 1945*, *Port Cities*, and *The Routledge Handbook of Planning History*. She currently works on the transmission of planning ideas among port cities and within landscapes of oil.

Henk van Schaik is an Ambassador of Water and Heritage of ICOMOS Netherlands. He was co-editor of the ICOMOS book *Water and Heritage: material, conceptual and spiritual connections*. He is working on creating an International Scientific group between ICOMOS and its partner organisations to raise awareness of water-related heritage for the future.

Diederik Six is a senior architect who specializes in restoration and reuse as well in new traditional architecture. He is the former president of the Royal Dutch Antiquity Society KNOB and president of ICOMOS Netherlands. He seeks to bridge disciplines and build on traditional wisdom for a sustainable future.

Tino Mager studied media technology and art history and the communication sciences. He received his Ph.D. in 2015 from the Berlin Institute of Technology where he was an Elsa Neumann fellow. After research terms and lectureships in multiple locations internationally, he was a research associate at the Chair of History and Theory of Architecture at the Technical University in Dortmund 2015–2017. Currently, he is a postdoctoral researcher at the Chair of History of Architecture and Urban Planning at Delft University of Technology.

Jan (J. C. A.) Kolen is an archeologist and specialist of the history and heritage of cultural landscapes. He is Professor in Landscape Archeology and Cultural Heritage at Leiden University (Faculty of Archeology) and is Director of the Centre for Global Heritage and Development of Leiden University, Delft University of Technology, and Erasmus University Rotterdam.

Maurits Ertsen is Associate Professor in the Water Resources group at Delft University of Technology, The Netherlands. His books include *Improvising Planned Development on the Gezira Plain, Sudan, 1900–1980* and *Locales of Happiness: Colonial irrigation in the Netherlands East Indies and its remains, 1830–1980*. His current work focuses on how short-term actions produce long-term changes in ancient irrigation systems.

Steffen Nijhuis is Head of Landscape Architecture Research, Director European Post-master in Urbanism (EMU), and Associate Professor in the Department of Urbanism, Faculty of Architecture, and the Built Environment, Delft University of Technology. His book publications include *The Polder Atlas of the Netherlands*, *Delta Urbanism, and Polder Landscapes of the World*. He currently works on landscape-based strategies for urban development.

Gerdy Verschuure-Stuip is an architect and assistant professor of landscape architecture at Delft University of Technology. She publishes and teaches on both landscape architecture and heritage issues in the Netherlands. Research topics include country estates and their connection to the landscape, planted avenues, and military heritage.

Part I
Drinking Water

Large-scale water infrastructure above the ground such as the Pont du Gard, France, Wikimedia, Benh LIEU SONG, released under a Creative Commons Share-Alike 3.0 license

Chapter 2
Silent and Unseen: Stewardship of Water Infrastructural Heritage

Meisha Hunter Burkett

Abstract Historic waterworks, including aqueducts and sewers, are civil engineering achievements with unique heritage management challenges. Often designed to be silent and unseen, the subgrade and inaudible infrastructure that delivers water and removes waste is frequently ignored by the public unless it stops working. Although these systems can benefit from official designation as heritage, they are infrequently given this benefit, as the water and wastewater management community that is responsible for them often remains disconnected from the heritage management community. I argue for the establishment of best practice guidelines on the stewardship of historic waterworks infrastructure. Further, I examine the need to evaluate historic significance and identify character-defining features as well as to promote rehabilitation, redundancy, and sustainability of active elements. This discussion is illustrated by the cases of the Aqua Virgo and the Cloaca Maxima of Ancient Rome, the *karez* of China's Turpan Province, the Jerome Park Reservoir and Water Tunnel No. 3 in New York City, and the *mamanteo* canals of Peru. In addition, I recommend strongly that water and wastewater managers and preservationists consider adaptive use of historic waterworks infrastructure after they are decommissioned from active use. Vibrant examples of repurposing are included from around the world, including subgrade sewers, cisterns, and weirs, as well as above-grade gatehouses, wastewater treatment plants, and pumping stations. The heroic civil engineering achievements of the past were often realized amidst the need for social change. In particular, water infrastructure systems, such as gravity-fed aqueducts, delivered reliable sources of potable water to communities while also consistently reducing outbreaks of disease and fire. Preservation of this heritage poses multiple challenges, owing, in part, to the inaudibility and invisibility of these systems, the general public's limited awareness of this heritage's civil engineering significance, underrepresentation in World Heritage listings, and limited funding from government agencies that often privilege utility over aesthetics. Although it is inextricably linked to cultural, agrarian, industrial, and maritime landscapes in urban and rural communities worldwide, this infrastructural heritage can suffer unsympathetic alteration, encroachment, obsolescence, demolition, and abandonment. Nonetheless, some waterworks heritage has

Meisha Hunter Burkett (✉)
Li/Saltzman Architects, New York, USA
e-mail: meishahunter@me.com

C. Hein (ed.), *Adaptive Strategies for Water Heritage*,
https://doi.org/10.1007/978-3-030-00268-8_2

been preserved within active systems, and some decommissioned sites have successfully been repurposed. All merit closer study by modern planners, engineers, and policymakers as they work to meet water delivery and wastewater removal needs today. In regulatory and political climates where heritage is increasingly threatened, there is a need for management guidelines for historic waterworks infrastructure, in order to consistently apply best practices to analysis, decision making, and modes of treatment. Conservation must be balanced with the demand for new construction and upgrades, in a process that (1) acknowledges significance and identifies character-defining features; (2) evaluates the choice between rehabilitation and replacement for prudence and feasibility; (3) views preservation as a pragmatic and cost-effective means to extend purpose-built service life; (4) replaces sacrificial elements; and (5) repurposes decommissioned elements.

Keywords Historic water and wastewater infrastructure · Heritage management · Stewardship guidelines · Rehabilitation · Sustainability · Adaptive use

What Is Water Infrastructural Heritage?

Civil infrastructure has been described as the "skeletal and vascular framework that supports social, economic, and cultural life in modern civilization" (Sparrow 2001). Water, wastewater, and sewer systems are frequently linear by design; that is to say, a linear relationship can be found between point A (which might be a dam) and point B (which could be a distributing reservoir) as well as any interconnective components (such as conduits) located between these points. Water infrastructure includes active water supply, wastewater, drainage systems, and decommissioned components. Drinking water systems collect source water from aquifers, groundwater, lakes, and rivers; remove pollutants; and distribute reliable and abundant sources of potable water to communities. Wastewater systems collect used water and sewage, remove contaminants, and discharge clean water back into rivers and lakes for future use. These systems include subgrade components, such as pipes, mains, conduits, weirs, cisterns, and underground as well as above-grade components, such as gatehouses, aqueduct arcades, spillways, reservoirs, pressure-equalizing towers, and bridges.

Most of the drinking and wastewater infrastructure we rely on is silent and unseen, whether they be pipes that deliver water to a faucet or conduits through which sewage and wastewater flow to a plant to be treated. The concept of invisible and inaudible infrastructure is illustrated in a *National Geographic* cross section of public utility and transportation facilities located under a typical Manhattan street (as shown in Fig. 1) where we see a complex, color-coded sequence of meticulously stratified underground layers representing electrical, phone, cable, Internet, and telecommunications lines as well as water and sewer pipes and subway tunnels. While this level of complexity is most typical of densely developed metropolitan areas, subgrade infrastructure is ubiquitous in communities large and small.

Fig. 1 Axonometric cross section of subgrade infrastructure under a typical Manhattan street. Image courtesy of *National Geographic*; all rights reserved

Unfortunately, most people in developed countries do not appreciate infrastructure until it is not working. The Columbia University economist Kate Ascher has argued in her book, *The Works: Anatomy of a City*: "Rarely does a resident of any of the world's great metropolitan areas pause to consider…the myriad systems that operate around the clock to support it…even the most mundane tasks would be impossible without the far-reaching, complex, and often invisible networks of infrastructure that support them" (Ascher 2005). Simply because conduits are not readily accessible or visible, or the water rushing through them is not audible, does not mean the infrastructure is less valuable or significant than a visible bridge might be.

Historian Rosalind Williams has grappled with the multivalent challenges of confronting underground infrastructure in her book, *Notes on the Underground.* She observed that "the subterranean environment is a technological one–but it is also a mental landscape, a social terrain, and an ideological map" (Williams 1990, p. 21). Since so much of our water and wastewater infrastructure is invisible and inaudible, an "out of sight, out of mind" predicament can result that inevitably affects public consciousness, diminishes awareness of engineering and architectural significance, dampens political will, and limits funding appropriations. Admittedly, it is not easy to bridge the gap between recognizing the intrinsic engineering, scientific, aesthetic, and social values of historic waterworks and allowing this heritage to perform a didactic, public service function amid the very real challenges of managing water systems as cultural heritage. It is difficult for the general public and elected officials to value water and wastewater infrastructure that is physically concealed, publicly inaccessible, or silent by design. This problem is compounded by the general public's perception that there is a physical disconnect between underground and above-grade components within active water systems. In addition, decommissioned elements are physically disconnected from active systems. Moreover, in heritage management jurisdictions, wide disparities within the same water system can be found in which some elements may benefit from heritage designation and others do not. Above-grade elements, such as reservoirs, spillways, and bridges, are most easily observed and inventoried for historic landmark designation and heritage protection. It is the underground elements, like water tunnels, that are less frequently seen and less often inventoried or protected.

Of course, heritage protection in and of itself does not guarantee funding for, or conservation of, the protected resource. The American Society of Civil Engineers (ASCE), Congressional Budget Office, the Environmental Protection Agency (EPA), the Harvard School of Public Health, and the Water Infrastructure Network have all prepared independent studies that forecast investments needed for future maintenance and expansion of our nation's water and wastewater infrastructure. ASCE's *2017 Infrastructure Report Card* awarded the US water infrastructure a D rating and estimated that one hundred fifty billion dollars will be needed by 2025 to address deficient conditions (Fig. 2). On World Water Day 2016, the EPA announced that two hundred twenty-seven billion dollars would be needed to replace thirty to sixty million miles of lead pipe. This state of affairs is in force while, every year, actual spending falls short of capital needs by many billions of dollars.

Since ongoing maintenance of infrastructure has often not been a priority, deferred maintenance has accelerated the decline of aging systems (Schrader 2011; Milman 2016). Elected officials have resisted committing funds to infrastructure projects that cannot be seen by constituents. Roy Sparrow, a professor of public management at New York University, posited that committing government dollars to improvements invisible and inaccessible to the public is politically unpopular: "I can't recall any politician who ran successfully on the issue of infrastructure" (Sullivan 1998). Despite endemic political unpopularity and funding shortfalls—along with existential and physical challenges–the need for prudent water infrastructure management remains.

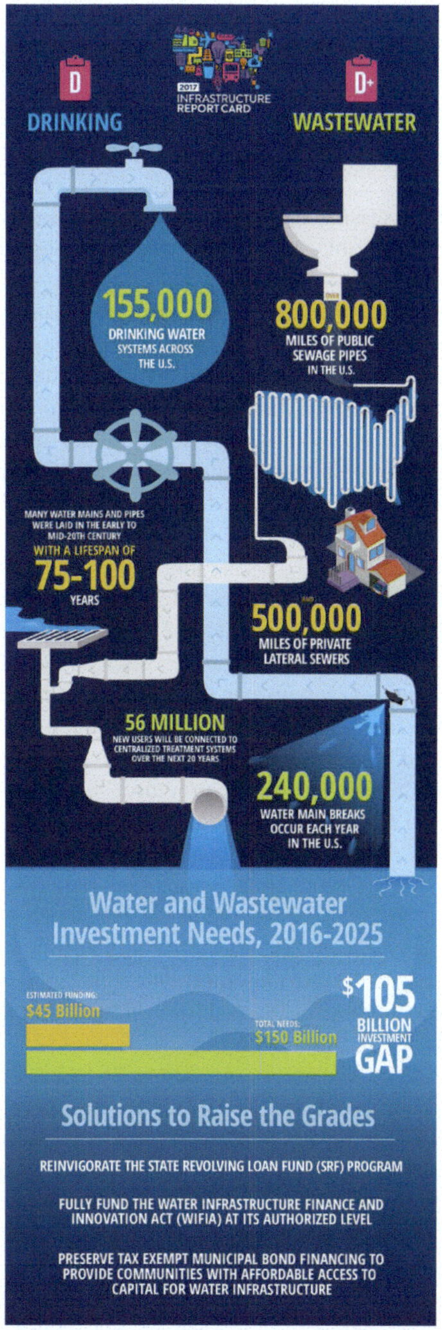

Fig. 2 2017 ASCE Infrastructure Report Card, Water and Wastewater Investment Needs 2016–2025. Image courtesy of the American Society of Civil Engineers; released under a Creative Commons Attribution-NonCommercial-NoDerivatives 4.0 International License

Historic Significance

The authors of the 2007 *Guidelines for Historic Bridge Rehabilitation and Replacement* state that, "the first step…that will underlie all planning and preliminary engineering assessments, is to understand why a bridge is historic" (Harshbarger et al. 2007, p. A-8). Understanding the significance of a historic resource is essential to assessing potential impacts to character-defining features; evaluating appropriate, feasible, and prudent treatment options to address those impacts; and making informed decisions that promote long-term stewardship.

It is instructive for twenty-first-century water and wastewater managers, among others, to consider the architectural, engineering, and social significance of ancient Roman potable water and sewer infrastructure, as well as the links between that infrastructure and politicians. Two thousand years ago, ancient Roman engineers designed aqueducts with gravity-fed delivery from source to terminus in cost-effective and militarily defensible underground conduits. Pragmatic engineering design and construction techniques, durable materials selection, and regular maintenance protocols of aqueducts and sewers were intended to offer maximum longevity and utility. Both aqueduct bridges carrying water aloft on monumental arcades and exuberant terminus fountains were often the dramatic, and purposefully visible landmarks, of a much longer, silent, and unseen underground delivery system. Likewise, visible outlets, provided by sewers such as the Cloaca Maxima, did not immediately reveal the extensive, unseen expanses of underground conduits that received and conveyed wastewater and sewage to the outlets.

The journalist Oliver Schwaner-Albright's thesis, that "the history of civilization is the history of infrastructure," underscores the fact that it was Rome's aqueducts and sewers that permitted the residential population of the ancient capital to swell to in excess of one million inhabitants–a number which was unsurpassed in the world until the nineteenth century (Schwaner-Albright 2008). Integrated infrastructure and urban planning provided the substantive building blocks for rational, high-density development (Boatwright 2000; Brunn 1991; Hodge 1992). Rome's aqueducts and sewers served the empire's capital as essential civic utilities that sustained a high quality of life, allowed for in-house plumbing, and fed numerous bathing complexes, fountains, and pools.

The Aqua Virgo (built from 25 BCE to 19 BCE), financed by the consul, statesman, general and architect Marcus Vipsanius Agrippa (63 BCE–12 BCE), exemplifies how an aqueduct could serve as a vehicle for driving political agendas (Nicolazzo 1998; Rinne, Katherine Wentworth. *Aqua Urbis Romae: The Waters of the City of Rome.* http://www.iath.virginia.edu/rome). The formal *termini* of the Virgo were the *Thermae Agrippae*, public baths, and the adjacent *Stagnum Agrippae*, an artificial pool, located roughly three blocks south of the Pantheon; all of these commissions were financed by Agrippa. The Virgo's terminus was strategically chosen: it was no accident that the aqueduct arrived at the massive bathing complex to which Augustus (63 BCE–14 CE) instituted free public access in 12 BCE and which shared the same piazza as the voting porticoes in the Campus Martius (Platner 1929; Aicher

2004). Water infrastructure in Rome was a powerful vehicle for promoting political ambitions, spreading propaganda, and obtaining political currency.

Today, the ancient Aqua Virgo (later, renamed the Acqua Vergine) and the Cloaca Maxima continue to operate in their purpose-built uses and are protected cultural heritage monuments (Frontinus 1925; Rinne, Katherine Wentworth. *Aqua Urbis Romae: The Waters of the City of Rome*. http://www.iath.virginia.edu/rome). The aqueduct and sewer heritage fall under the jurisdiction of the Ministry for Cultural Affairs (*Sopraintendenza per I Beni e l'Attivita Culturali*), which acts at both the municipal and regional levels. A ministry office, *Ufficio Vincoli*, administers right-of-way protection (*vincolo di rispetto*) for the subterranean aqueduct's substructure and superstructure. ACEA SpA (*Azienda Comunale Energia e Ambienti*) manages the aqueducts and sewers of the Municipality of Rome. Water, wastewater, and sewer operations were first entrusted to ACEA SpA in 1985 (for wastewater treatment service) and 2002 (for sewers), in accordance with the provisions of the Galli Law, national law 36/1994 (Drusiani et al. 2014). Considered together, these monuments are physical reminders of the former empire's engineering, construction, and maintenance capabilities, and vast geographical presence. While cultural heritage and water, wastewater, and sewer management fall under separate agency jurisdictions, the Cultural Affairs Ministry and the ACEA SpA water authority jointly collaborate and promote the preservation, utility, and longevity of the eternal city's water and sewer infrastructure.

Just as bridge engineers begin evaluating the significance of a historic bridge by asking "Why is this bridge historic?" water and sewer managers may ask, "Why is this aqueduct or sewer historic?". We can follow the trajectory of questioning and probe further, asking, "What is this site trying to teach us?". If ancient Roman aqueducts and sewers were a means for achieving political currency two thousand years ago, the preservation of this heritage amplifies our understanding of the actors who designed, built, and maintained these systems. Arguably, the propagandistic lessons of Ancient Rome can also motivate modern political actors to renew our infrastructure today.

Character-Defining Features

Identifying character-defining features, those visual aspects and physical features that individually and cumulatively help to define the distinctiveness of a historic resource, is a critical second step in heritage protection. The National Park Service's *Preservation Brief 17* discusses how to identify and evaluate character-defining features (Nelson 1988). These can include style, typology, structural system, materials, craftsmanship, finishes, details, spaces, site, and setting.

The ancient water canals and wells of Turpan, China, known as *karez*, merit consideration in a discussion of character-defining features and hydraulic heritage stewardship (Fig. 3). The *karez*, consisting of a two-thousand year-old, gravity-fed irrigation system dug into the Tianshan Mountains and is connected by underground tunnels and at-grade canals, have sustained grape growers, goat-herders, and silk

Fig. 3 Collecting water from a *karez* water channel in Turpan, China; ADAM DEAN/The New York Times/Redux; all rights reserved

road traders with essential glacial runoff in this harsh climate. Character-defining features of the *karez* include interconnected components such as underground channels, vertical wells, small reservoirs, and open-air canals. The underground channel is the principal watercourse of the system and links the glacial runoff with the Turpan Basin's inhabitants. Vertical wells allow removal of debris. Small reservoirs or outlets (known as dragon mouths) connect the underground channels with the open-air canals. Water empties into the small reservoirs before entering the canals.

The technology of the *karez* is not exclusive to the Turpan region; in fact, it may have been developed during the Islamic Umayyad Empire in Spain (where *karez* were known as *mayrit* or *galleria*) and subsequently exported to other countries where variants were adapted to the geology and hydrology of individual regions. While known as *karez* in Afghanistan, Central Asia, Iran, northwest China, and Pakistan, they are also known as *qanat* in Arabic, *falaj* in Oman, *foggara* in Egypt, Libya and Algeria, *ghayl* or *miyan* in Yemen, *khettara* in Morocco, and *qanat Romani* in Syria (Caponetti 2016).

The Water Resources Department of Xinjiang Uyghur Autonomous Region is responsible for maintaining the *karez*, controlling water development and usage, and regulating the construction of new wells (Hungshi Chao, personal communication, 2017). In 2006, the *karez* were designated a Major Historical and Cultural Site protected at the national level by the State Administration of Culture Heritage (SACH), the heritage administrative department of the State Council of the People's Republic of China. According to the department's Web site, the SACH's mission includes identifying and protecting relics as well as drafting and enforcing preservation legislation (http://www.english.cach.org.cn). In 2008, the *karez* of Turpan (in Xinjiang Uyghur Autonomous Region) were added to China's World Heritage watch list and have been enumerated on UNESCO's World Heritage tentative list (UNESCO).

Unfortunately, the *karez* are vanishing; the potable water itself is also disappearing from the area and its irrigation system is drying up (Jacobs 2016). Industrial scale agribusiness and petroleum companies in this arid region have met their demand for water by drilling their own wells, annually taking three million cubic meters (one hundred six million cubic feet) of water from the region's aquifer. These new wells, as well as electrical water pumps, have disrupted the historic balance of freshwater access, lowered the water table precipitously, and rendered seventeen hundred *karez* defunct since the 1950s. Cumulatively, these actions threaten the way of life in the region, because the glacial runoff cannot reach the communities who rely on it. While several hundred *karez* remain in operation, others continue to run dry each year or are contaminated by petroleum and abandoned. Small-scale farmers whose local *karez* have dried up must buy lesser quality water than the *karez* water; others fight over the limited water that is available (Sulaiman 2017; Jacobs 2016).

Several modern campaigns have attempted to renew the *karez*. In 2007, China's national government committed RMB 20M ($2.9 million) to renovate approximately 100 *karez*, which was approved by the National Development and Reform Commission of the State Council (NDRC). The NDRC's goals are enumerated in fifteen points, covering a broad spectrum of topics from the economic system's restructuring to environmental sustainability (http://www.en.ndrc.gov.cn). Two of the Commission's goals are relevant to the Tupan *karez* project: "to approve, authorize, and review key construction projects" and "to participate in the formulation of plans for ecological improvement and environmental protection" (http://www.en.ndrc.gov.cn). Unfortunately, the nationally-funded campaign used inappropriate concrete and metal materials, which compromised the *karez*' authenticity (Sulaiman 2017). In response, the Cultural Relics Department of Turpan City Government partnered with local craftspeople to restore the *karez* using traditional conservation methods. Between 2009 and 2016, SACH invested RMB 93 million to conserve one hundred sixty *karez*, which has increased the quantity of potable water (Hungshi Chao, personal communication, 2017). Despite this success, some Chinese officials have expressed resignation that the extinction of the *karez* system is inevitable (Jacobs 2016).

Guidelines

As the Turpan *karez* example demonstrates, even if historic water infrastructure is recognized and protected, tremendous obstacles to heritage stewardship can still exist. Heritage protection without preservation management can result in damage and decay to historic monuments, including waterworks infrastructure. Moreover, heritage protection in and of itself does not ensure the preservation of natural resources (as in potable water, for example) or the sustainability of purpose-built use. While designation represents a significant part of the preservation planning process, formal protection in and of itself does not address the fundamentally difficult topic of managing change to historic infrastructure systems in active use or planning for the adaptive use of decommissioned historic elements. When undertakings that affect historic waterworks are planned, and no preservation goals are set for the project, inappropriate treatment decisions and alteration campaigns may have far-reaching adverse consequences.

Since many of the world's water supply and drainage systems include aging and modern components, water managers must decide what to retain, what to replace, and what to decommission as they plan for the future sustainability of these systems. At present, there are no clear, consistent guidelines to assist water and wastewater managers in assessing the feasibility and prudence of rehabilitation versus replacement or in addressing the post-9/11 necessity of redundancy. Furthermore, there are no industry standards for guiding water managers in the potential adaptive use of water infrastructure that has already been, or is anticipated to be, decommissioned. While multiple reports on water and wastewater systems governance, historic bridge management, industrial heritage conservation, and infrastructure recycling are informative, few specifically tackle the issue of managing change to active and decommissioned historic waterworks infrastructure. Arguably, the most relevant document, authored jointly by water management and preservation agencies, is the *Memorandum of Agreement Concerning the Continued Operation of Jerome Park Reservoir* (MOA). The Commissioner of the Office of Parks, Recreation, and Historic Preservation of New York state, the New York State Historic Preservation Officer, and the New York City Department of Environmental Protection (DEP) Commissioner executed the Memorandum regarding the operation of the reservoir as a historic facility in active use for New York City's water supply. The MOA is a rare example of a management tool that balances water management needs with preservation concerns.

The Jerome Park Reservoir, which holds five hundred million gallons of water, was originally constructed between 1895 and 1906 at the southern confluence of the Old and New Croton aqueducts; it currently functions as a balancing reservoir for incoming water collected from the upstate Croton watershed, which supplies 10 percent of the City's water. In 1994, the New York State Historic Preservation Office (SHPO) determined it to be eligible for listing on the New York State and National Registers of Historic Places. After sixteen years of discussions between SHPO and DEP about the consequences of listing an operable reservoir within an active water supply system as a historic resource, DEP agreed to list the site on the State and National Registers in 2000. That year, New York State Office of Parks, Recreation, and Historic Preservation, SHPO, and DEP executed the Memorandum of Agreement for the Jerome Park Reservoir. Since 2015, water from the reservoir has been pumped upstream to the $3.2 billion Croton Filtration Plant, after which it is distributed to neighborhoods in the Bronx and Manhattan.

The Memorandum defines future DEP undertakings that will either (1) trigger or (2) be exempt from SHPO review and approval. Undertakings that do not require SHPO review will exempt the DEP from certain regulatory compliance requirements.

As enumerated in the MOA, exempt undertakings include essential repair and in-kind replacement activities, emergency response, selective demolition, upgrades, and new construction that will not compromise character-defining features of substructure and superstructure components.

As a working agreement between the three agencies over the Reservoir's future water and heritage governance, the MOA evidences a spirit of cooperation and pragmatism; it recognizes the DEP's current and future needs to "rehabilitate, modify, upgrade, expand and/or make additions" (MOA, p. 1) to the Reservoir complex while respecting the engineering and architectural significance of the site. After almost twenty years of the MOA's execution, engineers at the DEP's Bureau of Engineering, Design, and Construction involved with projects at the Reservoir and the Croton system had a positive assessment of the SHPO coordination during the design and permitting process, and of their working relationship in general. The DEP's Assistant Commissioner for Intergovernmental Affairs stated that, "Based on the MOA, the [engineers] would provide information on the work scope for these projects which would be endorsed by SHPO, and then follow the MOA to provide notifications on the progress of their design and construction work. The [engineers] followed guidelines for record keeping for these projects and provided this information as requested by SHPO. They did not regard the experience as burdensome in any way" (Mario Bruno, personal communication, 2017). The DEP's optimistic appraisal of the MOA and SHPO, expressed by the water management agency itself, could encourage water managers elsewhere to be more willing to allow heritage designation and inter-agency collaborative management agreements on the future stewardship of historic yet operable water infrastructure. The Jerome Park MOA and other examples of collaboration between water managers and heritage stewards are notable; however, there is a tremendous and persistent need for partnerships of this kind to become more commonplace in the future.

Rehabilitation and Redundancy

Investing in rehabilitation of historic water infrastructure can be a cost-effective means of tapping extant but deteriorated networks to meet modern needs, at the same time, extending service life and utility, as cases in Rome and Lima illustrate. With sufficient investment, water managers can acknowledge the vast portfolio of extant water infrastructure that exists, as well as the enormous investment in embodied energy that this heritage represents.

In the post-war decade between 1955 and 1965, Rome's ACEA SpA surveyed and repurposed the Aqua Virgo's twenty-one kilometer (thirteen-mile), gravity-fed, masonry-lined conduit or *speccus* (shown in Fig. 4). In the years following World War II, unregulated residential construction above the aqueduct–and the associated absence of wastewater utilities–resulted in soil contamination, environmental degradation, infiltration, and structural damage which degraded the aqueduct's water to non-potable status (ACEA SpA personal communication, 2007). In 1975, ACEA

Fig. 4 Inspection of tunnels (left) and section illustrating concrete intubation (*intubazione*) reinforcement in deteriorated segments of the *speccus* (right). Image courtesy of Roma Sotteranea and ACEA SpA; released under a Creative Commons Attribution-NonCommercial-NoDerivatives 4.0 International License

was entrusted with the task of expanding service to illegal settlements in peri-urban areas (the *borgate*), an important contribution to Rome's sustainable development (Lobina 2005). Instead of abandoning or demolishing the aqueduct, ACEA structurally reinforced failed segments of the aqueduct's subterranean tunnels by intubating them with cement tubes (a technique known as *intubazione*) and repurposed the aqueduct's water for ornamental fountain displays and irrigation of public parks (Nicolazzo 1998).

In Peru, scientists are working to restore the ancient, but largely abandoned, canal systems called *mamanteo* that are thought to pre-date the Incan empire, in an effort to address Lima's increasing water shortage (Collyns 2015; Schatz 2015). In the *mamanteo* system, the drainage system funnels hydrological runoff from highland streams into a mountain, where the water percolates through cracks and aquifers to emerge in springs and natural reservoirs (shown in Fig. 5). The *mamanteo* canals are protected heritage. According to the country's General Law on Cultural Heritage (Article 6 of Law 28296, 2004), the canals are archaeological heritage owned and managed by the Peruvian State. More specifically, the *mamanteo* canals are defined as archaeological landscape heritage and are protected both for their original purpose-built use and for their integrity (Brunke, personal communication, 2017). Peru's

Fig. 5 Detail view of ancient *mamanteo* canal, part of a network restoration project being undertaken by SUNASS in Peru. Image courtesy of Leah Bremer/The Natural Capital Project; released under a Creative Commons Attribution-NonCommercial-NoDerivatives 4.0 International License

constitution asserts national sovereignty over all sub-grade and above-grade heritage sites of any size, type, or age within the national territory (Silverman, 2006, p. 60).

Re-grouting, de-silting, and performing other maintenance on the canals to restore *mamanteo* to service may lead to weeks or months of long delays in the delivery of water to end users. However, allowing more time for wet-season moisture to percolate through the mountainous highlands (at a height of three thousand five hundred feet above sea level) is advantageous, particularly during the dry season when water supplies are scarce. By slowing water delivery during the dry season, user communities will receive reliable supplies of water for longer periods of time. Significantly, the proposed restoration project is also cheaper to realize than building a new canal.

The project, which began in March 2015, is funded by the national water regulator SUNASS as part of an investment of 7 million Peruvian Soles ($21.5 million) in green infrastructure. The work to be conducted also includes recuperating highland wetlands, adapting to climate change, and performing disaster risk mitigation. The campaign is anticipated to increase Lima's water rate by a least one cubic meter per second (Collyns 2015).

In addition to rehabilitation, sustainability upgrades to water infrastructure such as system-wide redundancy and resilient materials can address maintenance issues. They can also address security requirements. In New York, the completion of City Tunnel No. 3 will provide redundancy to the water system and cause the Department of Environmental Protection to temporarily close two existing tunnels (built in 1917 and 1936, respectively) for inspections and repairs for the first time (http://www.PlaNYC-NYC.gov). The Tunnel, the largest capital construction project in New York

City's history at a cost of $5 billion, is to be completed in the 2020s. In a post 9/11 era, actively protecting against terror attacks and integrating redundancy is essential for life-sustaining utilities such as water delivery and sewers.

Adaptive Use

Although many civic water infrastructure systems built by earlier generations were envisioned to provide many years of useful service, they are not, *sui generis*, immortal. At a certain point, water managers may indeed choose replacement and decommissioning over rehabilitation. In some cases, entire water and sewer infrastructure systems can be decommissioned from their purpose-built use.

Once they no longer serve their original use, decommissioned water infrastructure components can be adaptively, and creatively, reused, sometimes becoming iconic cultural heritage tourism destinations. Hydraulic infrastructural heritage poses unique challenges for repurposing. Not least among these are finding appropriate new users and programs, performing code-compliance upgrades, and balancing new interventions with respect for historic character-defining features. Several examples of adaptive use of below-grade components, such as cisterns, sewers, and waste weirs, illustrate how decommissioned heritage can be vibrant magnets for cultural heritage tourism.

The Basilica Cistern (dating from 542 CE), in Istanbul's Eminonu district, originally supplied water for the city during the reign of the Emperor Justinian I. It features three hundred thirty-six marble columns and cloister vaults, as well as a firebrick perimeter wall pointed with hydraulic mortar. After cleaning the site and constructing a walking platform above the water level, the city repurposed the cistern as an exhibition gallery. In Houston, the Cistern at Buffalo Bayou Park, (constructed in 1926 and decommissioned in 2007) was repurposed according to the design of architecture and engineering firm page and reopened to the public in 2016. The forest of columns in the underground space visually recalls the Basilica Cistern in Istanbul. Similarly, the Musée des Égouts de Paris occupies a segment of a decommissioned nineteenth-century sewer. This underground museum has informative exhibits about the city's urban planning history and the essential role played by its sewer system. In Vienna, where the chase scene in the 1948 suspense film, The Third Man (starring Orson Welles) was shot on location in the city's underground sewers, today, visitors can tour a segment of the sewers, known as the "city beneath the city." There, they can learn about Vienna's wastewater and sewer system (shown in Fig. 6). Finally, there is the case of the Egyptian revival style Ossining Waste Weir, built from 1881 to 1886 in Westchester County, New York and finally decommissioned in 1965. A weir chambers, formerly used to ventilate and divert water from the aqueduct when the water level was too high, have been turned into a publicly accessible tour site along the Old Croton Aqueduct State Historic Park.

Some instances of superstructure adaptive use include museums and performing arts venues in former gatehouses, wastewater treatment plants, and pumping stations.

Fig. 6 At grade view of the Vienna River adjustments (*Wienflußregulierung*), including a partial view of the subterranean passage. Image courtesy of Axel Föhl; released under a Creative Commons Attribution-NonCommercial-NoDerivatives 4.0 International License

The former Croton Aqueduct 135th Street Gatehouse in New York (engineered by Frederick S. Cook, 1887) sat empty for decades before Olhausen Dubois Architects and WASA Studio redesigned the building in 2007 as an award-winning performing arts venue known as Harlem Stage. Similarly, the Museum in Altes Wasserwerk, Berlin (built by Henry Gill and Richard Schultze from 1889 to 1893), once one of the largest and most modern waterworks in Europe, was repurposed as a water museum (Berliner Wasserbetriebe 1993). The former Wastewater Treatment Plant in Prague-Bubeneč (the work of William Henry Lindley and constructed from 1900 to 1906) offers tourists an opportunity to learn about Prague's sewerage and wastewater treatment history as well as its industrial architecture. Decommissioned in 1967, the plant previously served seven hundred thousand residents and is the oldest preserved facility of its kind in Europe. The former Ryhope Pumping Station (the work of Thomas Hawksley, 1868) previously supplied water to the Sunderland area of Great Britain. Decommissioned in 1967, the site was repurposed as the Ryhope Engines Museum and is a Grade II listed building (Föhl 1985). The former Metropolitan Waterworks' high-service pumping station in Boston (originated by Arthur Vinal, 1885–1887 and expanded by Edmund Wheelwright, 1897–98) was recently opened as the Waterworks Museum. The World Heritage listed Woudagemaal (built 1920) is the only steam-powered pumping station in the world that is still in use and features interactive exhibits (Fig. 7). The Fairmount Water Works in Philadelphia (built from 1799–1801; redesigned by John Davis and Frederick Graff in 1811) were repurposed

Fig. 7 View of the World Heritage listed steam pumping station (Woudagemaal) in the city of Lemmer, the Netherlands. Image courtesy of Daniel Hartog; all rights reserved

by Philadelphia's Department of Environmental Protection as an interactive water and wastewater history museum.

Future Currents

Our Common Future, the report of the United Nations World Commission on Environment and Development, defines sustainability as, "Meeting the needs of the present without compromising the ability of future generations to meet their own needs" (United Nations 1987). The examples described in this chapter have demonstrated that historic, gravity-fed delivery infrastructure can be a cost-effective means for meeting modern day consumption and irrigation needs; that investing in the rehabilitation of historic water infrastructure can extend its service life and utility; that upgrades to historic water infrastructure systems are viable; and that decommissioned waterworks can be adaptively and successfully used for cultural heritage tourism. As individual water management authorities and preservation agencies have enjoyed limited successes in managing change to their own water and sewer infrastructure, why can we not aggregate these examples, evaluate them, and discern best practices that can guide decision making in the future? The stewardship of water infrastructural heritage can be part of a cohesive, civic planning framework for new construction, preservation, and repurposing that ultimately promotes consistent decision-making, economic viability, environmental sustainability, infrastructural resiliency, and cultural authenticity.

Acknowledgements The author is indebted to Lisa Ackerman (Interim Chief Executive Officer, World Monuments Fund); Albert Appleton (Senior Fellow, The Cooper Union Institute for Sustainable Design); Mario Bruno (Assistant Commissioner, Intergovernmental Affairs, NYC DEP); Andrew Burdick (Architect and Design Strategist, Macquarie Group); Lorenzo Caponetti (Casa Caponetti); Hunghsi Chao (Program Associate for Chinese Projects, World Monuments Fund); Axel Föhl (retired State Officer for the Preservation of Industrial Monuments in Germany); Juan Pablo de la Puente Brunke (former Viceministro de Patrimonio Cultural e Industrias Culturales, Peru); Andrew Potts, Esq. (retired Partner, Nixon Peabody); Christina Rasmussen (US Army Corps of Engineers, New York District); Professor Ethel Sheffer (Columbia University GSAPP); my husband, Morgan, and my family for their encouragement and support. The author's participation in the Water and Heritage for the Future Conference was funded by US/ICOMOS; however, the views and opinions expressed herein do not necessarily reflect those of US/ICOMOS or any other person or entity.

References

ACEA SpA (2007) Personal communication with the author

Aicher P (2004) Rome alive: a source guide to the ancient city, vol 1. Bolchazy-Carducci

American Society of Civil Engineers Infrastructure Report Card: http://www.infrastructurereportcard.org/

Ascher K (2005) The works: anatomy of a city. The Penguin Press, New York

Basilica Cistern. Istanbul. http://www.ibb.gov.tr/SITES/KS/EN-US/1-PLACES-TO-GO/MUSEUM/Pages/basilica-cistern.aspx

Berliner Wasserbetriebe (1993) Wasserwerk Friedrichshagen, Industriedenkmal und Versorgungsbetrieb 1893–1993. Berlin/München

Boatwright MT (2000) Hadrian and the cities of the Roman Empire. Princeton University Press, Princeton

Brunke P (2017) Personal communication with the author

Brunn C (1991) The water supply of the ancient Rome: a study of roman imperial administration. The Finish Society of Sciences and Letters, Helsinki, Rome

Bruno M (2017) Personal communication with the author

Caponetti L (2016) The invisible landscape: The Etruscan Cuniculi of Tuscania as a determinant of present-day landscape and a valuable tool for sustainable water management. In: Christian, Stump D (eds) The Oxford handbook of historical ecology and applied archaeology, Isendahl

Chao H (2017) Personal communication with the author.

Collyns D (2015) Peru harnesses ancient canal system to tackle Lima water shortage. The Guardian, June 22

Department of Environmental Protection. City Tunnel Number 3 website: http://www.nyc.gov/html/dep/html/dep_projects/cp_city_water_tunnel3.shtml

Drittmanntour. Vienna. https://www.drittemanntour.at/de/der-wiener-kanal/geschichte

Drusiani R et al (2014) Evolution of sanitation services in the city of Rome between urban development and environmental quality. In: Angelakis A (ed) Evolution of sanitation and wastewater technologies through the centuries. IWA Publishing, London

Environmental Protection Agency Water Infrastructure and Resiliency Finance Center. https://www.epa.gov/waterfinancecenter

Fairmount Water Works Museum. http://fairmountwaterworks.org

Föhl A (1985) Die Industriegeschichte des Wassers, Düsseldorf

Frontinus SJ (1925) De Aqueductu Urbis Romae (Trans. C. E. Bennett). Loeb Classical Library, Boston
Harlem Stage. New York. http://harlemstage.org/about
Harshbarger JP et al. (Lichtenstein Consulting Engineers, Inc. in association with Parsons Brinckerhoff Quade & Douglas, Inc.) (2007) Guidelines for historic bridge rehabilitation and replacement, as requested by the American Association of State Highway and Transportation Officials, Standing Committee on the Environment
Hodge T (1992) Roman aqueducts and water supply. Duckworth, London. http://en.ndrc.gov.cn
Institute for Advanced Technology in the Humanities. University of Virginia. Rome Reborn: Thermae Agrippae. http://www.Archive1.village.virginia.edu
Jacobs A (2016) In a parched corner of Xinjiang. Ancient water tunnels are running dry. The New York Times, Sept 21
Lobina E et al (2005) D36: WaterTime case study—Rome, Italy. WaterTime. A research project supported by the European Commission
Metropolitan Waterworks' High Service Pumping Station. Boston. http://waterworksmuseum.org
Milman O (2016) White House enlists private sector for $4bn upgrade to water infrastructure. The Guardian, March 22
Musée des Égouts. Paris. http://equipement.paris.fr/musee-des-egouts-5059 and https://www.egouts.tenebres.eu/visite.php
National Development and Reform Commission. People's Republic of China
Nelson LH (1988) Architectural character—identifying the visual aspects of historic buildings as an aid to preserving their character. Preservation Brief 17. National Park Service, September
Nicolazzo V (1998) L'Acqua Vergine: E Suoi Acquedotti e le Fontane di Roma Attraverso I Secoli. ACEA, Roma
Ossining Waste Weir. Westchester, New York. http://aqueduct.org/events/old-croton-aqueduct-ossining-weir-tour
People's Republic of China, State Administration on Cultural Heritage. http://english.gov.cn/state_council/2014/10/06/content_281474992893400.htm
PlaNYC website. www.nyc.gov/html/planyc/downloads/.../full_report_2007
Platner SB (1929) A topographical dictionary of ancient rome. Thomas Ashby, Oxford (rev)
Rinne KW Aqua Urbis Romae: the waters of the city of Rome. http://www.iath.virginia.edu/rome
Ryhope Engines Museum, United Kingdom. http://www.ryhopeengines.org.uk/index.html
Schatz J (2015) Lima taps ancient Andean canals to help fight its water shortage. PRI, June 22
Schrader G (2011) Developing a new approach to water infrastructure. From the Yale program on strategies for the future of conservation as presented in the "How can conservation organizations help development go where it should?" Report of the 2011 Berkeley Workshop
Schwaner-Albright O (2008) Design challenge: fend off disaster. The New York Sun, July 22
Silverman H (2006) Cultural resource management and heritage stewardship in Peru. CRM J Heritage Stewardship 3(2) Summer
Sparrow R (2001) The evolving knowledge and skill requirements of America's civil infrastructure managers. Public Works Management and Policy, April 1
Sulaiman E (2017) Exploitation puts ancient well system at risk in Uyghur Region. Tuidang, May 7 (used by permission of Radio Free Asia)
Sullivan J (1998) Study puts a high price on fixing a crumbling city. The New York Times, Aug 25
The Cistern at Buffalo Bayou Park, Houston. https://buffalobayou.org/visit/destination/the-cistern/
UNESCO Tentative World Heritage Lists. http://whc.unesco.org/en/tentativelists/5347/
United Nations (1987) Report of the world commission on environment and development: our common future
Wastewater Treatment Plant in Prague-Bubeneč. http://www.prague.eu/en/object/places/623/old-wastewater-treatment-plant-in-bubenec-stara-cistirna-odpadnich-vod-v-bubenci?back=1
Williams R (1990) Notes on the underground: an essay on technology, society, and the imagination. MIT Press, Cambridge

Meisha Hunter Burkett is senior preservationist at Li/Saltzman Architects and a recipient of the NEA Rome Prize (Historic Preservation) from the American Academy in Rome. She has contributed to *Water Industry as World Heritage* (The International Committee for the Conservation of Industrial Heritage, 2018), *L'aque, le Pietri e Bronzi le Fontane Monumentali* (Commune di Roma, 2012), *City Secrets: Rome* (Fang, Duff Kahn, 2011), and *Sustainability and Historic Preservation: Toward a Holistic View* (University of Delaware Press, 2011).

An ice house, *yakhchal*, like this one in Meybod, was used to store ice in underground chambers throughout the year, Wikimedia, Ggia, released under a Creative Commons Share-Alike 3.0 license

Chapter 3
The Qanat System: A Reflection on the Heritage of the Extraction of Hidden Waters

Negar Sanaan Bensi

Abstract This chapter focuses on a traditional Iranian water infrastructure, the *qanat* system, a technical solution to the problem of accessing water for irrigation and urbanization that has shaped the landscape and organized the territory. The qanat was the basis for habitation, construction, and prosperity (*abadani*). It is also a key to understanding the culture and civilization of the Iranian Plateau and has evolved as a form of cultural heritage. Therefore, preserving this heritage is more than protecting an old technology. Rather, it requires a deeper understanding of the territory in which the qanat operated and of its limitations and possibilities. Discussing a historical work, *The Extraction of Hidden Waters* by Muhammad Al-Karaji (953–1029), this chapter explicates the multivalent role of the qanat system in managing and organizing the territory, society, life, and culture in the Iranian Plateau; this multiplicity of aspects and scales shapes its consideration of qanats' heritage today.

Keywords Iranian plateau · Territory · Qanat · Al-Karaji · Heritage

Introduction

The current severe scarcity of water in Iran and its surrounding region has necessitated a revision of the ways in which water resources are used, managed, and consumed, especially, as they appear in urbanization, planning, culture, and everyday life. Since the middle of the twentieth century, forces influencing the administration of water resources in Iran have included: excessive use of new pumping technologies and the abandonment of traditional ways of dealing with water, too many new dams, easy access to water for an increasing population and production sector, ongoing regional conflicts, and political pressures and instability. Indeed, these forces led to the current

---

Georg Gerster and the world of his photography.

N. Sanaan Bensi (✉)
Architecture Department, Delft University of Technology, Delft, The Netherlands
e-mail: N.SanaanBensi@tudelft.nl

C. Hein (ed.), *Adaptive Strategies for Water Heritage*,
https://doi.org/10.1007/978-3-030-00268-8_3

water crisis, shifting the culture of water management and, more largely, how we inhabit our territories.

Controversially, several scholars have recently argued that a disorganized bureaucracy is to blame. For example, Kaveh Madani, a civil and environmental engineer and water resources expert who has himself worked in Iran's government, said that "Iran's water problems are not due to a lack of access to technology or technical expertise, as some decision makers claim. Indeed, Iran is suffering from disintegrated decision making and problem solving by knowledgeable experts who act independently" (Madani 2014). Isa Kalantari, the recently appointed head of Iran's Department of Environment, points to the Iranian Force Ministry's dam-making and vast transition projects as the main reasons for today's water problems. If continued, he observes, this mismanagement will destroy the country's major water resources within the next 15 years. In fact, the technocratic approach has ignored the significance of the vital and scarce resources of water in human-made, cultural landscapes and the heritage values of water infrastructure like qanat. This amnesia has led to an undesirable governance structure and to a practice of ignoring the effects of development on the environment in favor of short-term benefits (Madani 2014; Ardakanian 2005).

The current water crisis forced scholars in various disciplines to rethink and reconsider ways of managing and inhabiting territories. In this regard, it is helpful to look critically at water heritage: its infrastructure, its spatial and physical dimensions, its culture, and the complexity inherent in its managerial systems. This chapter focuses on the qanat system as cultural heritage. The qanat is a traditional technique for accessing and managing underground water; it shaped the Iranian landscape and was the basis for inhabitation, construction, and prosperity (*abadani*). If we look at the qanat system purely as a technique, it might appear to be an obsolete system that is unable to keep up with the increasing speed of urbanization. But the qanat system is more than a technical solution to the problem of accessing water for irrigation and urbanization; it is a key to understanding culture and civilization in the Iranian Plateau. The heritage value of the qanat system is related to the preservation of an ancient technology; more importantly, it can provide a deeper understanding of the limitations and possibilities of the territory and its cultural, social, political, and legal complexity.

This transition to a holistic understanding of the qanat is well expressed in the work of Henri Goblot, a French geologist. Goblot went to Iran in 1940 at the invitation of the king at the time to prepare a plan for the underground water sources and give advice on modern irrigation technologies. At the beginning of his stay, he spoke against using qanats, as he considered them an obsolete technique; he suggested instead using deep wells and new pumping technologies (Goblot 1992). However, he came to appreciate the complexity of the qanat system and, slowly became so fascinated by qanat that he stayed for 20 years in Iran and wrote a book on the subject. His extensive and valuable treatise, *Les qanats: une technique d'acquisition de l'eau* (1992), is an important contribution to the study of the qanat system—its history, technology, and geography. Goblot understood the context in which the qanat could operate, where it was historically placed, and what it does.

As a technique, qanats have been used throughout history in different parts of the world—in the Middle East, around the Mediterranean, in the Americas, and even in west China (Yazdi and Khaneiki 2017). Although its exact history and origin is disputable, it is believed to be an Iranian invention and has been in use in Iran for thousands of years.

Modern studies of the qanat system first focused on the functional outline of its technology and the justification for continuing to use it to manage water (Jomehpour 2009). In the last three decades, scholars have also addressed its history and importance in different regions (Hu et al. 2012; Lofrano et al. 2013; Martínez-Santos and Martínez-Alfaro 2014) and expressed concern about its preservation and protection (Yazdi and Khaneiki 2017; Ghasemi et al. 2013; Jomepour 2009; Harandi and de Vries 2014). More recently, scholars have proposed the importance of developing a wider cultural understanding of the qanat system and the role of its heritage values in the planning and design of cities (Agah 2014; Safi Nezhad 2017; Beheshti and Najar Najafi 2017).

The chapter is structured to transit in scale and content: between landscape and object, between surface and depth, and between territory and thing: the qanat. According to Elizabeth Grosz, "the thing is the precondition of the living and the human, their means of survival, and the consequence or product of life and its practical needs. The thing is the point of intersection of space and time, the locus of the temporal narrowing and spatial localization that constitutes specificity or singularity." (Grosz 2001) To discuss the qanat as a thing, then, means to consider it in a reciprocal relation with the territory which accommodated it. In this way, the text pays simultaneous attention to a multiplicity of levels.

In order to understand the fundamental role of the qanat system and the management of the underground hidden waters in the formation of life in Iranian Plateau, it is important to state that the city—and any settlement within the Iranian Plateau—had to establish an intimate relationship with its territory, landscape, and geography. This study opens its analysis with a look at that relationship, reflecting on the archeological works and aerial photography of Erich Schmidt. Next, the specific geographical and geological conditions of the Iranian Plateau are briefly described, with a focus on the qanat system. Drawing on Muhammad Al-Karaji's historical treatise, *The Extraction of Hidden Waters*, further reflection follows, on the multiple aspects of the qanat which should be considered when its heritage value and related preservation strategies are discussed. Heritage considerations can also shape responses to the current and future water crisis, not necessarily to solve them, rather to avoid short-sighted solutions and, as Madani suggests, in that way, to focus on and identify the causes of particular problems by formulating better questions (Madani 2014).

Flights Over Territory: Between Landscape and Object

One of the most fascinating studies on the relationship between the city and its territory was carried out by Erich Schmidt, a German scholar who used aerial photography, drawing, and mapmaking to conduct archeological surveys in Iran in the

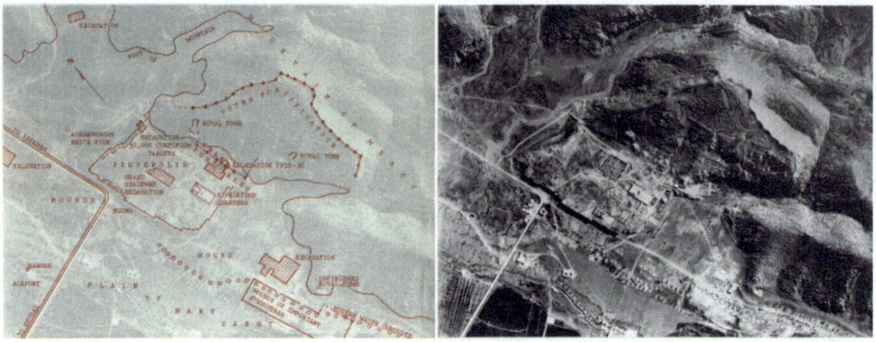

Fig. 1 Aerial photograph at the right side shows a vertical view from the site of Persepolis; "In the form of a plastic map the system of fortifications with its towers, the complex of palaces, and the physical environs of the royal site are spread below." Aerial photograph by Erich Schmidt, 27 September, 1935. The left map is the edited version of the vertical view photograph in which the relation between elements—i.e., the topography, fortification, water infrastructure (qanat), road and other elements—are highlighted on a tracing paper. *Source* Schmidt, Erich Friedrich. *Flights over ancient cities of Iran* 1940. PLATE 2: Vertical View of the site of Persepolis; April 20, 1936; 7:39 A.M.; altitude, 2440 meters; 1/100 se.; no filter. Courtesy of the Oriental Institute of the University of Chicago; released under a Creative Commons Attribution-NonCommercial-NoDerivatives 4.0 International License

1930s. *Flights over Ancient Cities of Iran* (1940) is his most well-known and fascinating publication; it combines descriptions of his flights over a large landscape (noting the relation between topography, roads, water infrastructure, agriculture, and settlements) with description of architectural elements and archeological objects found in each excavation site. This attention to the concurrence of issues is present in Schmidt's literal act of tracing in the drawings of excavation sites over his aerial photographs (shown in Fig. 1).

Schmidt's text constantly shifts between scales and between history, geography, and anthropology. Schmidt was genuinely interested in more than what he saw; he looked into the relations between things. He wished to know something about the ways that the territory was inhabited in a general sense. "Schmidt's publication elevated archeological surveying to a new art form", employing aerial surveying, photography, and prose to look at archeological excavations and discern relationships between objects, landscape, as well as history in their "enigmatic story" (Sobti and Hosseini 2016).

Schmidt also relied on some important geographical histories of the Medieval Islamic period–such as Le Strange's *Lands of the Eastern Caliphate*, Al-Muqaddasi's (945–1005 CE) *Ahsan al-Taqasim* and al-Tabari's (838–923 CE) *Tarikh al-Tabari* (Sobti and Hosseini 2016)—to render the multiplicity and interconnection of settlements with their immediate landscape and the larger region. For example, he captures the historical settlement of Band-e Amir within its territory in relation to the larger irrigation system and the *band*, or, dam, built over the Kur River in the Fars region in Iran during the Buyid dynasty (934-1055) (Fig. 2). Schmidt offers Le Strange's observation that: "Ten great water wheels raised the water to such a high level that

Fig. 2 Historical settlement of band-e Amir. *Source* Schmidt, Erich Friedrich. *Flights over ancient cities of Iran* 1940. PLATE 43: Bustam (Bastam), A place of pilgrimage in Northern Iran; September 23, 1935; 7:27 A.M.; altitude, 610 meters; 1/140 se.; yellow filter. Courtesy of the Oriental Institute of the University of Chicago; released under a Creative Commons Attribution-NonCommercial-NoDerivatives 4.0 International License

three hundred villages could be supplied with the water, so precious in Iran; and at each wheel there was a flour mill" (Schmidt 1940). His aerial photograph of Turut (or Torud), a small village south of the Salt Desert of Damghan, captures a clear border between inhabited territory and the desert that absorbs the hidden underground water from the heart of the mountains. In the caption of this photograph (Fig. 3), Schmidt writes: "Turut is one of these places at the rim of the infernal *kavir* [desert]. It is actually wedged between two salt deserts, the *kavir* of Damghan and the 'great *kavir*'. It owes its existence to a threat of sweet water breaking from low range of hills, which partly separates the lifeless plain" (Schmidt 1940).

This multiplicity of scales and relation between architecture and territory in Schmidt's writings and photographs goes beyond a purely formal and visual presentation of landscape. He depicts a complex relation that is present at various social, cultural, economic, and legal levels. Schmidt was well aware of the delicate geographical and geological condition of the Iranian Plateau and its determinant role in the formation of its civilization, culture, and built environment.

B. TURUT, A TOWN IN THE SALT DESERT

On a tongue of solid land, south of the Salt Desert of Damghan, lies this strange little town.

Fig. 3 Turut (Torud), a town in the Salt Desert. The rows of *qanats* are visible on the top of the photograph. *Source* Schmidt, Erich Friedrich. *Flights over ancient cities of Iran* 1940. PLATE 46 B. September 23, 1935; 8:10 A.M.; altitude, 610 meters; 1/140 se.; orange filter. Courtesy of the Oriental Institute of the University of Chicago; released under a Creative Commons Attribution-NonCommercial-NoDerivatives 4.0 International License

An Insight into the Geographical and Geological Condition of the Iranian Plateau: Between Surface and Depth

Geologically, the Iranian Plateau was formed and shaped by the uplifting and folding of three giant plates: the Arabian Plate, the Eurasian Plate, and the Indian Plate pressing against each other (Harrison 1968; Fisher 1968) (Fig. 4). As a result, a series of chain mountains, primarily in the north and southwest, enclose the interior basin of Iran. This central basin includes the central deserts, *Dasht-e-Lut* and *Dasht-e Kavir*, currently two of the driest and hottest spots in the world.

The uniqueness of Iran is the result of the encounter of two extreme geographical conditions, as historian Heinz Gaube points out. On one hand, the country is part of the Eurasian mountain belt that "runs from the Iberian Peninsula, through the Alps, the Balkans, the Carpathians, the Taurus and Pontus, and the Iranian highlands rims of the Elburz and Zagros" (Gaube 2008). On the other hand, Iran is also part of "the arid belt of the Old World which stretches from the Sahara in the west across

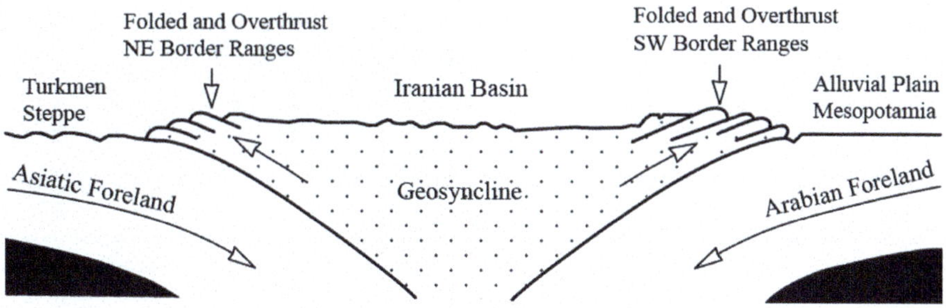

Fig. 4 Diagram of the Iranian Orogen by Harrison, J.V. *Source* Harrison, J.V. 1968. Geology. In the Land of Iran, vol. 1 of The Cambridge history of Iran. ed. W.B. Fisher. Cambridge: Cambridge University Press: 183. [© Cambridge University Press, 1968]; This figure cannot be reproduced, shared, altered, or exploited commercially in any way without the permission of Cambridge University Press, as it is copyrighted material and therefore not subject to the allowances permitted by a CC-BY license; all rights reserved

the Arabian Peninsula and the Iranian Plateau to the deserts of Central Asia in the east" (Gaube 2008). Although the main part of the plateau is covered by dryland, a considerable amount of water is stored between the layers of folded ground and faults. Hence, most Iranian settlements have been sited in a piedmont zone between mountain and desert in large alluvial fans (Kheirabadi 2000).

Water is the source and precondition of life. With few permanent surface water sources in this harsh environment, subsurface water reservoirs made human settlement possible here. People devised an underground water infrastructure, the qanat system, to access fresh water, using it both for irrigating farms and gardens and for drinking. This was territorial management operating on spatial, physical, legal, social, and cultural levels.

On the Qanat: Between Territory and Thing

The qanat system is a subterranean infrastructure that gave access to the hidden water at the foot of mountains. Vertical shafts of successively increasing depth were connected by a horizontal underground tunnel (*dehliz*), which directed the water from subterranean water sources down a slight slope to gardens, farms, and settlements (see the images in Fig. 5). However, the vertical shafts are not themselves used for accessing water. Rather, they are important, first, for calculating the right direction and proper angle of slope for the horizontal tunnel. Later, during excavation, they are used for faster removal of dug materials, as well as for regulating pressure and oxygen for workers. Finally, after the completion of construction, these shafts are used for maintenance, providing workers an easier way to get at the underground horizontal tunnels for repair.

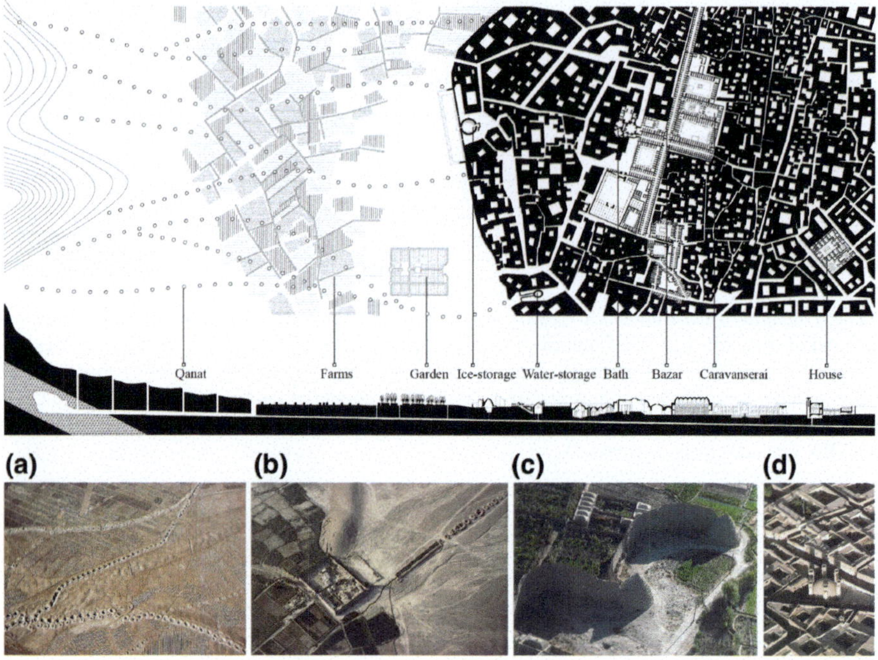

Fig. 5 **a** Drawing of a hypothetical territorial section showing how the *qanat* system and various architectural constructs such *bagh* (garden), *ab-anbars* (water storage), and *yakhchal* (ice storage) provided water for settlements and farms. It established a system that supported life in the territory. Courtesy of Negar Sanaan Bensi and Raul Forsoni; released under a Creative Commons Attribution-NonCommercial-NoDerivatives 4.0 International License. **b** Aerial photograph by Georg Gerster, 1976–1978 at Yazd, Iran. "A palimpsest of qanats crisscross an area of recently formed fields." The aerial photograph depicts different types of qanat structures: rows of several qanat with a single row of shafts at various distances and qanat with double shafts to facilitate cleaning and maintenance. It also exhibits faint traces of much older qanats, now replaced by new ones. *Source* Mousavi. Ali et al., *Ancient Iran from the air* 2012: 26; released under a Creative Commons Attribution-NonCommercial-NoDerivatives 4.0 International License. **c** Aerial photograph by Georg Gerster, 1976-78 at Golpayegan, Iran. The Point at which "a qanat comes to surface, having carried pristine water deep beneath bare, uncultivated *dasht* [desert]". It also depicts "an initial walled stretch of the surface stream, three separate irrigation channels then serve to direct the water toward the adjacent fields. *Source* Mousavi. Ali et al., *Ancient Iran from the air* 2012: 176; released under a Creative Commons Attribution-NonCommercial-NoDerivatives 4.0 International License. **d** Aerial photograph by Georg Gerster, 1976-78 at Yazd, Iran. Depiction of an *ab-anbar*, water storage feature, ringed by six tall wind-catchers designed to capture wind from various directions so as to ventilate and cool the water. At its foundation, a perimeter wall protects the base of the dome of the water feature and after occasional heavy downpour guides run-off into the cistern. *Source* Mousavi. Ali et al., *Ancient Iran from the air,* 2012: 168; released under a Creative Commons Attribution-NonCommercial-NoDerivatives 4.0 International License. **e** Aerial photograph by Georg Gerster, 1976-78. at Sirjan, Iran. The picture shows a *yakhchal*, an ice making and storing structure. Two tall curved walls shade shallow over pools located in front "where, in winter, qanat water was turned into ice, chopped out, and then stored in deep pits beneath the stepped, domed ice houses". *Source* Mousavi, Ali et al., *Ancient Iran from the air* 2012: 163; released under a Creative Commons Attribution-NonCommercial-NoDerivatives 4.0 International License

The location of the motherwell, distance between vertical shafts, proper slope for the horizontal tunnel, and length of the qanat were carefully examined during the process of construction. According to Goblot, the slope of the underground tunnel must be around 0.5/1000 to minimize the erosion of the inner surface of the horizontal tunnel. This is an important technical issue, as slope directly influences the point at which water emerges on the surface called *farhang*. In Farsi *farhang*, means ‘culture’. Both *culture* and *farhang* in their etymological roots refer to the cultivation (of land), the training or improvement of the faculties, care (of a monument), upkeep and the cultivation of the acquaintance (of a person). (*Oxford Latin dictionary*, 1968, “cultūra” and *Dehkhoda Dictiony* under فرهنگ). Such an implicit interwoven relationship between *farhang* (or culture) and water thus is not by chance (Najar Najafi 2015). The farhang affects the exact placement of a settlement, farm, or garden. Hence, the slope of the ground has a direct impact on how the water is managed. Where the slope is steep, water moves in linear fashion; it is then difficult to direct the water through the network of water systems. Gardens and distributing ponds or *moqassem*, directed the water into networks toward the city. In sum, qanats formed a low-tech system based on the interaction of gravity, minimal waste of materials, sources of water, and labor. Knowledge of its construction was embedded in daily life and taught, in practice, from master to pupil. It was also a long-term measure which created a constant flow of water within a harsh, dry environment.

This system can be presented in a hypothetical territorial cross-section through a city, including various architectural elements: garden, *bagh*, for distribution; water storage, *ab- anbar;* or ice storage:—or *yakhchal*—literally the ice-hole—along with the settlements and farms that were where the water was destined to be consumed (Fig. 5a). The importance of the qanat system lay in the way in which it organized territory through a process of revealing water to surface and, thereby, providing the possibility of habitation.

The multivalence and complexity of this territorial management become clear when we refer to treatises that accumulated and organized the knowledge of water management and the construction of qanats historically. These treatises are important objects of cultural heritage, though they have not yet been thoroughly investigated. Inbat al-miyah al-khafiya, by the Iranian Muslim mathematician and engineer of the late tenth and early eleventh centuries Abu Bakr Muhammad Al-Karaji, is one of the oldest surviving manuals on hydraulic and water supplies. In his treatise, Al-Karaji demonstrates his familiarity with contemporary concepts and principles of the hydrological cycle, classification of soils, description of aquifers, and the search for groundwater (Abattouy 2014). He observed these processes and practices during his lifetime and organized them scientifically and according to his educational background in the form of a treatise.

The title itself contains a word that is worthy of some relevant comment. According to Mohammed Abattouy, “The inbat, like istikhraj, means precisely ‘extraction’ of underground water, to show what is hidden and to extract ground and hidden waters for economic and social benefit. The term may have to do with the mathematical concept of istinbat, meaning ‘deduction by reasoning’. If this is verified, the link between the two is natural, as Karaji would have coined the term in the aftermath of his long experience as a mathematician" (Abattouy 2014). Karaji himself wrote, in the introduction, "I know no profession more beneficial than the extraction of hidden water, as it gladdens the earth and makes life possible." (Al-Karaji 1994, 22). He further explains that he wrote this book after his return from Baghdad to the iqlim of Jabal–mountain region–on the heights full of farms and villages, with cold, clean, and tasty water from several sources.

Opening first with a contextualizing of hydrology within "the larger field of natural science and geology" (Abattouy 2014), the book compiles a vast knowledge on the modes of finding and treating water as well as the methods and instruments used in building and preserving qanats (shown in the images in Fig. 6). The first few chapters introduce water as a system that is fluid: transmitting a liquid substance from one place to another and able to change its form in the water cycle. This understanding of water as a system is fundamental to the development of qanat as a non-pervasive technique that works with water itself as a renewable resource. Qanats facilitate water transit from source to destination, even as, since the level of the underground water table controls the flow, the qanat system does not drain an aquifer (English 1998).

Al-Karaji describes how to find underground water sources. This survey ranges from the geological conditions that influence where the water is naturally stored underground; the topography, type, and color of soil and stone; vegetation in the proximity; and other indicators. In the next chapter, he describes kinds of water, their sources, use, and ways of cleaning and purifying them. He, then, addresses miscellaneous topics related to water, including how earthquakes influence water underground or how to determine whether a well still contains water. The relationship he brings forth between the location of a qanat and earthquake lines is particularly important, as many Iranian cities are characterized by intertwining of a seismic landscape with arid and semi-arid areas—to form landscapes of risk and resilience (Ibrion et al. 2014).

In the last chapters of his volume, Al-Karaji addresses the legal dimensions of water management, as a techno-scientific discipline closely related to society and economy (1994, 67–87). In these chapters, he establishes a legal argument according to a discourse on the various Islamic schools of law, *fìgh*, concerning the construction, characteristics, and use of qanats. Ownership of qanats is complex, as the surface of the land and the underlying section which holds the infrastructure can belong to different people. In other words, he unfolds a legal condition where various ownership regimes overlay and superimpose each other. According to Karaji, these legal principles differ if the hydro-infrastructure is a qanat, well, or canal and if they are constructed for the use of a farm, animal, or a city. What is the legal and ownership status when a qanat constructed by someone for the use of a city or settlement passes underneath the farm belonging to someone else? And what happens if the qanat system needs to be repaired or even accessed? These are the more fundamental questions. Moreover, different territorial ownership principles apply to different geological conditions, for example, if a qanat or well is constructed in a soft porous soil or muddy or hard soil.

In this way, geometry measures not only the surface or the landscape, as its etymology literally says (*geo,* earth + *metry*, measure); it also encompasses the whole complexity and contradiction of the act of measuring. Indeed, measurement is tightly related to the management of land. As James Corner puts it, measurement is not an autonomous and instrumental part of human technologies, used to dominate and control the world; it is a way "to reveal culturally significant forms of order" (Corner and Alex 2000). He adds, critically, that modern technologies, steeped in efficiency and utility, characterize only an instrumental aspect of measurement or of the means taken to secure a particular end. This explication of the concept is close

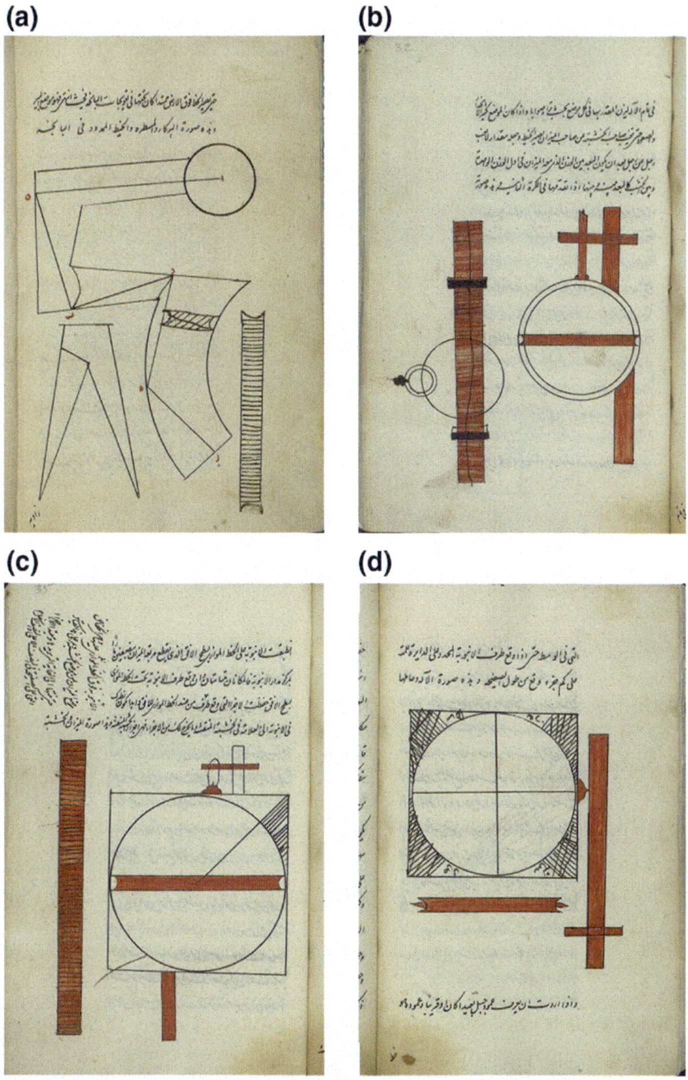

◂**Fig. 6** Pages from the book *The Excavation of Hidden Water* by Al-Karaji. **a** (indicated in the archive as page 45v) is related to a chapter discussing the tools and methods of digging the underground tunnel connecting the wells in the construction of a qanat. The drawing is a diagrammatic plan of this underground tunnel. It shows how to project the underground tunnel on the surface of the ground to be able to site and dig the well. The drawing shows a compass and a ruler. **b** (indicated in the archive as page 32r) and **c** (indicated in the archive as page 35r) are illustrations of leveling tools. These figures are included in a chapter which explains the leveling tools and techniques for surveying the ground. This surveying process for understanding the topography and height differences along the qanat's path is crucial in the construction process. **d** (indicated in the archive as page 37v) is an illustration of a tool for measurement of topographical distances and it is part of a chapter which describes the ways and tools for defining the height of a mountain, the distance between a mountain and where an observer stands and as well the distance between the summit of different mountains. *Source* Title: [Inbāṭ al-miyāh al-khafīyah] [manuscript]. Kitāb Inbāṭ al-miyāh al-khafīyah; Origin: [Iraq or Persia], A.H. 1084 (1674); Physical Description: 49 leaves: paper, col. ill.; 193 × 125 (138 × 70) mm. bound to 193 × 128 mm. Manuscript Location: LJS 399, Rare Book & Manuscript Library University of Pennsylvania LJS 399; Available online: http://hdl.library.upenn.edu/1017/d/medren/9948256513503681. Courtesy of Lawrence J. Schoenberg Collection of Manuscripts, Kislak Center for Special Collections Rare Books and Manuscript, University of Pennsylvania; released under a Creative Commons Attribution-NonCommercial-NoDerivatives 4.0 International License

to the Heideggerian understanding of technology as a way of revealing and grasping the world (Heidegger 1977). And qanat as infrastructure stem from this geometrical knowledge of the management and measurement of the territory.

One of the important legal notions relating to the qanat system is the law of *harim or* borders, which has given an owner protection over a territory surrounding a qanat and prohibited the sinking of new motherwells or any other construction within a defined distance of an existing qanat (English 1968). The *harim* law does not itself fix this distance. As it is clear in Al-Karaji's discussion, that distance was contingent, depending, as ownership did, on the type of hydro-infrastructure, type of qanat, its length and depth, the geological condition of the soil, the location of the motherwell and so on. Thus, the importance of *harim* is more than just the protection of the right of ownership; it defines frames for protecting underground water resources.

These are important issues to be considered for territorial and city planning—and even for a city's architecture. Historically, they are the dynamics by which the water network and the qanat system directly influenced the formation and structure of the city and the position of agricultural and farm lands (Bonine 1979; Aboutorabian et al. 2009). The lack of attention to such important issues as these in the last decades in Iran has caused it serious problems. It seems curious, for example, in Tehran, the urbanization of the northern areas during and after the 1960s often interrupted the practice and application of *harim* to older qanats: the deep foundations of many buildings blocked the flow of underground waters (Khosravi et al. 2016). One of the most problematic issues in Iran's current water crisis is the improper distribution of industry. It does not, for example, make sense to put steel mills in the central desert, which has limited water resources. Komeil Sohani, a young Iranian researcher, has recently directed a documentary that shows that the water crisis in Iran is rooted in decades of mismanagement and misunderstanding of the actualities of Iran's position.

The qanat system was also involved with the social structures of local communities and cities (English 1966; Bonine 1982). For instance, wealthier families often had private and direct access to water networks, or they lived closer to the outlet or upstream of it. At the same time, the rest of a city's inhabitants used the collective and public services available in each *mahalleh*—which might include that neighborhood's water storage. Sharing and maintaining the common basic resources helped form close social relationships within a neighborhood. Furthermore, the main public institutions such as the bazaar, mosques, baths, and schools usually had direct access to freshwater. This means that the economic structure of the city had a close relationship with the qanats and the water network in general. In short, a multiplicity of interwoven factors has shaped how people construct and maintain qanats throughout the history of their use, creating specific ways of living and inhabiting the territory.

Conclusion

The word territory derives from the Latin word *territorium*, meaning the land around a town, and terra, the dryland and earth, as opposed to sea, proximate to "specifically a Roman or a provincial city" (Online Oxford Dictionaries, s.v. "territory" and "terrain"). The term *terra* itself is derived from the Indo-European word *ters,* dry or to dry and points to the very early exercise of city making, and the possibilities of life, within the wet marshy land in Europe. In Avestan *ters* became *taršna*, or thirst, in modern Iranian it transformed into *tešneh*, or thirsty (Nourai 1999). Both signal the lack of water that substance whose absence jeopardizes life itself in the arid plateau of Iran. In this sense, *thirst* both literally and metaphorically has constituted the urge and motive for people to search for ways and means to inhabit the dryland, and to extract its hidden waters. It propelled people to seek, to imagine, and to create, and to reverse obstructions to new possibilities throughout their history. The qanat system is a clear example of this life-giving dynamic. To build a qanat, one must have a holistic understanding of where and when one lives. Goblot considered the qanat system to be one of the most imaginative engineering works in the history of humankind. In fact, the extraction of hidden waters and construction of oases and gardens, the life-filled *Pardis* or Paradise within a landscape that is a seeming tabula rasa, could not occur without the presence of rich imagination.

During World War II many qanats were replaced by deep wells, introduced to Iran by the Allied Armies. The premises of this introduction were multiple that: it was more efficient to get water from a deep well than from the qanat; one could get more water faster; deep wells were separate matter from topography and soil conditions; and wells could be built without using local materials or local labor (English 1997). In essence, deep wells involve fewer constraints. However, for these very reasons, they might tax water sources and empty aquifers. Notably, the shift from qanats to deep wells represented a complete shift in the management of the territory; it was a shift from a collective legal act to exploitation according to individual interest.

As several scholars have noted, it is important to strategically conserve qanats as cultural memories. In "The Qanats in Yazd: The Dilemmas of Sustainability and

Conservation" Bharne introduces the qanats of Yazd and their decline, then puts forth several solutions: He, for example, suggests using the existing system of qanats as a gray water recycling network or renovating them as sites and vessels of cultural memory. Although these proposals seem feasible and interesting on first view, they focus mainly on an immediate alternative functional question rather than on the larger cultural and historical value of the qanat.

Some old qanats are in use still today in Iran, principally in villages and smaller towns. Qasabeh Qanat, in Gonabad, a city in the province of Razavi Khorasan at the northeast of the current political boundary of Iran is one of the oldest remaining qanat. Even though most are abandoned–and only a few are available for tourist visits–they remain complex systems that influence and are being influenced by various scales and levels of geography, society, administration, economy, and culture. The heritage value of systems like qanat consists exactly in the interwoven understanding that they can give us, not in their expression as, purely, a technical solution for the exploitation of water, nor as museums.

Perhaps the restrictions and principles embedded in systems like the qanat are necessary for the delicate geography of the Iranian territory. And perhaps, it is exactly these principles and restrictions that could establish a base for new managerial structures and decision making, planning, construction, and production. This would certainly be a difficult shift, necessitating a cultural understanding of the issue of water as well as the involvement of all levels of sectors, communities, and citizens.

Last, but not least, we preserve water infrastructure such as the qanat system, and must do so, not only for the safe-keeping of a past heritage, but in order to take the right steps toward the future. We require a deep understanding of the logic and complexity of such a system, of what exactly it was doing as a system and in its parts, and how it was doing it, the limits it was creating, and how it made a certain way of life possible and sustained a territory. Such an understanding aids us to move into the future as we harbor and live with the past.

References

Abattouy M (2014) Muhammad Al-Karaji: A mathematician engineer from the early 11th Century. Muslim Heritage, Foundation for science, technology and civilization. http://www.muslimheritage.com/article/muhammad-al-karaji. Accessed 10 Nov 2014

Aboutorabian B, Najafi EN, Latifian T (2009) The Inter-relation between the physical structure of the historical city of Shiraz with water. Paygah-e siasat gozari va miras-e farhangi-e bafte tarikhri-e Shiraz, Shiraz

Agah M (2014) The influence of qanat on the culture of Iran. Pesteh 100:56–59

Al-Karaji, Abū Bakr ibn Muḥammad ibn al Ḥusayn (1994) Estekhraj-e Ab-ha-ye Penhan or The Extraction of Hidden Waters. Arabic Edition: Al-Karaji Muhammad (trans. Hoseyn Xadiv Jam). Iranian National Commission for UNESCO PRESS, Tehran

Al-Karajī, Abū Bakr ibn Muḥammad ibn al Ḥusayn. 1674. Kitāb Inbāt al-miyāh al-khafīyah. Rare Book & Manuscript Library University of Pennsylvania LJS 399. Courtesy of Lawrence J. Schoenberg Collection. http://dla.library.upenn.edu/dla/medren/pageturn.html?id=MEDREN_9948256513503681. Accessed 1 Nov 2014

Ardakanian R (2005) Overview of Water Management in Iran. In: Water Conservation, Reuse and Recycling: Proceedings of an Iranian-American Workshop. National Research Council, The National Academies Press, DC, pp 18–36

Beheshti SM, Najar Najafi E (2017) Teshnegi Servat-e Mast! or Thirst is our wealth!. Andisheh iranshahr, 24–25(1396): 8–14

Bharne V (2011) The qanats in Yazd: the dilemmas of sustainability and conservation. In: Sang L(ed) Aesthetics of sustainable architecture, 010 Publishers, Rotterdam, pp 152–167

Bonine ME (1979) The morphogenesis of Iranian cities. Ann Assoc Am Geogr 69(2):208–224

Bonine ME (1982) From qanat to kort: traditional irrigation terminology and practices in central Iran. IRAN. British Inst Persian Stud 20:145–159

Calder N (1999) Feqh. In: EncyclopaediaIranica, vol IX, Fasc. 5, pp 504–511. Online edition available at: http://www.iranicaonline.org/articles/feqh

Corner J, Alex SM (2000) Taking measures across the American landscape. New Haven, Conn. [u.a.]: Yale Univ. Pr

English PW (1966) City and village in Iran: settlement and economy in the Kirman basin. University of Wisconsin Press, Madison

English PW (1968) The origin and spread of qanats in the old world. Proc Am Philos Soc 112(3):170–181

English PW (1998) Qanats and lifeworlds in Iranian plateau villages. Yale F&ES Bulletin 103:187–205

Fisher WB (1968) Physical geography. In: Fisher WB The land of Iran, vol 1 of The Cambridge history of Iran, pp 3–110. Cambridge University Press, Cambridge

Gaube H (2008) Iranian cities. In: The city in the Islamic world, Vol. 2, ed. Salma K. Jayyusi, Renata Holod, Attilio Petruccioli, Andre Raymond. Leiden: Brill

Gerster G, Sachs M (2008) Paradise lost, persia from above. Phaidon, London

Ghasemi H, Valipour E, Morad DH (2013) Application of traditional architectural structure as sustainable approach to mitigation of shortage water supply in desert regions. Acad J Sci 2(1):125–132

Goblot H (1992) Les qanats: une technique d'acquisition de l'eau, Trans. A. Sarvqad Moqadam and M.H. Papoli Yazdi. Astan-e Qods, Mashhad

Grosz EA (2001) Architecture from the outside: essays on virtual and real space. Cambridge, MIT Press, Mass

Harandi MF, de Vries MJ (2014) An appraisal of the qualifying role of hydraulic heritage systems: a case study of qanats in central Iran. Water Sci Technol: Water Supply 14(6):1124–1132

Harrison JV (1968) Geology. In: Fisher WB (ed) The Land of Iran, vol. 1 of the cambridge history of Iran. Cambridge University Press, Cambridge, pp 111–185

Heidegger M (1977) The question concerning technology, and other essays. Harper & Row, New York

Hu W, Zhang J, Liu Y (2012) The qanats of Xinjiang: historical development, characteristics and modern implications for environmental protection. J Arid Land 4(2):211–220

Ibrion M, Lein H, Mokhtari M, Nadim F (2014) At the crossroad of nature and culture in Iran: the landscapes of risk and resilience of seismic space. Int Proc Econ Dev Res 71:38–44

Jomehpour M (2009) Qanat irrigation systems as important and ingenious agricultural heritage: case study of the qanats of Kashan, Iran. Int J Environ Stud 66(3): 297–315

Kamalvand, Sara. Hydocity, http://www.hydrocity.ca/

Kheirabadi M (2000) Iranian cities: formation and development. Syracuse University Press, Syracuse

Khosravi H, Djalali A, Riedijk M, Marullo F, Frausto S (2016) Tehran life within walls. Hatje Cantz Verlag GmbH, Berlin

Lofrano G, Carotenuto M, Maffettone R, Todaro P, Sammataro S, Kalavrouziotis IK (2013) Water collection and distribution systems in the Palermo plain during the middle ages. Water 5(4):1662–1676

Lombard M (2009) The golden age of Islam, Trans. Joan Spencer. Princeton, Markus Wiener Publishers, NJ

Madani K (2014) Water management in Iran: What is causing the looming crisis. J Environ Stud Sci 4(4):315–328

Martínez-Santos P, Martínez-Alfaro PE (2014) A priori mapping of historical water-supply galleries based on archive records and sparse material remains. An application to the Amaniel qanat (Madrid, Spain). J Cult Heritage, 656–664

Matricide (2015) Film: Komeil Sohani. https://vimeo.com/173872128

Meidaan (2017) Interview with Isa Kalantari, 13 September 2017. http://meidaan.com/archive/41558. Accessed 10 Oct 2017

Mousavi A, Stronach D, Gerster G, Beazley E, Harverson M, Huff D, Sumner WM, Wilkinson TJ (2012) Ancient Iran from the air. Philipp Von Zabern Verlag, Darmstadt

Muqaddasi. Muḥammad ibn Aḥmad Shams al-Dīn (1981) Ahsan al-taqasim fi ma'rifat al-aqalim. Arabic Edition: Muqaddasi (trans, Alinaghi Monzavi). Tehran: Sherkat-e Mo'alefan va Motarjeman-e Iran

Najar Najafi, E (2015) To build and to be built: revisiting the relation between culture and architecture. Ph.D. diss., Shahid Beheshti University of Tehran

Nourai A (1999) An etymological dictionary of persian, english and other indo-european languages. Xlibris, Bloomington

ONLINE ETYMOLOGY DICTIONARY, s.v. "terrain". http://www.etymonline.com/index.php?allowed_in_frame=0&search=terrain&searchmode=none. Accessed 16 Sep 2014

Online Oxford Dictionaries, s.v. "territory" and "terrain". http://www.oxforddictionaries.com/definition/english/territory. Accessed 16 Sep 2014; Online Etymology Dictionary, s.v. "terrain". http://www.etymonline.com/index.php?allowed_in_frame=0&search=terrain&searchmode=none. Accessed 16 Sep 2014

Online Oxford Dictionaries, s.v. "territory". http://www.oxforddictionaries.com/definition/english/territory. Accessed 16 Sep 2014

PERSIAN DICTIONARY DEHKHODA, s.v. "قلمرو", http://www.etymonline.com/index.php?l=t&p=12. Accessed 16 Sep 2014

Qanat Qasabeh Gonabad Documentry. Saeed Tavakolifar (2015) https://www.youtube.com/watch?v=OHN0fRiKdNo

Safi Nezhad J (2017) Kariz dar Iran va shive-haye sonati-ye bahregiri az an or Kariz in Iran and the traditional ways of its using. Pouyeh Mehr Ishraq Cultural Institute, Tehran

Schmidt EF (1940) Flights over ancient cities of Iran. Ill: University of Chicago Press, Chicago

Sobti MP, Hosseini S (2016) Re-examining 'Persianate Civitas': networked urbanities and suburban hinterlands in erich schmidt's flights. In: Mohammad G (ed) The historiography of Persian architecture. Routledge, London, New York

Yazdi AAS, Khaneiki ML (2017) Qanat knowledge construction and maintenance. Springer, Dordrecht

Negar Sanaan Bensi is a researcher and tutor at the Architecture Department, Delft University of Technology. Her research offers a critical reading on the architecture of the "bazaar" as part of exploring complex spatial and territorial mechanisms in contemporary cities and their histories. She is part of the editorial board of the Footprint Journal and an editor of issue Footprint #23: The Architecture of Logistics.

The UNESCO World Heritage Site Yuanyang rice terraces in Yunnan, China, Wikimedia, Yulin Jia, released under a Creative Commons Attribution 2.0 Generic license

Chapter 4
Studying Ancient Water Management in Monte Albán, Mexico, to Solve Water Issues, Improve Urban Living, and Protect Heritage in the Present

Araceli Rojas and Nahuel Beccan Dávila

Abstract In the past, between the sixth century BCE and the ninth century CE, the Zapotec people managed rainwater in Monte Albán, in the state of Oaxaca, south of Mexico, through terraces, canals, dams, and wells. Water was a keystone of their worldview and ritual practice. Today, this knowledge is in oblivion. Rapid but irregular urbanization threatens the remnants of these water control systems, still hidden on this archaeological hill site. Our ongoing interdisciplinary project, Parque Monte Albán, has centered on the water that flows down the hill and offers new strategies to increase the value and quality of water by revitalizing and redesigning ancient hydraulic technology. In the short and long term, our solutions can restore the natural environment, improve the quality of urban living, and help protect archaeological heritage.

Keywords Water management · Urban planning · Monte Albán · Heritage · Archaeology

Introduction

Monte Albán is an archaeological site in the south of Mexico, posed at the top of one of the three hills in the vicinity of Oaxaca City (shown in Fig. 1). The three-hundred-meter-long ceremonial plaza; dozens of temples, palaces, and residential areas; hundreds of tombs and artificial terraces visible on the slopes below; splendid ceramic effigy vessels; fine engraved stelae; and masterly worked jewelry have earned its status as a UNESCO World Heritage Site. In the last few decades, after twelve hundred years of decline, Monte Albán has faced a second round of rapid and uncontrolled

A. Rojas (✉)
University of Warsaw, Warsaw, Poland
e-mail: araceli.rojas@wn.uw.edu.pl

N. Beccan Dávila
Beccan Davila Urbanismo, Leiden, The Netherlands

C. Hein (ed.), *Adaptive Strategies for Water Heritage*,
https://doi.org/10.1007/978-3-030-00268-8_4

urbanization, marked by a new informal city springing up primarily on its southern and eastern sides. Fences have been installed to protect the archaeological heritage of Monte Albán, but they have not stopped people from building houses and reaching the top. Lamentably, this unplanned urbanization has been unable to guarantee even basic necessities to the new settlers.

The natural environment is also under serious threat. Deforestation is causing unstoppable erosion. Without trees, rainwater is not held in their root systems, and instead runoff on the hill is exponentially increased, raising the number of flood events. New houses, streets, and infrastructure obstruct the runoff water that could otherwise flow into streams or come out of springs as it moves to reach Río Atoyac, the main river at the base of the valley. Moreover, the new inhabitants contaminate the water with their waste. Furthermore, climate change has caused this area to suffer ever heavier storms and harsher droughts. Residents who have settled at the very top of the hill are more vulnerable than others, since urban services cannot reach them or are simply unavailable because of the greater concentration of archaeological remains.

Our work explores solutions to these problems of water, heritage, and urban life in Monte Albán. Our aim is to recover the flow of water and improve the quality of life of the new informal city by using archaeological knowledge, cultural values, and social participation. We build on earlier projects conducted from 2012 to 2014 that studied the condition of the Río Atoyac and confirmed its deterioration–caused mainly by pollution and disposed sewage from Oaxaca City (Visión del Río Atoyac, Oaxaca 2012; Duurzaam Stedelijk Waterbeheer Oaxaca 2013–2014). These studies proposed several pathways to end contamination of the water. However, short terms in office of local authorities and lack of funding have obliged us to focus on smaller subprojects and segments of the larger project designed to rehabilitate the river. We find it useful to draw on the concept of acupuncture to develop strategies for more effective water management, envisioning that small pinpoints of improvement can help heal the overall body of the landscape and its arteries of water. One of the areas where it is viable to apply this healing acupuncture is the hill of Monte Albán and the streams and waters running from it to the Río Atoyac.

The stunning archaeological setting and great cultural and historical value of Monte Albán invited us to look closely at and learn from the ways that the site's ancient inhabitants, the Zapotec people, controlled rain and groundwater. This past culture of respect for water, rain, and lightning inspired us to bring back, or to remember, that culture through archaeological research. Thus, our design project *Parque Monte Albán* was born.

From the moment of contact, Europeans began the massive destruction and alteration of indigenous water control systems. Revealing how these systems worked, and in some cases critically assessing them, as well as recovering this knowledge and improving it with modern urban and hydrological expertise and technology, allows us to address current water problems. Solutions for contemporary challenges can be found in local knowledge rather than in importing foreign technologies, or, even worse, by imposing them–as it happened five hundred years ago. We investigated the technology Zapotec people used to manage water along with the ways in which

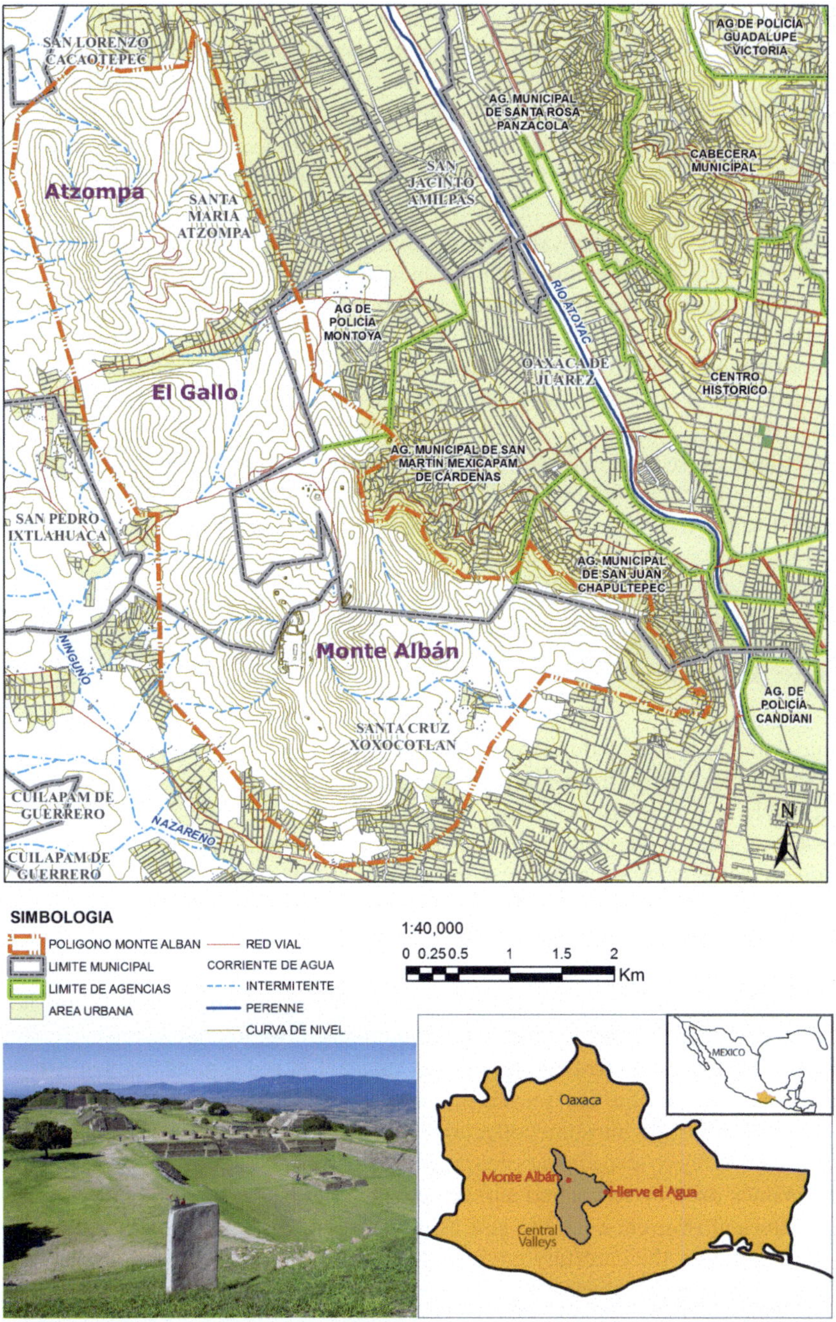

Fig. 1 Monte Albán, perimeter of protection, Río Atoyac, Oaxaca City, and municipal division (maps courtesy of Beccan Davila Urbanismo, photograph courtesy of S. Kerkhof; released under a Creative Commons—Attribution-Noncommercial-NoDerivatives 4.0 International license)

they valued water culturally. While conducting our work, we proposed a range of subprojects to improve urban life, actively involving local people to develop strategies so that awareness of the vulnerable archaeological heritage of Monte Albán and the crucial role of water there could be developed.

By continuing to both investigate the history of control of the water at the site and to learn more about both effective and failing strategies, we can elicit ideas and actions to prevent floods and provide freshwater to current occupants of the site. Moreover, there is an urgency to change how these problems are viewed. Rather than considering the new houses on the ancient site as an illegal invasion, we have to consider the deep social problem–which is rooted in economic marginalization and social rejection–behind it. This project conceives of long-term solutions as well as, more importantly, solutions that can be sustained and maintained by society: by those who live daily with the current problems of water, heritage destruction, and social exclusion. Rather than perceiving local people as the "enemies" that cause these issues, we see them as the best allies possible in the effort to protect the site. The same goes for rain: We should stop seeing it as the hostile party. Rainwater is good fortune today, as it was for the ancient Zapotecs, a blessing and benefit to society.

Monte Albán and Water

Monte Albán was founded in five hundred BCE, the earliest urban center in Middle America. At its height, from 450 to 600 CE, it had more than two thousand terraces inhabited by nearly thirty thousand people (Blanton 1978, p. 30). With temple pyramids surrounding its large plaza, Monte Albán is also considered the most important religious center of its time. It sat at the core of the Mesoamerican cultural region and was visited by thousands of people from the surrounding areas and far away alike (Joyce 2004). Monte Albán's main deity, for its more than one thousand years of active use, was Cociyo, the deity of water and rain whose name literally means lightning in the Zapotec language (Cruz 1946). Cociyo was responsible for breaking the clouds with his "serpent of fire"—lightning—and bringing rain to the agricultural communities that depended on him to grow maize. Cociyo is most frequently represented by the shape of effigy vessels that are found interred in tombs. Most of these are now in private and museum collections (Marcus and Flannery 1994; Sellen 2002; Urcid 2009). Cociyo also appears engraved on stelae, as his name was apparently one of the names used by rulers and authorities (Urcid 2001; Jansen and Oudijk 1998). This historical reverence for the god of thunder and lightning at Monte Alban, in combination with the carefully engineered systems, stands in stark contrast to the disrespectful attitudes toward water today, a time when people pollute and obstruct watercourses.

Although the religious side of water seems well understood, scholars have overlooked the water control systems at Monte Albán. A prominent interpretation is that the site is a hill with no water. Scholars who have studied the evolution of the site, from its foundation to its decline, have suggested that Monte Albán was created as a neutral "disembedded" capital for a confederation which used the site to defend itself from enemies and launch military attacks to control the region (Blanton et al. 1999, p. 65). This new center was strategically established on marginal, sinuous, and unproductive soil, without natural sources of water, and at a distance of more than four kilometers from the main river, Río Atoyac (Blanton 1978, p. 36; Flannery and Marcus 1983, p. 81; Wolf 1959, p. 97). This image of a "hill with no water" contradicts what our present project has shown on maps and has already been asserted: the presence of multiple streams and runoffs. In 2014, we were able to prove that ancient water control technology existed in some of these streams. Indeed, the Monte Albán site accords with the Mesoamerican concept of *altepetl*: the sacred hill that is a source of sustenance and, in particular, of water (Rojas 2017).

Preliminary Results of the Zapotec Water Technology Study

In 2014, we conducted a brief non-intrusive survey of Monte Albán (Rojas 2015). The aim was twofold: to locate water and to seek out ancient water management technology. We first mapped some of the streams from their origin and traced them into the city, inspecting conditions (we gauged level of pollution, the presence of obstruction, and, where redirection had taken place, whether water still ran) along with potential for being included in the improvement project. We targeted densely urban areas including Mexicapam, Chapultepec, and Montoya on the eastern side of Monte Albán, since financial support for this project came from Oaxaca City, of which these areas were a part. To identify the ancient technology that had managed the watercourse downhill, we surveyed several streams for evidence of human intervention.

Using satellite imagery and geographic information systems (GIS), we verified the existence of four streams on the eastern side (streams A, B, C, and D) which we had previously identified as originating at the top of Monte Albán, continuing down into the city (where they are obstructed by garbage, houses, and streets), and eventually flowing into concrete outfall structures (shown in Fig. 2). Stream C originates at the top of the hill, where it is fed by rainwater. Water then percolates into the ground, surfacing farther down, beneath the main road (at the Carretera Monte Albán), in the form of a spring. Here, neighbors have built a well, sheltering it from inclement weather with a cage; water comes through a hose out of the well to flow freely along the street (shown in Fig. 3). This spring, located only 1.5 km from the main plaza, confirms that the hill provides freshwater.

We documented canals and small reservoirs nearby of approximately thirty centimeters width in another stream (in Stream D), excavated from the bedrock. These bodies of water are similar to those found at another Zapotec site, Hierve el Agua (Doolittle 1989; Kirby 1973). Both sites prove that the surface has been modified by human hand to control and obtain water (shown in Fig. 4).

Fig. 2 Streams verified on field whose origins are on top of Monte Albán and go through Mexicapam and Montoya; released under a Creative Commons—Attribution-Noncommercial-NoDerivatives 4.0 international license

We also surveyed within the protected zone of Monte Albán, a terrain which is much less disturbed by recent human intervention. We walked the north and west sides of the hill, where other studies had previously found water management features (Neely 1967, 1972; Neely and O'Brien 1973; O'Brien et al. 1980). There, along approximately four hundred meters of what we called the north and south streams, we found thirty-four features that would have been placed there to manipulate the flow of water, that is to say, to stop, hold, drain, or divert water to adjacent terraces. The associated ceramic material, though small in sample size, dated to circa 500 BCE to 200 CE (Epoch 1 and Epoch 2 or the Danibaan and Niza phases) (Caso et al. 1967, pp. 24–25, 219 [Figs. 127, 185, 186c]).

Many of these features were wells that inhabitants had excavated in the course of the stream, either using the natural banks of the stream as walls or building walls from stones or a wattle-and-daub mixture (shown in Fig. 5b, d).

Canals, which were probably constructed to bring in water, were connected to some wells at the upstream edge (shown in Fig. 5a). Other wells had walled, V-shaped steps (at the point facing upstream). Some of these were cut from the bedrock, while others were made from stones and slabs in order to spread water over the terraced fields next to the stream.

A long wall of more than fifteen meters in length and 0.6 m in height crossed through one of the streams from north to south (shown in Fig. 5c). At the point

Fig. 3 Spring in Colonia Monte Albán, Mexicapam. (Photographs by S. Kerkhof and A. Rojas)

Fig. 4 **a–b** Canal and small reservoirs in stream D; **c** canal and "pocito" in Hierve El Agua (photographs by A. Rojas and T. Lovett); released under a Creative Commons—Attribution-Noncommercial-NoDerivatives 4.0 International license

at which this outstanding feature intersected with the stream–inside the stream gorge–there lays another V-shaped stepped feature made of modeled stones and slabs (shown in Fig. 5d). This formation seemed to be a sort of dam and a well; at its lowest point, it was of more than one meter in depth, and seems in its entirety to have functioned to spread water to terraces immediately.

Another feature worth noting was almost at the base of the hill: A large reservoir that seemed to be excavated into the bedrock was also modeled into a stepped shape

Fig. 5 **a** Canal and reservoir in stream north; **b** circular well in stream south; **c** wall that intersects stream north; **d** stepped well next to wall and inside stream north; **e** large reservoir at the end of streams north and south; released under a Creative Commons—Attribution-Noncommercial-NoDerivatives 4.0 International license

(shown in Fig. 5e). It was at least two meters deep (making it similar to other reservoirs reported by O'Brien et al. 1980). This feature collected the water coming from both north and south streams, and it is probably connected to the so-called defensive wall some scholars have mentioned to be very close to this area (Blanton et al. 1999). The reservoir indicates that this presumed wall is more likely to have been a water management feature than a work used for military protection.

These archaeological features demonstrate that Zapotec people developed technology to control running waters on the hill, probably already in use since the site's foundation. According to previous studies, water was stopped, retained, deviated, channelized, drained, and transferred to terraces for agricultural and possibly domestic purposes. These are the features that have inspired us to emphasize water flow in our initiatives for the Parque Monte Albán project. In fact, we will try to recover and rebuild some of these works.

Thinking of Solutions

During fieldwork in the urban areas of Mexicapam, Chapultepec, and Montoya, we were able to see that today's inhabitants use water inefficiently. Rather than keeping the streams clean and sourcing their water from them, families throw garbage, sewage, and soapy water into the streams. They in addition buy water from companies that send tanker trucks to fill large containers (called *tinacos*) which are on top of or next to their houses (Fig. 6).

Overall, neighborhoods in this area are informal, marginalized, and lack urban services, including light, electricity, gas, and sanitary water. Most of the houses near the fence that protects the archaeological site have neither a supply of freshwater nor drain pipes. As their status is illegal, inhabitants have to also build their houses with perishable and fragile materials on steep slopes (Fig. 7). These factors make living there a constant risk. Even the maps clearly show the lack of open and green areas; our walks, however, revealed directly that there were no public places where people residents can sit, enjoy the shade of trees, exercise, or simply walk safely away—on sidewalks, for example—from passing vehicles. This stands in contrast to the Historic Center of Oaxaca City and its Zócalo. Public lighting is also lacking, which makes these places dark and at risk to crime–a fact which was confirmed by the Comunidad Segura (2014) report, prepared by a civic organization that measures crime rates statistically in Oaxaca City, noting the Monte Albán neighborhoods as most dangerous.

However, the area does have an abundance of water. And water is the principal axis of this project: Wherever there is water, there is also probably a suitable location for natural, urban, and heritage intervention. Our strategy for improvement focuses precisely on such sources of water, not only to defend them but also to use them as a trigger for improvements in domains like environmental recovery, heritage protection, and reconstruction of social bonds. Indeed, we propose to build parks featuring water.

Fig. 6 Condition of streams and runoffs at Monte Albán; released under a Creative Commons—Attribution-Noncommercial-NoDerivatives 4.0 International license

Fig. 7 Conditions of houses in the eastern neighborhoods of Monte Albán; released under a Creative Commons—Attribution-Noncommercial-NoDerivatives 4.0 International license

Our proposal for the design of these parks has five parameters. *Healthier water* is the principal axis. Thoughtfully incorporating the technology of the ancient Zapotecs, improved by modern hydraulic expertise, the parks, as we envision them, will feature healthy and clean streams and will allow rainwater to infiltrate into the subsurface soil and aquifers. Local people will be able to collect water from streams or rainwater reservoirs. *Greener areas* will help make those streams healthy: The aim here is to generate more vegetation so as to stabilize the banks of the streams and enhance the absorption of rainwater. With the help of specialists, we will plant species appropriate for the type of soil and elevation on site. The parks will constitute *living archaeology* that incorporates Zapotec technologies of water management such as those we find in our survey and other researchers have reported. Wherever possible, we will revitalize the Monte Albán water management system that is in place and make it visible in the parks. We hope in this way to raise awareness of the cultural value of the site.

Along with these living connections to the past, the parks will have *smart connections* to the present. Sites will be well connected to roadways that make it easy to get to the archaeological site, the riverside of Río Atoyac, or even the historical center of Oaxaca City. The goal here is to enhance paths so as to allow safe transit for walking, running, and cycling.

Having transformed unattractive spaces into green, dignified, equipped places, we expect residents to appropriate them and start turning them into public areas with *social appeal*: areas that light up the community and are unattractive to criminals. This greater access to sports and recreation will improve families' lives.

We believe that the only way to make this project succeed is to sustain it from the very bottom, that is, with the participation and *social engagement* of the people who day by day suffer from floods and scarcity of water, live next to or even on top of the archaeological heritage site, and experience urban, economic, and social exclusion. Their participation in finding solutions to the challenges of living on this hill: Developing concepts, making decisions, and even building will ensure that the projects not only materialized but also are maintained in the future. The parks themselves can help foster this engagement, as a *new form of education:* a new way of learning about the ecological, hydrological, and archaeological importance of this area. These green, hydrological, and public spaces can become a living solution for a variety of intertwined problems.

The Natural, Urban, and Social Environment

In order to sketch a plan for parks that deploys old and new technology to help water, people, and heritage, it is necessary to build on a wide range of information from different disciplines. We have already looked at the geographic features of the area (Proyecto Parque del Agua Monte Albán 2016). Moreover, we have built on others' documentation of flora and fauna and have consulted expert botanists on how to restore the ecology of the area. We have mapped the type of soils, ground altitude, land tenure (communitarian, state-controlled or *ejido,* and private property), and the use of land (habitational, commercial, agricultural, as well as for urban development such as schools, sports or recreational areas, public buildings, churches, and the like) in order to understand the overlapping of spaces and the possibility of finding free ones. In this regard, we have considered other issues: population size, political division (according to varying jurisdictions of municipality, agency, and neighborhood), standing legislation regarding urban settlement, and the system of roadways. The final variables in the plan were current management of archaeological sites, the existence of archaeological remains in the area, and other cultural traits such as local traditions and festivities.

A critical layer in this matrix of information has been the hydrological features of the landscape at Monte Albán. It has been thoroughly studied by Deltares, a Dutch agency specializing in the research for developing water technology (Duurzaam Stedelijk Waterbeheer Oaxaca 2013–2014). Besides revealing the high number of water sources that pour down from the hill, Deltares has studied the features of the surface, its elevation, the Río Atoyac, and the great aquifer underneath the Great Valleys of Oaxaca, where Monte Albán sits. It has identified four different zones on the hill, documenting the problems in each and proposing solutions. The water originates in the upper area where it runs quickly down steep slopes and erodes soil. To halt erosion, Deltares proposes reforesting the slopes and building barriers such as terraces or dams. Urban obstacles block the streams and runoff on the two intermediate areas, in the high and low piedmonts. In addition to restoring vegetation and slowing down the flow water from steep slopes with terraces and walls, the proposed solution would allow water to flow cleanly through side canals and build wells to help rainwater infiltrate to the groundwater. The last area is closest to the valley's bottom; it is also closest to the Río Atoyac. Here, the problem is flooding into urban areas. The solution is to allow water to follow its course to the river and drain into it.

All the data from these studies has been analyzed using a strengths, weaknesses, opportunities, and threats approach (see other examples of SWOT analysis in urban planning in Prestamburgo et al. 2016). To put this approach into practice, it was necessary to use these variables and set some criteria by which we could assign proper places to design and build the parks. First, we determined to establish limits and define a target area. We selected the area of five hundred meters around the protected site and decided to treat it as a buffer zone. Within this area, we searched for sources of water, such as a stream, the presence of runoff, or a spring, and an absence of urban infrastructure such as a house, buildings, road, or power and water supply. We preferred locations with trees—which meant there was a likelihood of running water nearby—or barren sites that would have room to replant vegetation. Given these considerations, we looked at the maps to find free spaces to site, design, and eventually build the parks. Next, we considered several additional variables: the presence of roads, commercial centers, population density, and the land as either private or communal property. Areas with low urban density, low population, and flexible land use, *soft urban tissue*, were identified as suited for locating the parks. Figure 8 shows the graphical results of the SWOT analysis, along with our proposals for park sites.

Social participation has been a key component of this research process. From the beginning, in this bottom-up project, we have invited the people who live along Río Atoyac to share ideas, problems, and needs with us. To date, in three separate meetings, local people have participated in talks and workshops with experts on water, urbanism, and archaeology. Local people have worked with authorities from neighborhood, agency, municipal, and state entities, to voice their views on how to solve water issues and to map out new strategies. These workshops have been

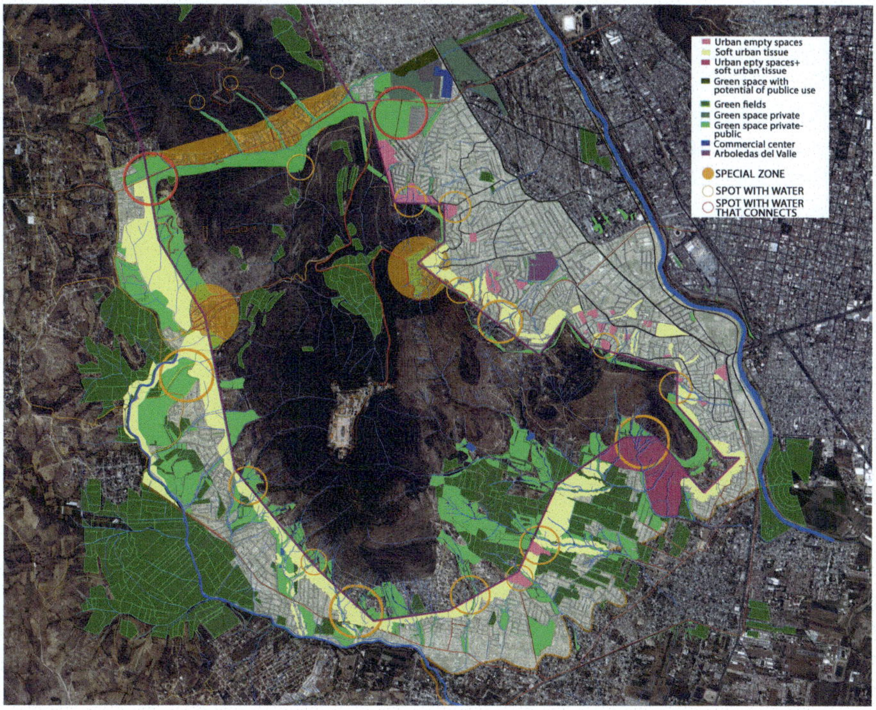

Fig. 8 Results of SWOT analysis (map courtesy of Beccan Davila Urbanismo); released under a Creative Commons—Attribution-Noncommercial-NoDerivatives 4.0 International license

very fruitful in creating a connection between all stakeholders. However, our main concern has been to center the people living on Monte Albán and, particularly, on the site itself in all our discussions and decisions.

Designing Solutions

From the beginning of this water and urban space study, we realized that it was no longer possible to create an ideal buffer park: a green zone with no construction around Monte Albán due to the dense urbanism and effective absence of free space on the eastern side of the site. Moreover, such an effort would be extremely costly and time consuming. Our acupuncture strategies, in which delimited projects modify a small space and improve its hydrological features, create slightly better conditions while being cost-effective. These may be as large as a planting three trees—or two, or, even, one—at a street corner, adding greenery along a sidewalk, or building small canals that could permit the flow of water. These small projects embrace out guiding principles: healthier water, greener areas, living archaeology, smart connections, social appeal, social engagement, and new form of education. In the long term, if placed well, in key spots, we expect them, in their turn, to trigger larger transformations. Our vision of them all together is as a green dotted line or strand of pearls.

To date, we have proposed eight of our acupuncture projects. Each one imagines new green space, revitalizes some part of the Zapotec water control systems, and adds urban and recreational infrastructure. We describe two of the more advanced projects below.

Ojito de Agua

The uniqueness of the Ojito de Agua site, and its great value, relies on the existence of a natural spring (shown in Fig. 3). The particular acupuncture strategy to be applied begins with respecting the water in the spring, the well that collects it, and the surface path that remains without concrete cover. One other particularity of this site is the presence of washtubs which people of Mexicapam neighborhood use for laundry. We propose making minor repairs to dramatically improve how the area looks and works. In particular, we would raise the washtubs, reinforce them with local materials–which could make them more comfortable to use– and add stones to the river bed to strengthen and protect them. We also propose adding entirely new elements to the site. The smallest of these would be benches, streetlights, and more plants (such as agaves and bougainvillea). More broadly, new terraces in the walls could prevent them from eroding, while also enclosing the spring; structures with a stepped shape, like those found in archaeological record, could keep the water on course while broadening its flow. A green floor and new reservoirs made from pots, following the historical model, can help water infiltrate into the ground. It may be possible to construct a canal along the nearby street, visible to the inhabitants (Fig. 9). These embellishments can also work to prevent crimes in the area, a principal concern of inhabitants and passersby.

La Crucecita

The value of the La Crucecita site lies in its social use. It is a space that has already been appropriated by the residents of Chapultepec: Every May 3, they use it to celebrate the Holy Cross Festival, a Catholic festivity that probably has roots in the precolonial period. The proposal to modify this area into a green space for families and to meet social and communal needs was an idea that raised strong enthusiasm among inhabitants. In this case, as with the other acupuncture projects, there is no need to make massive changes. Although it can already be seen that some erosion challenges will have to be met, the main aim here will be to reforest the area and put in place urban infrastructure to improve the atmosphere for the recreational, cultural,

Fig. 9 Acupuncture proposal for Ojito de Agua (render courtesy of Beccan Davila Urbanismo); released under a Creative Commons—Attribution-Noncommercial-NoDerivatives 4.0 International license

and religious activities already being held. This acupuncture intervention is also a great opportunity to connect the city and the archaeological park.

One peculiarity of the area is its undulating terrain. One task to be dealt with is to level these, restoring the historic defined terraces with green and permeable blocks to prevent erosion. This will add not only more space but beauty to the area, and it will be useful, as we will also add streetlights and urban furniture made from local materials. The terraces could include canals to bring water down the hill; an added

Fig. 10 Acupuncture proposal for La Crucecita (render courtesy of Beccan Davila Urbanismo); released under a Creative Commons—Attribution-Noncommercial-NoDerivatives 4.0 International license

wall could help control the water flow and allow local people to collect it for local use. This technology is similar to that of archaeological features found uphill.

One key element of the site is the large cross around which the people gather. Here, they are welcomed ceremonially, receive food, hold religious services, dance, and hold games. We propose to add a playground for children nearby whose floor should be made of some form of permeable cobblestone so that rainwater does not run off but filters into the subsoil. In some areas, these could be of local brick–which would result in a level surface. Trees such as pine, *ocote*, would cast shade to protect the children from the direct and often intense sunlight (Fig. 10).

Conclusion

Water can be the factor that triggers solving heritage problems at Monte Albán, at the same time resolving quality of living problems as well as those of water itself—addressing both consumption and harvesting needs. Our findings are consistent with those of previous studies (Neely 1967, 1972; Neely and O'Brien 1973; O'Brien et al. 1980; Sansores 1992) which lamentably have been neglected. We believe that this technology which has been documented in archaeological research may, to a certain extent, be revitalized. At the very least, it will be worthwhile to critically investigate and learn from it, since once—for more than a thousand years—it proved to be useful to a city of thirty thousand people.

Today, finding sustainable strategies to combat worldwide water crises is a crucial task (Millennium Development Goals of the United Nations 2014; World Water Council 2014). If we take no action, freshwater resources will soon become scarce while cities and their massive demands for energy will continue to bleed natural resources from the landscape. Climate change itself will only worsen the problem of flooding. Monte Albán constitutes a unique opportunity to learn from worldviews and water control technologies that deeply respected water. This case study offers an innovative methodology which may also be applied to solve water problems in regions and sites worldwide to achieve engineering and urban designs that are uniquely local and inspired by archaeological knowledge.

The acupuncture projects we have presented are low-cost strategies for bringing an urgently needed healing process to the urban environment. This model of working might be replicated in other urban areas where water, heritage, and unchecked development collide.

In the long term, our study also seeks to improve the connection between the archaeological site and the Historic Center of Oaxaca City, which is itself a UNESCO World Heritage Site because of its sixteenth-century Spanish architecture, gastronomy, and living culture. We ultimately expect to enhance synergy between the archaeological and contemporary parts of the city and, in this way, to soften the barriers between the formal and informal cities, that is, to integrate marginal areas into the formal city.

Finally, we hope that by once again learning ancient techniques of water management, we might recover and remember the respect ancient people had for water. We hope that the inspiration we get from the archaeology of ancient Zapotec peoples can be transmitted to people living in the present. This work is our attempt to honor the ancestors who built a magnificent city in the past and to bring a part of their memory to the present, to live among their heirs, who hopefully will defend it with a sense of pride.

Acknowledgements We would like to thank Deltares, in the Netherlands, and Arquinistas, in Mexico, for their work and support in the development of the project "Parque del Agua Monte Alban". Our gratitude to the faculty of Archaeology of Leiden University, for the facilities given to make the archaeological survey in Monte Alban possible, as well as to Svenja Kerkhof, Nienke Verstraaten, Philippa Jorissen, and Tom Lovett for participating and involving in this project. Thanks to Instituto Nacional de Antropología e Historia, in particular Dirección de la Zona Arqueològica de Monte Albàn and Coordinación Nacional de Arqueología, for conceding the permits required to conduct survey in Monte Alban. Our gratitude to Ana Osante for her skills in making maps and Sarah Kellar for the English revision of this text. This contribution was made possible by the support from the European Union's Horizon 2020 research and innovation programme under the Marie Skłodowska-Curie grant agreement No. 800253.

References

Blanton RE (1978) Monte Albán: settlement patterns at the ancient Zapotec capital. Academic Press, New York

Blanton R, Feinman G, Kowalewski S, Nicholas Linda (1999) Ancient Oaxaca. Cambridge University Press, Cambridge

Caso A, Bernal I, Acosta J (1967) La cerámica de Monte Albán. INAH, Mexico City.

Comunidad Segura (2014) Report of the Oaxacan organization affiliated to the safe community model (World Health Organization, Stockholm, Sweden). http://www.oaxacacomunidadsegura.org.mx/blog/?p=310. Accessed 30 November 2016

Cruz W (1946) Oaxaca recóndita. Linotipográficos Beatriz Silva, Mexico City

Doolittle WE (1989) Pocitos and registros: comments on water control features at Hierve el Agua, Oaxaca. American Antiquity 54(4):841–847

Duurzaam Stedelijk Waterbeheer Oaxaca (2013–2014) Project developed by Dutch agencies Deltares, Arcadis, and MAP Urban strategies, funded by the Oaxacan State government, the municipality of Oaxaca and Partners voor Water. Document in MAP Urban strategies's archive

Jansen M, Oudijk M, Kröfges Peter (1998) The shadow of monte Albán: politics and historiography in postclassic Oaxaca, Mexico. CNWS, Leiden

Joyce A (2004) Sacred space and social relation in the valley of Oaxaca. In: Hendon Julia A, Joyce Rosemary A (eds) Mesoamerican archaeology. Blackwell Publishing, Malden, pp 192–216

Kirby AVT (1973) The use of land and water resources in the past and present valley of Oaxaca, Mexico, memoirs of the museum of anthropology, number 5, vol 1. University of Michigan, Ann Arbor

Marcus J, Flannery K (1994) Ancient Zapotec ritual and religion: an application of the direct historical approach. In: Renfrew Colin, Zubrow Ezra B (eds) The ancient mind: elements of cognitive archaeology. University of Cambridge Press, Cambridge, pp 55–74

Millennium Development Goals of United Nations (2014) http://www.un.org/millenniumgoals. Accessed 10 Oct 2014

Neely JA (1967) Organización hidráulica y sistemas de irrigación pre-históricos en el Valle de Oaxaca. *Boletín del Instituto Nacional de Antropología e Historia* 27: 15–17

Neely JA (1972) Prehistoric domestic water supplies and irrigation systems at monte Albán, Oaxaca, Mexico. Paper presented at the 37th annual meeting of the society for American archaeology, Instituto Welte archive: 1–15

Neely JA, O'Brien MJ (1973) Irrigation and settlement nucleations at Monte Albán: a test of models. Paper presented at the 38th annual meeting of the society for American archaeology, Instituto Welte archive: 1–19

O'Brien MJ, Lewarch DE, Mason RD, Neely JA (1980) Functional analysis of water control features at monte Albán, Oaxaca, Mexico. World Archaeology 11(3):342–355

Prestamburgo S, Premru T, Secondo G (2016) Urban environment and natura. A methodological proposal for spaces' reconnection in an ecosystem function. Sustainability 8(407):1–13

Proyecto Parque del Agua Monte Albán (2016) Memoria del proyecto Parque Metropolitano Ecológico del Agua, Área Monte Albán. Descriptive memory and studies done for the after mentioned project elaborated by MAP Urban Strategies and Arquinistas, manuscript in State Government of Oaxaca and authors' archives

Rojas A (2015) Informe del recorrido en escurrimientos, arroyos y manantiales de Monte Albán (Verano 2014). Report in INAH, Mexico City, archive

Rojas A (2017) El agua en el cerro del Rayo: Nueva evidencia sobre la presencia y manejo del agua en Monte Albán. Revista Española de Antropología Americana 47: 15-42.

Sansores FJ (1992) El control del agua en Monte Albán, nuevas evidencias. Cuadernos de Arquitectura Mesoamericana 18:19–28

Sellen A (2002) Storm-God Impersonators from Ancient Oaxaca. Ancient Oaxaca 13:3–19

Urcid J (2001) Zapotec hieroglyphic writing. Dumbarton Oaks, Washington D.C

Urcid J (2009) Personajes enmascarados. El rayo, el trueno y la lluvia en Oaxaca. Arqueología Mexicana 16(96):30–34

Visión del Río Atoyac, Oaxaca (2012) Project developed by Dutch agencies MAP Urban strategies, Urhan and Deltares, funded by the government of the municipality of Oaxaca City. Portfolio in MAP Urban Strategies's archive

Wolf E (1959) Sons of the shaking earth. University of Chicago Press, Chicago

World Water Council (2014) http://www.worldwatercouncil.org. Accessed 10 Oct 2014

Araceli Rojas is a postdoctoral researcher at the Institute of Iberian and Iberoamerican Studies, University of Warsaw, Poland, and recently a visiting fellow at the International Consortium for Research in the Humanities at Erlangen University, Germany. Her book, *El tiempo y la sabiduría*, documents the ongoing use of a 260-day calendar among the Ayöök people of Oaxaca. She currently participates in multidisciplinary projects to restore the streams and protect the heritage at the archaeological site of Monte Albán, Mexico.

Nahuel Beccan Dávila is Urban Planner and Director of Beccan Davila Urbanismo. He has helped design and implement the strategies for efficient use of water and space in cities in the Netherlands, Argentina, Peru, and Mexico. He is a proponent of interdisciplinary work, planning grounded in local culture and knowledge, and participatory workshops in which stakeholders work together to solve problems.

The oldest watertoren of the Netherlands (constructed 1678–1683), served the fountains of the gardens of Paleis Soestdijk, a former palace and hunting lodge of the Dutch Royal Family. The water reservoir was supplied by means of a windmill and a horse drawn mill. *Source* Beeldbank Rijksdienst voor het Cultureel Erfgoed; released under CC BY-SA 4.0 Photographer Jan van Galen

Chapter 5
Thirsty Cities: Learning from Dutch Water Supply Heritage

Suzanne Loen

Abstract Cities worldwide currently face freshwater shortages. Forecasts predict that demand will outstrip naturally renewable and available water supplies by 40% by 2030. This poses a serious threat to livability in cities and urban areas that are already struggling with water-related issues like floods and land subsidence. Water insecurity, in particular, now intensified by climate change, calls for integrated and creative solutions. The Dutch heritage in freshwater management, sometimes overlooked and undervalued due to its utilitarian and often modest orientation, is able to provide knowledge and inspiration toward developing water-secure, water-sensitive cities. Three lessons can be learned from Dutch water supply heritage. First, the quest for clean drinking water has sometimes driven the development of valuable urban greenscapes, waterscapes, and nature and landscape conservation areas. Second, those who initiated and managed water-related innovations were often private and commercial parties which collaborated with public entities. Third, although Dutch water supply heritage has become invisible or is not recognized, it embodies valuable systems and practices, particularly with regard to the multisource water supply, private and collective rainwater harvesting, that today could benefit both water supply and wastewater systems. Reducing the need for clean drinking water, generating less storm water runoff, needs to be done while engaging the public in building water-sensitive, safe cities.

Keywords Dutch water supply heritage · Urban drinking water management · Rainwater harvesting · Water-sensitive cities · Water cistern

---

S. Loen (✉)
Department of Urbanism, Section Landscape Architecture & LILA Living Landscapes, Delft University of Technology, Delft, The Netherlands
e-mail: suzanneloen@lilalandscapes.nl

C. Hein (ed.), *Adaptive Strategies for Water Heritage*,
https://doi.org/10.1007/978-3-030-00268-8_5

Introduction

The Netherlands, with its polders and flood defense systems, is well known as a model nation in sea and polder-boezem (that is, outlet-pool) water management and water heritage. Slightly less famously, it is also an exemple in freshwater supply management. In 1853, the country's first drinking water company delivered piped water in Amsterdam (Leeflang 1974). By 1968, more than 99% of the Dutch population had access to piped drinking water, and today all households are connected to piped water. One consequence of this transformation is that the country decreased its use of surface water for drinking and, today, 60% of its freshwater supply comes from groundwater (De Moel et al. 2006). Climate change is expected to intensify the negative effects of groundwater exploitation in both cities and the agricultural hinterland worldwide: depleting water tables, damaging the quality of surface water, and increasing salination.

Dutch systems for accessing drinking water developed from decentralized off-the-grid systems, which included rainwater harvesting, into large-scale centralized urban utility networks with a civil engineering orientation. In the densely populated urban and industrial, western part of the country, an area of low-lying deltas, polders, and coastal regions—more particularly, the cities of Rotterdam and Amsterdam—groundwater sources were often scarce or of poor quality due to brackish conditions, salination, as well as domestic and industrial pollution. Locals met these problems with resourceful public and private innovations and initiatives. In this chapter, such historic local developments, taking place between the Middle Ages and the introduction of centralized piped water utilities networks during the Industrial Revolution, are first presented then placed into national and international perspective. Exploration of historic tangible infrastructure in spatial, architectural, and landscape architectural follows. These interventions were based on drinking water type—whether it was groundwater, surface water, or rainwater. The intangible dimensions of this heritage, that is how the systems were organized, are then addressed. Here, the focus is on the three possible organizational arms: public, communal, and private involved in water management. As governing bodies and private parties require the participation of citizens in developing resilient and water-secure cities, this focus provides insight into how such collaborations were coordinated in the relatively highly urbanized and industrialized areas of the Netherlands. The objective is to investigate and identify lessons to be learned from intangible and tangible Dutch water heritage that can be applied to building water-secure and water-sensitive cities in the future in areas that face overexploitation and pollution of water resources.

Water Heritage from the Middle Ages to the Nineteenth Century

Dutch cities were relatively late in implementing centralized piped drinking water facilities compared to other European cities (Leeflang 1974). The cities of London and Paris, for example, introduced piped riverine surface water supply systems during the sixteenth and early seventeenth centuries (Leeflang 1974; Wijmer 1992; Wijntjes 1982). Citizens of Dutch cities generally relied on three types of water for consumption: groundwater, surface water, and harvested rainwater. Residents of cities and villages obtained their drinking water from the surface water in shallow wells, rivers, canals, and brooks (Wijmer 1992). (Figure 1) The urbanized western part of the country, consisting of South Holland, North Holland, Utrecht, and North Brabant, relied mainly on surface water (Vogelzang 1956). Most shallow wells were fed, vertically, with infiltration-storm water runoff, or, horizontally, with surface water from nearby rivers and streams (Rapport aan den Koning van de commissie tot onderzoek van drinkwater 1868). Shortage of water in general was not a problem as the Netherlands had an abundance of rivers, lakes, and canals.

Fig. 1 Detail of a nineteenth-century drawing of a woman fetching water from the canal in the town of Delft by Petrus Augustus Beretta. *Source* Rijksmuseum; released under a Creative Commons Attribution-NonCommercial-NoDerivatives 4.0 International License

In the fifteenth century, many Dutch cities began to grow in economic activity. Highly polluting industries within them, such as textile industries, tanneries, and beer breweries, required large quantities of clean water as they flourished and expanded (Wijmer 1992). Along with growth of the urban population, this change affected the self-cleaning capacity of the surface water system (Groen 1978). Securing urban freshwater supply for domestic and economic purposes in cities became very challenging. The quality of urban surface water deteriorated substantially, forcing cities and their inhabitants to build more rainwater reservoirs and drill groundwater wells. In the western part of the country, the drilling of wells often proved unsuccessful (Groen 1978).

In 1868, a commission advised the king of the need for a centralized piped water system (Rapport aan den Koning 1868). In the process, it described the conditions of the water supply system that had characterized Dutch life since the sixteenth century. The commission's research showed that the percentage of deaths due to the waterborne disease cholera was significantly higher in areas where surface water was the dominant source of drinking water than in areas where groundwater was the predominant source. Areas where rainwater predominated as source were the least affected by cholera. The report divided surface water into three subtypes: riverine water from the main rivers; water from the polder system (a closed and controlled water system in contrast to the open riverine system); and water from open surface bodies like shallow wells (*putten*), ponds, lakes, ditches, canals, smaller rivers, streams, and water from filtration of storm water, *zakwat*er. The water quality in this last category varied substantially. Cities with an open canal system subject to the tides of its main rivers were better off than villages and cities and parts of cities with a closed urban canal system that was not flushed regularly. Villages and cities in polders, in turn, were worse off than these (Rapport aan den Koning 1868).

Water provision in Rotterdam exemplifies the difficulties. The city was serviced by three different water systems: a) the polderwater system, b) the innercity grachtensystem, and c) the riverine system of the River Maas. (Figure 2) The inner city of Rotterdam was divided from the waterfront extension by a dike and surrounded by two large *grachten* (known also as *stadsvesten;* these canals were closed off from the river by a sluice). Originally, a defense system—but mostly used as docks and harbors—, the canals did not profit from the influx of freshwater during high tide (Van den Noort and Blauw 2000). Water quality deteriorated substantially over time due to lack of sufficient water flow and domestic and industrial pollution. In the ditches, streams, and canals in the polder surrounding the city, the surface water was affected by stagnation and agricultural pollution. The water quality in the polder was therefore not much better than in the inner city. It was worse than in the waterstad that benefited from the relative clean water from the river Maas.

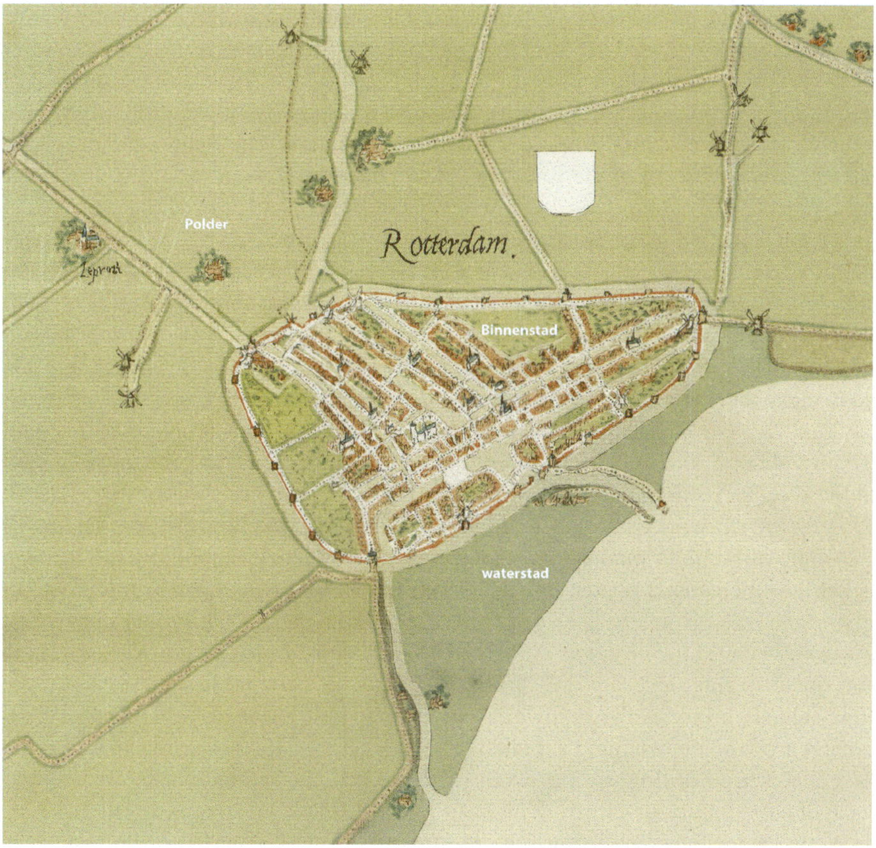

Fig. 2 Map of the inner city of Rotterdam in 1550 by Jacob van Deventer. *Source* Nationaal Archief; released under a Creative Commons Attribution-NonCommercial-NoDerivatives 4.0 International License

The seventeenth-century extension of the city in the Waterstad, with open harbors, canals, and docks, was subject to river tides. Water was consequently of relatively good quality. There, water was taken directly from surface water bodies and indirectly from water reservoirs, or *vloeiputten,* a system of arched openings in quay walls and cisterns (Vogelzang 1956): When water rose, surface water entered these openings and flowed through pipes to cisterns located in building or to pumps that then made it available to the public (Vogelzang 1956). City architect W.N. Rose promoted this system in 1840, as he thought the quality of the Meuse water was quite good (Dijkstra 1974). He pointed to one such *vloeiput* which was already providing water to the public at the Stadstimmerhuis, a public works building located on the outer quay of the waterfront (Dijkstra 1974).

With the nineteenth-century expansion of the city, the *stadsvesten* became part of the new polder city, closed off from the open riverine water system. The surface water in this new polder city deteriorated substantially (Van den Noort and Blauw 2000). Dutch polders' inferior water quality can be explained by the traditional attitude of the Dutch water boards to polder water management. The main concern of the water boards was to keep the low-lying peat polders dry, so they set up the water management system to mill as much water out of the polders as possible, rather than letting water in. Maintaining water quality was of secondary importance. The city of Rotterdam could therefore get rid of its dirty water during low tide; however, the water board did not let in freshwater from the Meuse to flush the canals during high tide. In 1840, to tackle this urgent lack of clean drinking water, W.N. Rose proposed building eight more *vloeiputten* to collect water from the river Meuse at high tide and to carry the water in metal pipes to the dwellings and buildings of the inner city. (Figure 3) A year later he expanded the plan to nineteen. This *Waterproject* would also provide new and improved sluices, public green water ways [*parksingels*], and public water wells to both the inner city and the polder city. In addition, it proposed filling up the *stadsvesten* to turn them into roads. (Dijkstra 1974) (Figure 4).

In spite of the poor quality of surface water, many citizens continued to throw a bucket into the canals, ponds, or ditches to retrieve water for their needs. The city did provide water from city wells and also permitted citizens to drill their own wells. However, because the city was constructed on a layer of peat which was separated from the deep sand layer by only a very thin layer of clay, public and private wells often yielded a mixture of groundwater and polluted infiltration water (Dijkstra 1974). To meet growing demand and expand the supply of good quality drinking water, the city council tried drilling deeper wells, often to find that groundwater was beyond reach, of poor quality, or simply of an insufficient amount (Vogelzang 1956). In 1874, the Water Project was finally carried out, accompanied by a centralized piped water supply plan by Scholten. This plan, in fact, laid the foundation for the city's drinking water company (Van den Noort and Blauw 2000).

The situation of Amsterdam was slightly different. At the beginning of the sixteenth century, with nearly sixty-five thousand inhabitants, the city of Amsterdam was forced to economize on freshwater. A hundred years earlier, the water in the canals had a sufficient self-cleaning capacity, so that inhabitants had only to throw a bucket into the canal for clean water for drinking, washing, and cooking (Groen 1978). But water quality deteriorated: increasing salination of the IJ and Amstel rivers matched by increasing pollution (Leeflang 1943) as the urban population grew and required more water. To clean the water in urban water bodies and canals, the city introduced sluices and mills to the urban water system. The logic was: If polluted canal water could be flushed out of the city, cleaner water could be let in. Around 1675, the city of Amsterdam constructed the *stadsvuilwatermolens* (wastewater mills)—three horse-drawn mills, each drawn by nine horses—to move dirty water out of the city into the IJ when water levels outside the city rose higher than the city's water level. (Figure 5) The mills proved ineffective and were soon replaced by windmills (Groen 1978; Jurgens and Van der Kaay 2008).

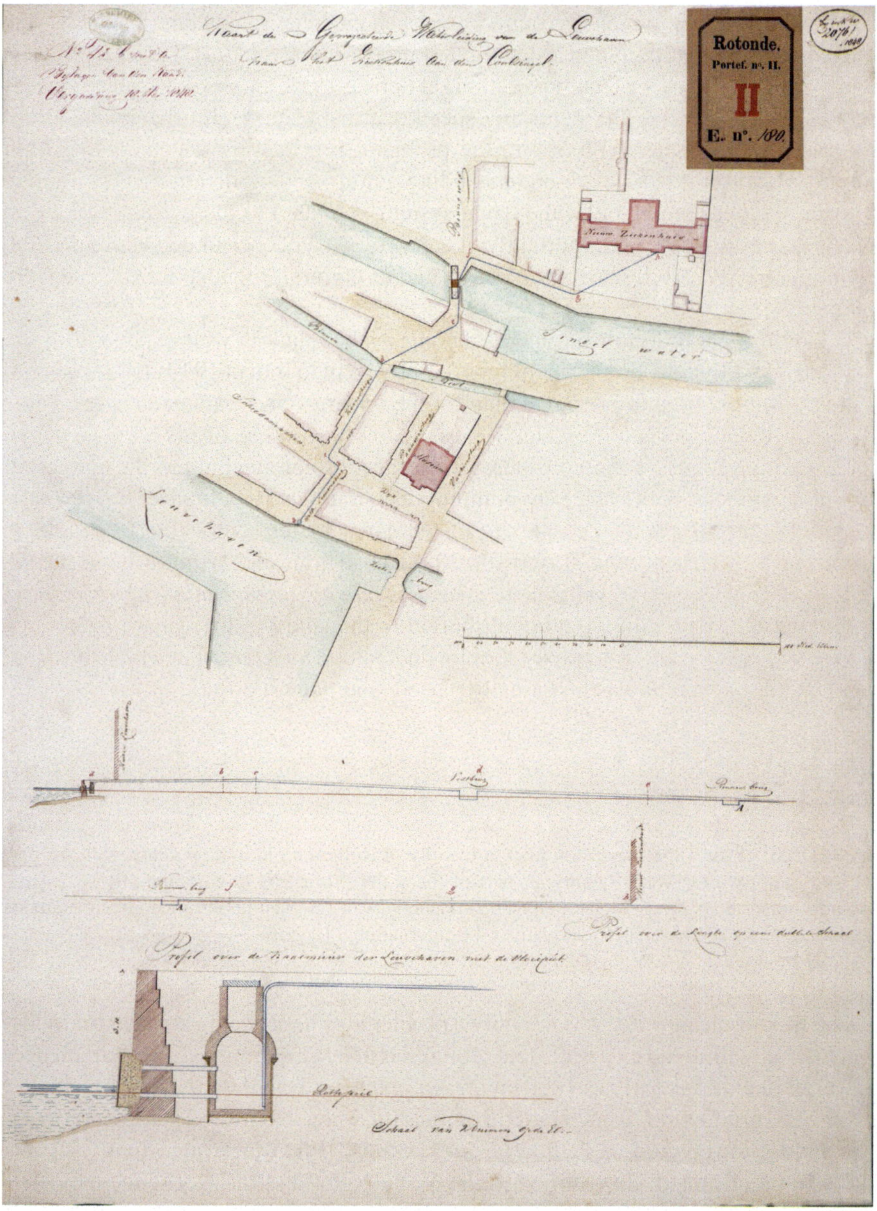

Fig. 3 Detail of plans and profile drawings of a *vloeiput* at the quay walls of the Leuvehaven and water pipe connecting it to a hospital in the city center of Rotterdam (1848). *Source* Stadsarchief Rotterdam; released under a Creative Commons Attribution-NonCommercial-NoDerivatives 4.0 International License

Fig. 4 Detail of the 1858 expansion plan of the city of Rotterdam into the polders, with the new singels to improve the water quality, as part of the water project by city architect Rose. *Source* Hoogheemraadschap van Schieland en de Krimpenerwaard; released under a Creative Commons Attribution-NonCommercial-NoDerivatives 4.0 International License

The lack of clean water was not only a problem for households but also for industries located within the city walls that required it in large quantities. Amsterdam beer breweries decided as early as 1480 that water obtained from the canals was no longer fit for beer production. They started bringing in water from the Haarlemmermeer, a lake close to the city, in shallow sailing vessels. When the water quality of this lake also deteriorated, they turned to the Vecht River and the Kockengenpolder, a polder forty kilometers from the city (Leeflang 1974). The construction in 1639 of the Weespertrekvaart, a canal with towpaths near the town of Weesp between Amsterdam and the Vecht where drinking water was collected, was of great importance to the urban water supply, improving and simplifying the route (Zondergeld-Hamer 2010).

Fig. 5 Drawing of the inner and outer stadsvuilwatermolens (waste water mills) in Amsterdam ca. 1765 by Jan Spaan. *Source* Stadsarchief Amsterdam; released under a Creative Commons Attribution-NonCommercial-NoDerivatives 4.0 International License

By 1658, a group of Amsterdam brewers collectively invested in a shallow horse-drawn vessel that could break through ice in the waterways to secure water even in harsh winters. If necessary, as many as thirty-six horses could pull the vessel through the thick ice, assisted by men with ice picks. The brewers' water barges supplied the brewers and sold water to the public. Due to its importance to the urban water supply, the icebreaker became the property of the city council in 1786, and a distinction was made between barges selling water to the public and barges supplying the brewers. This protected the public from soaring water prices during droughts and harsh winters. The water vendors organized themselves in the Verschwatersocieteit, or freshwater society (Leeflang 1974). Freshwater from the water barges was transported from boats to smaller vessels that could navigate the smaller canals or to floating platform, where the water was pumped up from beneath the vessel and sold to the public.

Along with riverine water, water stored in the western coastal dune massif was sought after by the beer brewing industry, bleachers, and the urban population (Jelles 1968; Oneindig Noord-Holland 2017). Over the course of centuries, rainwater was stored in the dunes, sometimes to a depth of one hundred meters, pressing down salt

Fig. 6 Drawing of the Brouwerskolk with sluice (1798) by Franciscus Andreas Milatz. *Source* Noord-Hollands Archief; released under a Creative Commons Attribution-NonCommercial-NoDerivatives 4.0 International License

water deposed there. The groundwater level in the dunes, at above sea level, appeared on the surface in valleys and brooks that flowed in a west to east direction. This fresh and clean water was extracted from the sixteenth century until the eighteenth century on a small scale in shallow wells, then transported to cities, and sold to the public (Leeflang 1974). Textile bleachers, dyers, and paper factories had also settled early in this area in the old dunes behind the young dunes to make use of the clean and readily available water (Jelles 1968). Fresh dune water appeared at the surface of the Brouwerskolk, a manmade pool located at the end of the Brouwersvaart canal in the older dunes west of the city of Haarlem. (Figure 6) Water vendors obtained water here and sailed with it back to the city's brewers. The Brouwersvaart, earlier known as the Santvaert, was initially dug to transport sand from the dunes to the cities of Haarlem and Leiden for construction but became both an important transportation route for freshwater and a source of freshwater (Hoekstra 1947; Regtdoorzee Greup-Roldanus 1936). In the nineteenth century, rich bankers from Amsterdam bought land, which included the Brouwerskolk, to build an estate. In 1928, the Brouwerskolk was sold to the city, which converted it into a public park, in that way, protecting it from building development.

Fig. 7 Detail of map of the town of 's-Gravenzande by Coenraet Oelenzn (1566) with clearly visible public *stadswaterputten* (wells operated and constructed by the city counsels for its inhabitants) with sweeps. One sweep was located near the harbor, and drinking pools providing both ground and surface water to its inhabitants. *Source* Nationaal Archief; released under a Creative Commons Attribution-NonCommercial-NoDerivatives 4.0 International License

Along with surface water, the water well was a common typology in water management in the Netherlands. In villages and cities, public wells, often located on squares, were small but important landmarks. (Figure 7) Two types of wells were in use in cities and in rural areas: the rain well for storing and obtaining rainwater and the *welput* for groundwater. (Rapport aan den Koning 1868) (Fig. 5) A *welput* reached below groundwater level and was built in a circular shape, from clay bricks which filtered impurities from groundwater seeping into the well. Such structures were usually located on a farm or dwelling but could also be found inside buildings, conveniently located in a stable, for example (Maris 2010). Public versions of these were centrally and strategically located near churches and markets. (Figure 6) Sometimes, the water obtained from them was free of charge, but sometimes a water vendor would charge customers a fee for it. (Vogelzang 1956).

By the eighteenth century, densely populated villages and cities needed more and more public wells to meet the demand for drinking water (Groen 1974). These wells were usually drilled and managed by the local government. By 1871, six thousand wells were in use by a population of sixty thousand in the city of Utrecht. Sixty of those wells were reserved for public use (De Bruin et al. 2000); a civil servant in the city of Utrecht, the *boormeester*, was in charge of drilling and maintaining them ('T Hart 2005).

In the Netherlands, rainwater and surface water were stored in cisterns both large and small (Vogelzang 1956; Zondergeld-Hamer 2010). Larger cisterns were sometimes filled with water brought from the Vecht River (Zondergeld-Hamer 2010). In the province of Zeeland, in parts of West Friesland, Groningen, and Overrijssel, and in cities like Amsterdam, rainwater was traditionally the first or second most important source of water, as surface water was too saline or polluted and groundwater was brackish (Vogelzang 1956; Groen 1978). Rainwater pools or ponds, each with a clay base and walls, can be found in the northern province of Friesland (Leeflang 1974; Baas et al. 2005).

A common type of small rainwater harvesting cisterns was directly connected to a farm, dwelling, or building. This rectangular, lidded, low well, built from bricks, collected rainwater from roofs and sent it through pipes, often made of lead and sometimes a sand filter, into a small underground reservoir. (Figure 8) The water also hosted *tench*—edible fish—to keep it clean. In the eighteenth century, people became aware of the risk of lead poisoning when using rainwater from large cisterns (Groen 1974). When this was addressed, rainwater was one of the safest water sources available, as it did not spread water-borne diseases (Rapport aan den Koning 1868). In rural and sparsely populated areas, small cisterns were in use into the 1950s. A regular one-family cistern had a volume of about three to five hundred buckets.

To secure the public urban water supply, larger cisterns were located below squares, larger dwellings, public buildings, churches, or courtyards (Fig. 9). Cisterns were also constructed beneath commercial and industrial sites like harbors or tanneries (Gawronsky and Veerkamp 2007). In 1505, the city council of Amsterdam built nine such cisterns to harvest rainwater, as the quality of freshwater from the canals had deteriorated increasingly due to domestic and industrial pollution (Groen 1978). The size and volume of cisterns vary considerably (Gawronsky and Veerkamp 2007). A cistern with a volume of twenty-eight thousand liters was discovered in the foundation of a hothouse—dating back to 1715—at the Amsterdam Botanical Gardens. Research by the city's archeologists suggests this cistern was part of a larger system of interconnected underground water reservoirs both in and outside the grounds of the garden. Rainwater stored in this system was probably used not only for watering the plants but also as a public water reserve in case of drought or fire (Gawronsky and Veerkamp 2007). Construction of a cistern of considerable size was a costly affair and was therefore only affordable for wealthy citizens or businesses (Vogelzang 1956).

The beer brewers of Amsterdam also took an interest in harvesting rainwater. In 1784, Amsterdam brewer Isaac Decker proposed a plan to the city council to build fifty-two freshwater tanks, evenly distributed through the urban territory within the city walls, to secure the water supply for both brewers and inhabitants. Between 1790

Fig. 8 Two *regenputten* (rainwater wells) connected to a rainwater pipe. *Source* Rijksdienst voor het Cultureel Erfgoed; released under a Creative Commons Attribution-NonCommercial-NoDerivatives 4.0 International License

and 1824, thirty-three of the fifty-two were built; twelve of these were exclusively reserved for the breweries (Groen 1978). (Figure 10) However, these rainwater catchment systems (regenputten) actually provided relatively little water. Water imports and surface water therefore remained important sources of urban water.

Outside cities, cisterns were constructed at estates, castles, and fortifications to secure the water supply in times of siege, drought, or natural disaster. Most forts on the Defense Line of Amsterdam and the New Holland Waterline caught rainwater on their sloping, earthen roofs. The water then infiltrated through the sand or earth layers for purification and dripped through small pipes into the cistern (Van Ginkel 2004). This means of diverting and purifying water was much safer and more sustainable than the use of lead pipes.

The water utilities and infrastructure from the Middle Ages to 1853 examined in case studies turn out to be very rich and diverse. The practice of building large-scale rainwater harvesting stands out in particular, not only because of the relative obscurity of these systems but also because the practice was so widespread. In fact, both public and private parties initiated, lobbied, and invested heavily in the expen-

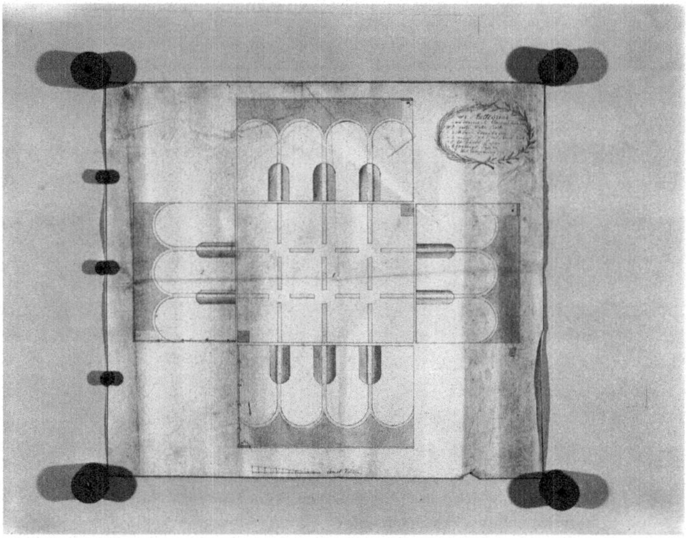

Fig. 9 Drawing of a *waterkelder* or water cistern for an orphanage in Amsterdam. *Source* Rijksdienst voor het Cultureel Erfgoed; released under a Creative Commons Attribution-NonCommercial-NoDerivatives 4.0 International License

sive construction of cisterns. And while the literature does not mention storm water buffering, the actual or potential storm water buffering capacity of these rainwater harvesting cisterns may have been substantial. Another striking characteristic of the historical water supply systems is that they were sourced from multiple locations and were fit for different purposes. Rainwater, for example, was used only to produce costly drinking water, while surface water and groundwater were used for agricultural or industrial activities. The water supply system was multilayered, featured both small and larger infrastructure and utilities, and was funded and operated by public as well as private parties. This created an awareness and sense of shared responsibility, exemplified in the participation of citizens and private parties in decision making, and the number of private initiatives and investments to secure urban water supplies. From a spatial point of view, the examples of the parksingels of Rotterdam, the Weespertrekvaart of Amsterdam, and the Brouwerskolk in nearby Haarlem were all part of a drinking water supply scheme, are today valued for their contribution to spatial quality but not recognized as integrated components of the water supply systems as such.

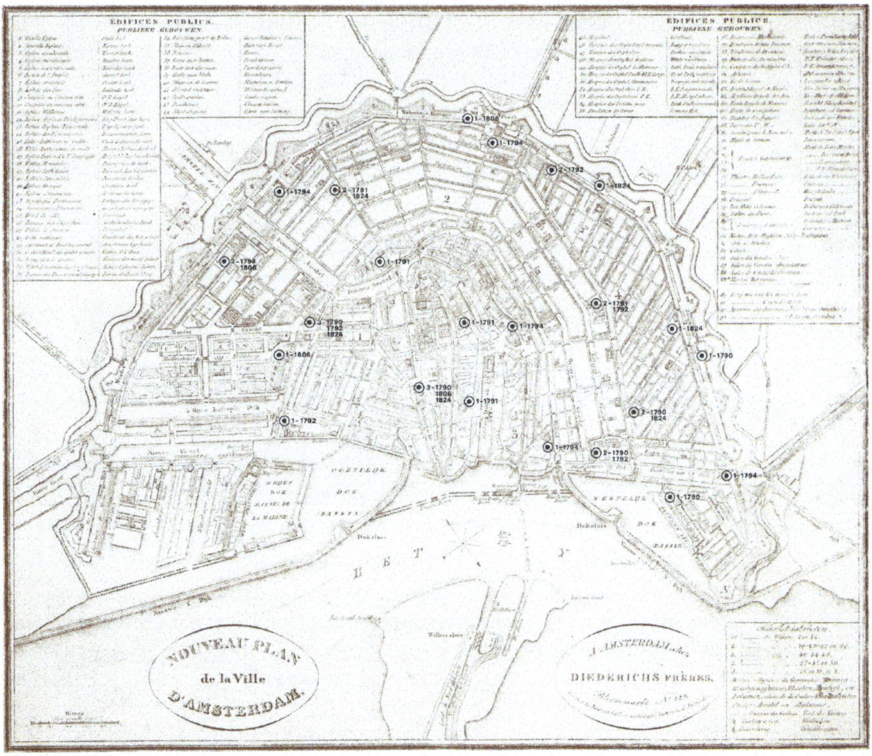

Fig. 10 Map with cisterns in the city of Amsterdam after the plan of beer brewer Isaac Deckers built between 1790 and 1824. *Source* Groen 1978; released under a Creative Commons Attribution-NonCommercial-NoDerivatives 4.0 International License

Piped Systems as Drivers for Nature and Landscape Conservation (1853—Present)

Dutch cities were relatively late in implementing centralized piped drinking water facilities compared to other European cities (Leeflang 1974). The cities of London and Paris had introduced piped riverine surface water supply systems in the sixteenth and early seventeenth centuries. Finally, on December 12, 1853, citizens collected the first buckets filled with water from the coastal dunes at the Willemspoort fountain in Amsterdam, paying one cent per bucket (Groen 1978). Amsterdam was the first Dutch city with a centralized, piped drinking water supply. However, other cities soon followed.

The Dutch coastal dunes, stretching along the North Sea at its western border, have always been exploited on a smaller scale for their high-quality freshwater. The dunes have also been exploited for sand, wood, and agriculture. There the nobility

Fig. 11 Canal in the dune water catchment area today. Photograph: Joop Hilster. *Source* Stichting Waternet; released under a Creative Commons Attribution-NonCommercial-NoDerivatives 4.0 International License

and rich citizens built their estates, with attendant hunting grounds and gardens. Urban water demand led to a private initiative to extract water on an industrial scale from the dunes for the city of Amsterdam. In 1851, the Oranjekom, a water body six meters deep, was dug parallel to the coast in the former wet dune valleys and connected to two canals for phreatic shallow water extraction (Baeyens and Duyve 1992)—that is, to get at water below the water table. This water was first transported by a pipe to a fountain at the Willemspoort in Amsterdam. Soon, more public water taps for distributing piped drinking water, or standpipes, were installed. By 1854, the canals measured three thousand five hundred fifty meters; by 1863, they were at seven thousand, six hundred twenty meters (Baeyens and Duyve 1992). By 1866, Amsterdam had fifty-six standpipes in use (Groen 1978). Tapping the dune water was so successful that the drinking water company, *Duinwater-Maatschappij,* struggled to meet demand from Amsterdam. In 1870, a law was passed forcing landowners to sell their dune properties to the Duinwater-Maatschappij so that it could dig more canals to safeguard the urban water supply (Groen 1974). These canals, parallel to the coastline, are distinctive cultural elements that contrast with the coastal dune landscape. (Figs. 11 and 12).

To meet growing demand and to recharge the aquifer and push back the salt water, the Duinwater-Maatschappij decided to inject surface water from the Rhine River into the dunes (Figure 9). They started the process in 1957. The effort turned the

Fig. 12 Infiltration point of surface water from the lake IJsselmeer into the infiltration channels in coastal dunes area of Castricum. Photograph: T. Kisjes

groundwater extraction site into a water purifying plant for riverine surface water; and typical channels were reversed from structures for extraction to infiltration. As unfiltered riverine water threatened the local ecology and the ecosystem of the dunes, riverine water today is first purified before it is put into the dunes (Baeyens and Duyve 1992). Its vital role in the water supply of the city of Amsterdam and the water conservation area's status has protected the area against urban and recreational development.

A unique type of drinking water production is peat lake water extraction in the Bethunepolder. In 1850, the Belgian Marquis Bethune tried to transform a peat lake into a polder by milling water out of it. Unfortunately, seep water continued to appear at the surface of the lakebed, seeping in from a nearby peat lake as well as from the Gooise moraine (Kosman 1988). Because Bethune could not get it fully dry, the lakebed was unfit for agricultural purposes. Plans were then developed to turn the peat lake into a waste dumping area. Luckily, in 1930, the city of Amsterdam decided to exploit the polder for drinking water. Today, one-third of Amsterdam's drinking water comes from the Bethunepolder (Kosman 1988). Seep water from the polder runs through the *Waterleidingkanaal* (that is, the drinking water canal)

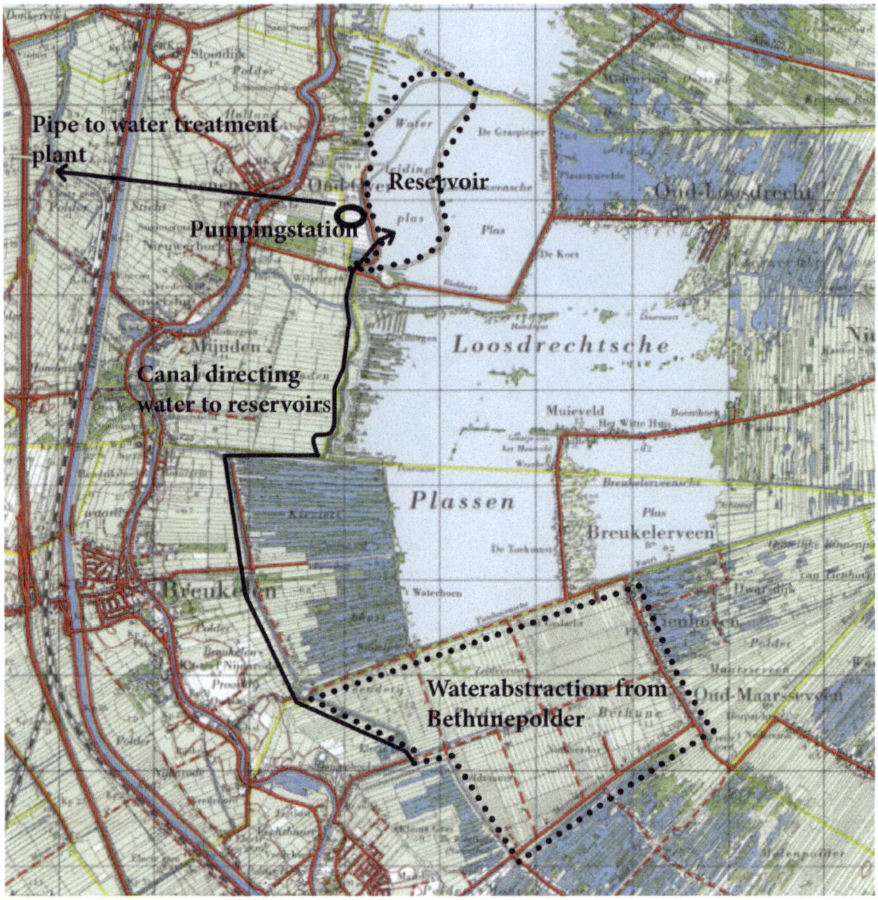

Fig. 13 Map of the peat lake water by ca. 1960. *Source* Nationaal Archief; released under a Creative Commons Attribution-NonCommercial-NoDerivatives 4.0 International License

to Loenderveense Lake, in a nature conservation area. (Figs. 13 and 14) There, the water is purified before being pumped at the water plant in Weesperkarspel, a district of Amsterdam. The Loosdrechtse peat lake, today an important recreational area, has also been preserved to protect the water supply from the Bethunepolder. The water reservoirs, the *Waterleidingkanaal,* and treatment plants form a spatial ensemble of both landscape architectural and civil engineering.

Finally, it is important to note that river water has been important to the water supply since the late nineteenth century. Populated areas are often located in river deltas, where surface water was and still is often available, commonly, in large amounts and is easily used. But surface water varies substantially in quality and often requires extensive treatment. In 1870, the Department of Public Works built a surface water

Fig. 14 Drinking water canal (*Waterleidingkanaal*) transporting seep water from Bethunepolder to the Loenderveenseplas for natural purification (ca. 1988). *Source* Het Utrechts Archief; released under a Creative Commons Attribution-NonCommercial-NoDerivatives 4.0 International License

treatment plant equipped with a monumental water tower, filter building, and deposit basins east of the city of Rotterdam. The plant extracted water from the Meuse River (Dijkstra 1974). By 1961, the local drinking water company had built a water reservoir and water treatment plant on Beerenplaat Island in the river De Oude Maas with a capacity of eight million cubic meters to meet the growing demand of the city and the water-intensive petrochemical industry in the harbors. However, water quality and quantity still required improvement, and so, the company moved even further away from the city to secure its freshwater demands.

In the early 1970s, the company built three water reservoirs to purify and store water from the Biesbosch, extensive freshwater tidal wetlands located further upstream, southeast of the city. From these reservoirs, water goes to five drinking water treatment plants, from which the water is distributed as drinking water and water for industry (Van den Noort and Blauw 2000). Due to its importance to the water supply, the Biesbosch has become a protected nature conservation area and an important recreational area (Figs. 15, 16, 17, and 18).

Modern centralized water supply systems did not only herald the beginning of a new era in urban drinking water supply in the Netherlands. It also resulted in the water and landscape conservation of a large area of the coastal dunes, Biesbosch,

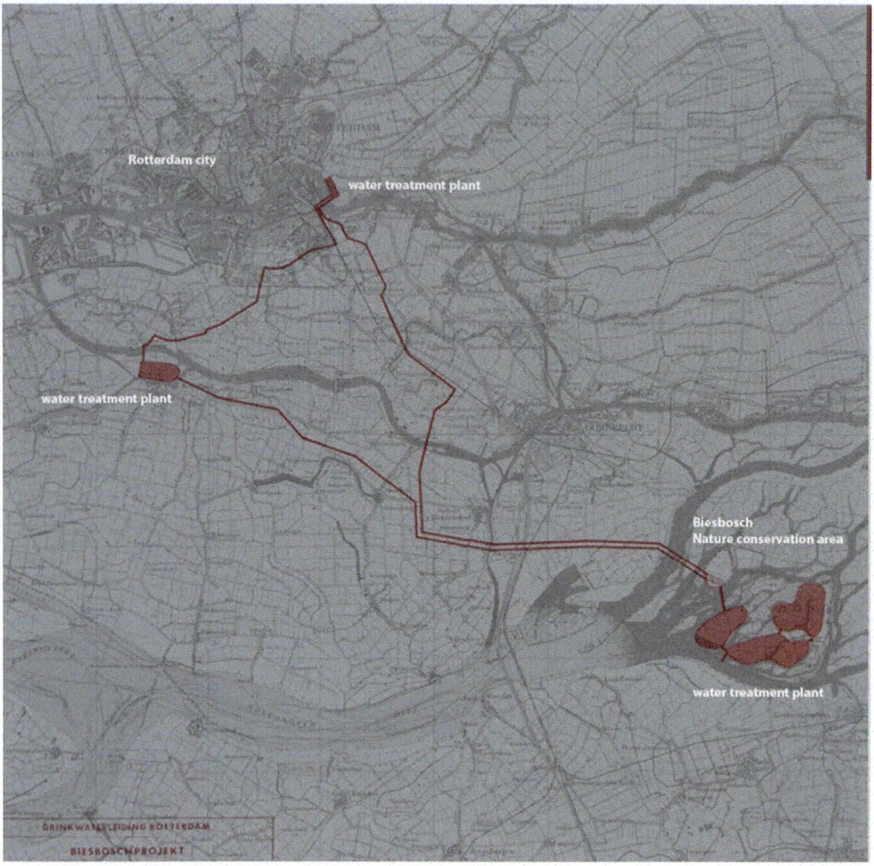

Fig. 15 Map of the Biesbosch project by the water company of Rotterdam in 1967. *Source* Stadsarchief Rotterdam; released under a Creative Commons Attribution-NonCommercial-NoDerivatives 4.0 International License

and peat lake areas, as relatively unspoiled nature reserves and recreation areas. At the same time, drinking water supply utilities like public wells and rainwater cisterns disappeared from the city, out of sight and out of the hands of individual citizens and private parties. Harvesting rainwater, a common practice up until then, was forgotten.

Today, 60% of drinking water in the Netherlands comes from groundwater (De Moel et al. 2006). Early on, drinking water companies extracted either groundwater or surface water, but the majority of the Dutch drinking water companies now extract groundwater from both shallow and deep confined layers. In the western part of the Netherlands, infiltrated surface water is the main source of drinking water. And along the Rhine, Meuse, Waal, Lek, and IJssel rivers, riverbank groundwater is extracted for this use (De Moel et al. 2006).

Fig. 16 Map of the freshwater tidal wetlands of the Biesbosch (1961) before the construction of the *spaarbekkens* (reservoirs). *Source* Regionaal Archief Dordrecht; released under a Creative Commons Attribution-NonCommercial-NoDerivatives 4.0 International License

Conclusion: Learning from Dutch Water Heritage

Along with its cultural, spatial, and historical value, Dutch water supply heritage is valuable for its potential to create public awareness on water in and around cities in such ways that this knowledge can be used creatively and effectively to build water-sensitive cities and sustainable water utilities systems and networks today and in the future. In sum, the examination of the case studies in this study from the point of view of spatiality raises a few points that are worth noting.

First, the quest for clean drinking water—sometimes deliberately but often unintentionally—was a driver behind the development of urban green and water areas as well as of nature and landscape conservation areas. This fact is not always recognized. Many people in the densely populated urban areas of Rotterdam and Amsterdam enjoy these landscapes and are unaware of their role in the drinking water supply system. Second, the typologies and case studies described in this chapter show that, up to the introduction of piped water and sewer systems, city councils and churches as well as citizens and the private sector cooperated, invested in, and sometimes even exploited rainwater harvesting utilities and water import companies. The challenge is to promote and develop policy and practices that visibly integrate water heritage and historical rainwater harvesting systems more prominently in the public realm. This may help to educate and to engage both citizens and private parties to collaborate in building water-safe and water-secure cities.

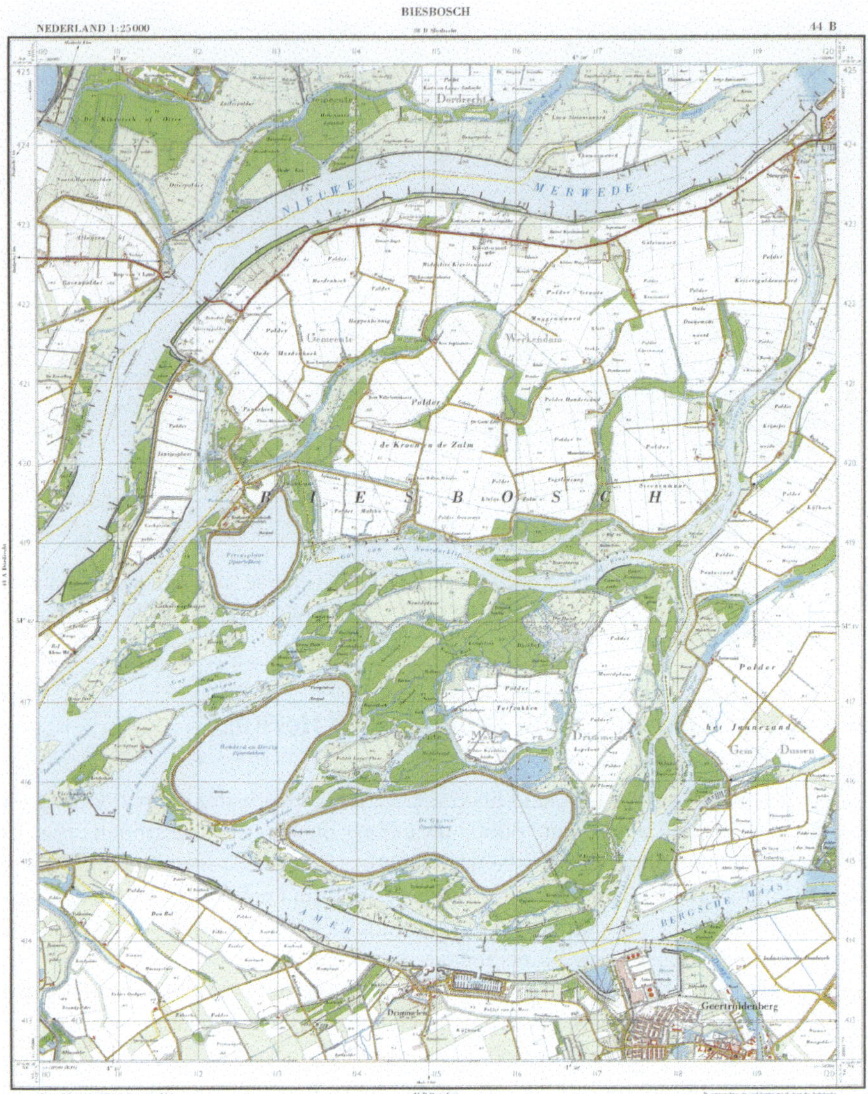

Fig. 17 Map of the freshwater tidal wetlands of the Biesbosch (1988) after the construction of the *spaarbekkens* (reservoirs). *Source* Regionaal Archief Dordrecht;

Fig. 18 Three *spaarbekkens* (water reservoirs) in the Biesbosch national park. *Source* Google Maps; all rights reserved

Third, Dutch water supply heritage has become invisible or is not recognized, although it embodies valuable solutions and practices such as multisource water supply, private and collective rainwater harvesting, and a combination of off-the-grid, centralized systems that today could have substantial benefits to both water supply and wastewater systems. They could reduce the need for clean drinking water and generate less storm water runoff while engaging citizens and private parties in building water-sensitive and safe cities.

To conclude: Drinking water supply systems and artifacts like public wells, water pumps, infiltration canals, pools, canals, and parksingels transcend the purely utilitarian. As modest but important landmarks they have given meaning to and enhanced urban public space. With the introduction of centralized piped water systems, the urban water supply became an invisible and utilitarian affair. At the same time areas of centralized water extraction and treatment quietly triggered the conservation and development of large nature reserves and recreational areas outside cities. Many enjoy these nature areas but few realize they are navigating the very source of their tap water.

References

Baas H, Mobach B, Renes H (2005) Leestekens van het landschap: 188 landschapselemen-ten in kort bestek. Landschapsbeheer Nederland, Utrecht, NL

Baeyens G, Duyve J (1992) Lezen in het Duin. Stadsuitgeverij Amsterdam/Gemeentewaterleidingen Amsterdam, Amsterdam, NL

De Bruin RE et al (red) (2000) 'Een paradijs vol weelde'. Geschiedenis van de stad Utrecht, Matrijs, Utrecht

De Moel P, Verberk J, Van Dijk J (2006) Drinking water: principles and practices. World Scientific Publishing, Singapore

De Vos P (2016) Ergoed Leiden en Omstreken (ELEO). https://www.erfgoedleiden.nl/nieuws/vondst-van-de-week/692 Retrieved (21 sep 2017)

Dijkstra J (1974) De watermakers: honderd jaar drinkwaterleiding rotterdam. Ge-meente Drinkwaterleiding Rotterdam, Rotterdam

Gawronsky G, Veerkamp J (2007) Water uit de kelder: De verdwenen cisterns van Am-sterdam. Monumenten Archeologie Amsterdam 6, Amsterdam

Groen J (1978) Een cent per emmer: het amsterdamse drinkwater door de eeuwen heen. Gemeentewaterleidingen Amsterdam, Am-sterdam

'T Hart (2005) Leven in Utrecht 1850-1914. Groei naar een moderne stad, Uitgeverij Verloren, Hilversum, blz. 83

Hoekstra P (1947) Proeve ener streekgeschiedenis. Wormerveer

Jelles G (1968) Geschiedenis Van Beheer en Gebruik van het Noord-Hollands Duinreservaat. Instituut voor toegepast biologisch onderzoek in de natuur, Arnhem

Jurgens N, Van der Kaay H (2008). http://www.molendatabase.org/molendb.php?step=details&nummer=7184. Retrieved (26 Mar 2018)

Kosman H (1988) Drinken uit de plas 1888-1988: honderd jaar Amsterdamse plassenwater-leiding. VEWIN, Rijswijk

Leeflang K (1974) Ons drinkwater in de stroom van de tijd. VEWIN, Rijswijk

Maris A (2010) Van wel- en waterput naar waterleiding op Schiermonnikoog, Cultuurhistorische vereniging 't Heer en Feer', Schiermonnikoog

Oneindig Noord-Holland (2017) Schoon duinwater uit de Brouwerskolk. https://onh.nl/verhaal/schoon-duinwater-uit-de-brouwerskolk. Retrieved (21 sep 2017)

Rapport aan den Koning (1868) Van de commissie tot onderzoek van drinkwater in verband met den verspreiding van cholera en tot aanwijzing der middelen ter voorziening in zuiver drinkwater. Van Weelden en Mingelen, 's Gravenhage

Regtdoorzee Greup-Roldanus S (1936) Geschiedenis der Haarlemmer bleekerijen. 's-Gravenhage

Van den Noort J, Blauw M (2000) Water naar zee: Geschiedenis van Waterbedrijf Europoort. NV Waterbedrijf Europoort, Rotterdam

Van Ginkel H (2004) Het rijke verleden van Vestingstad Muiden. Waanders Uitgeverij, Zwolle

Vogelzang I (1956) De drinkwatervoorziening van Nederland vóór de aanleg van drinkwaterleidingen. Rijksuniversiteit te Utrecht, Geografisch Instituut, Utrecht

Wijmer S (1992) Water om te drinken. Rijswijk: Vereniging van Exploitanten van Waterleidingbedrijven in Nederland (VEWIN)

Wijntjes W (1982) The water supply of the medieval town, in: Rotterdam Papers IV: a contribution to medieval archeology 1982

Zondergeld-Hamer A (2010) Van drinken uit de gracht tot kraanwater: De drinkwatervoorziening in Weesp van circa 1850 tot 1907. In: Vrienden van't Gooi & Tussen Vecht en Eem. 2010 nr. Water Geschiedenis & Actualiteit. Vereniging van Vrienden van het Gooi Stichting Tussen Vecht en Eem

Suzanne Loen is a Landscape Researcher, Architect, and Guest Tutor at the Department of Urban Planning and Landscape Architecture at Delft University of Technology. Her publications include *Land inSight* and *Water inSight*. Her research includes Dutch water management and the sustainable redesign of industrial kampungs in the peri-urban landscape of Bandung, Indonesia.

Part II
Agricultural Water

Stone bridge across brook in Lodge Park water meadow, Wikimedia, Celuici, released under a Creative Commons Share-Alike 4.0 International license

Chapter 6
Water Meadows as European Agricultural Heritage

Hans Renes, Csaba Centeri, Sebastian Eiter, Bénédicte Gaillard, Alexandra Kruse, Zdeněk Kučera, Oskar Puschmann, Michael Roth and Martina Slámová

Abstract From the Middle Ages until the twentieth century, water meadows in Europe were primarily irrigated to improve their productivity and to lengthen the growing season. They were water management systems designed to collect and use water and to discharge it: water had to be kept moving. This chapter presents a general overview and a history of research on European water meadows. It also examines examples from the sandy landscapes of northwestern Europe, from Slovakia, and Norway. Three main types of water meadows are distinguished: simple dam systems, more elaborate catchworks, and highly developed bedworks. Of these, bedworks were technically and organizationally the most complex; they were also the most costly in construction and maintenance. Most water meadows were abandoned in the twentieth century; in many places, however, their traces can still be

H. Renes (✉)
Faculty of Geosciences, Utrecht University, Utrecht, The Netherlands
e-mail: j.renes@uu.nl

C. Centeri
Institute of Nature Conservation and Landscape Management, Szent István University, Gödöllö, Hungary

S. Eiter · O. Puschmann
Department of Landscape Monitoring, NIBIO, Aas, Norway

B. Gaillard
Brandenburg University of Technology, Senftenberg, Germany

A. Kruse
insitu World Heritage Consulting, Buc, France

Z. Kučera
Department of Social Geography and Regional Development, Charles University, Prague, Czech Republic

M. Roth
Faculty of LandscapeArchitecture, Environment, and Urban Planning, Nürtingen-Geislingen Polytechnic, Nürtingen, Germany

M. Slámová
Department of Landscape Planning and Design, Faculty of Ecology and Environmental Sciences, Technical University in Zvolen, Zvolen, Slovakia

C. Hein (ed.), *Adaptive Strategies for Water Heritage*,
https://doi.org/10.1007/978-3-030-00268-8_6

recognized in the landscape. They are both an interesting part of European agrarian and landscape heritage and a carrier of regional identity. In recent years, a number of water meadows have been restored, for ecological, water management, tourism, and heritage purposes.

Keywords Water meadows · Irrigation · Landscape · Europe · Slovakia · Norway

Introduction

Well-functioning meadows were essential to the survival of farm animals; European farmers, therefore, managed them meticulously, partly with systems of irrigation. Irrigated meadows are known in historical and landscape literature by the general term water meadows (Fig. 1). They must have been successful, as they appeared all over Europe for many centuries (Leibundgut and Kohn 2014a, b; Leibundgut and Vonderstrass 2016). They were still an important feature in the European landscape in the early twentieth century, then fell into disuse, and disappeared from the landscape. Water meadows are now a rare sight, little known to modern farmers and water managers. Even most handbooks on irrigated agriculture do not mention them (see, e.g., Achtnich 1980).

Fig. 1 In 1820 or 1829 John Constable (1776–1837) painted water meadows near Salisbury. This painting is one of the oldest pictures of water meadows. Wiltshire is one of the regions in which extensive traces still exist nowadays. In fact, this particular complex, dating from the seventeenth century, is well preserved and is still managed by the Harnham Water Meadows Trust (Cook 2008; Cook and Cutting 2008). Original: Victoria and Albert Museum, London; released under a Creative Commons Attribution-NonCommercial-NoDerivatives 4.0 International License

Today, ecologists, landscape historians, and, to a lesser degree, water managers are paying more attention to water meadows. Traces of former water meadows have been rediscovered, mapped, restored, or reconstructed. They are valued as heritage and in some regions serve as a resource in small-scale tourism. They also have ecological potential and offer possibilities to local water management.

Water meadows were created by flooding meadows. There seem to have been two main reasons for irrigation of meadows. The first was to have an earlier harvest in the spring. At the end of the winter period, animal fodder became scarce (hence, the hunger gap); with flooding, which kept temperature at a constant and relatively high level, spring grass could start to grow and animals could begin to graze weeks earlier. The second reason for water meadows was its role in gaining higher productivity. When hay was harvested, meadows normally continuously lost organic material and minerals; however, irrigation kept conditions wet during summer and meadows—to which silt and lime were sometimes added—were kept productive (Williamson and Cook 2007, pp. 1–2; Cook et al. 2003). Dutch research from the end of the nineteenth century shows that the harvest from water meadows was more constant than that from regular meadows, particularly in dry years (Thissen and Meijer 1991).

The chemical effect of flooding is more complicated. The Swedish authors Emanuelsson and Möller (1990) have stressed that it can act as a sink to phosphorus and nitrogen, thereby reducing the need for artificial fertilizers. English research has confirmed the effect on phosphorus but has found no evidence for that on nitrate (Cook et al. 2003). Flooding may also have had the effect of diminishing the numbers of mice and grubs (Ewald and Klaus 2010, p. 118).

Some authors note additional reasons for the use of water meadows and make clear that regional differences were present. Whereas in temperate parts of Europe, flooding was used to avoid frost, in the North of Sweden winter irrigation was used to create an ice crust on a wet meadow, which prevented bushes and trees from springing up (Emanuelsson and Möller 1990, p. 136). In parts of the Alps, the amount of rainfall affected the use of water meadows. Drainage channels, *waale*, in South Tirol were concentrated in the driest places (Hallmann and Peters 1995); in the Mediterranean, a long tradition of irrigation made productive agriculture possible in dry regions (Emanuelsson 2009, p. 283). Most Mediterranean irrigation systems are primarily applied to the production of food crops and are not considered here—in this chapter, we focus on the use of irrigation in meadows.

In recent years, the growing interest in water meadows has led to much local research as well as to a number of survey studies. Notably, the Swiss geographer Leibundgut (Leibundgut and Kohn 2014a, b; Leibundgut and Vonderstrass 2016) describes meadow irrigation as an element of a larger group of activities under the banner of traditional irrigation, focusing on Central Europe. In this chapter, we offer additional information, particularly on Atlantic Europe, and look more closely at the heritage of this form of water management. After a section on the history of research on water meadows, we present a typology and history, followed by three regional case studies. In the final section of this study, we address heritage explicitly.

History of Research

Until the early twentieth century, most research and writing on water meadows came from agronomists, who were interested in how they functioned and were maintained.

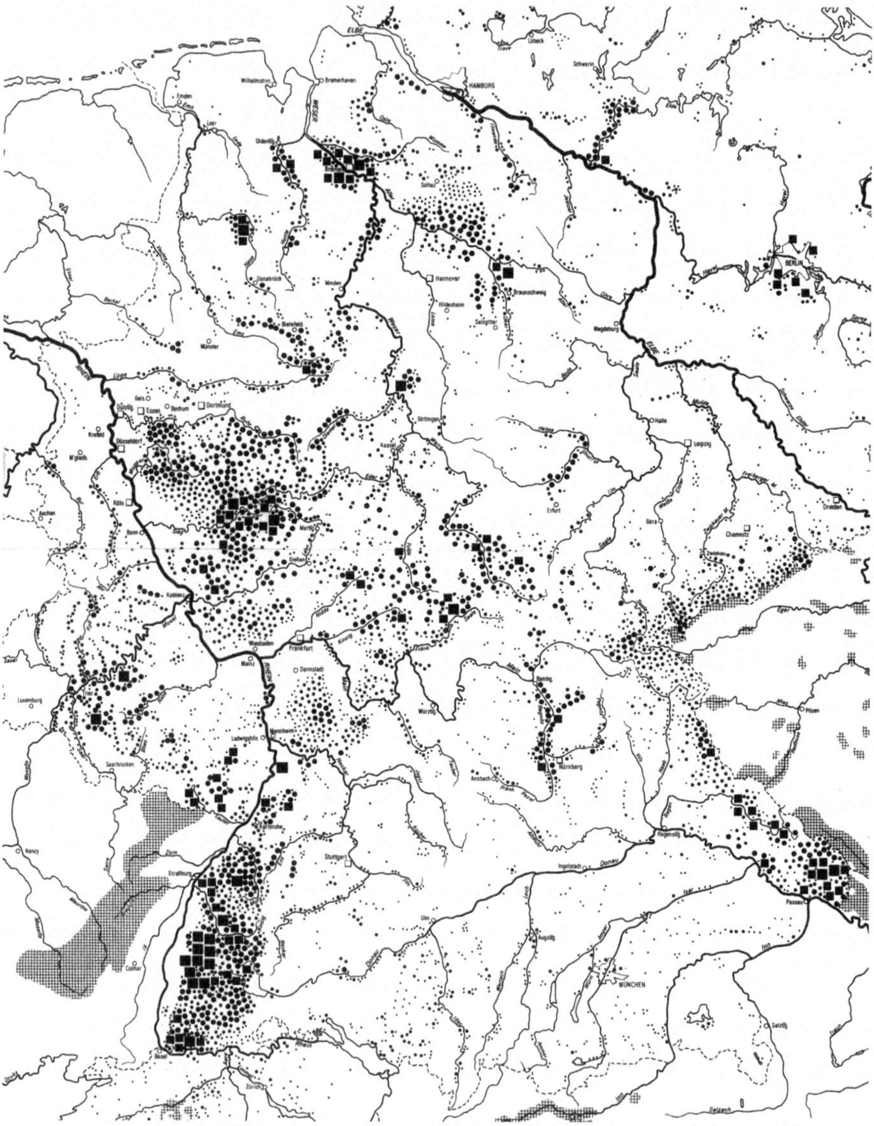

Fig. 2 Troll's map of water meadows ('Rieselwerke') in Central Europe in 1937. From: Böhm 1990; released under a Creative Commons Attribution-NonCommercial-NoDerivatives 4.0 International License

More than one hundred sixty references can be found in German libraries, ranging from 1743 to the present. In the second half of the twentieth century, when only a few water meadows were still functioning and the system was largely a relic, new groups became interested in the subject, among them historians, ecologists, and water managers. At that time, agricultural historians in the UK began studying water meadows (e.g., Kerridge 1954). Later, landscape historians and landscape archaeologists discovered the subject; their research, more highly developed in the UK than elsewhere, combines detailed and well-documented fieldwork with old maps and other archival sources along with the personal experiences of farmers who developed their knowledge of the workings of water meadows from direct experience (Cook and Williamson 1999, 2007; Brown 2005). In recent years, many publications in Germany and elsewhere in Central Europe have come from ecologists, who have often developed their studies as components of restoration projects.

There are few detailed maps of the historic distribution of water meadows. A map of water meadows in a large region of Germany was published by Böhm (1990) and was based on data collected by the famous German ecologist Carl Troll between 1939 and 1946, the last period in which a substantial number were still functioning (shown in Fig. 2). Troll was aware of the human influence on landscapes and advocated a geographical science that combined physical geography with human geography and ecology. Other distribution maps have also been constructed. Among others, the ecologist Baaijens and collaborators recently mapped five hundred systems identified in Dutch nature reserves (Baaijens et al. 2011, p. 21), and Emanuelsson and Möller (1990) mapped the nineteenth-century systems of Scania, in the South of Sweden. These maps use different criteria, classifications, and sources which makes them difficult to compare. A further map of European water meadows was produced by Leibundgut (Leibundgut and Vonderstrass 2016, pp. 1–53). Although it is rather simplified, this map does show the widespread occurrence of these systems in mountains and lowlands.

Types of Water Meadows

To understand the landscapes of water meadows, it is essential to keep in mind that the water had to be kept moving. Therefore, installations had to be deployed to flood meadows and to drain them. The diagrams exhibited in Fig. 3 show the complex annual program of some English water meadows. More such time schedules are available; together, they show a huge variety in approaches to management. Flooding could occur in different seasons, although emphasis was on winter. In one Swiss case, meadows were flooded between November 11 and the beginning of May (Ewald and Klaus 2010, p. 120).

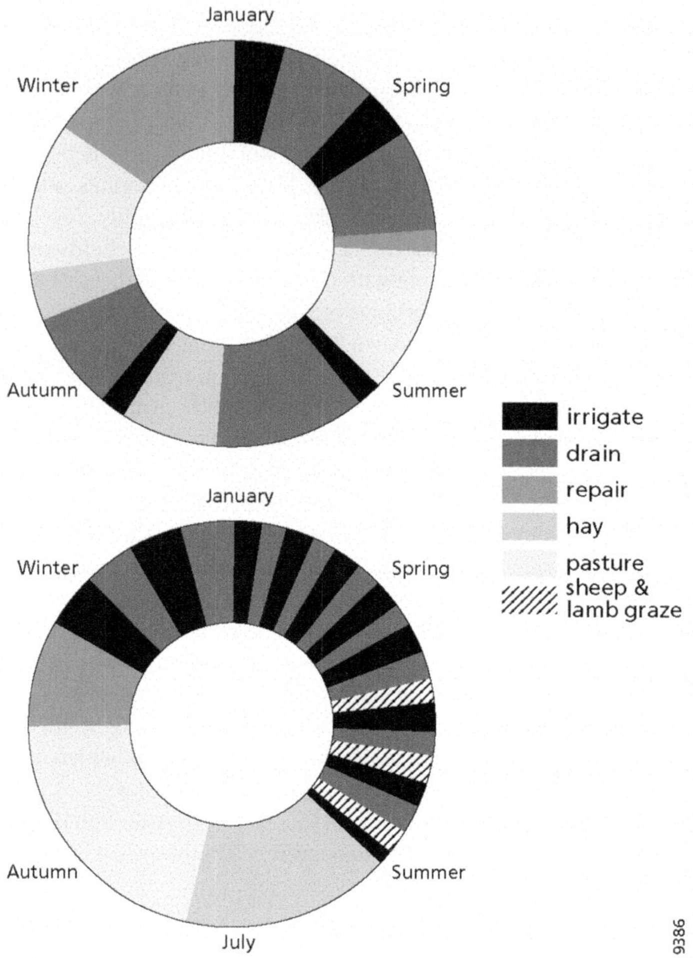

Fig. 3 Two yearly cycles from English sources. Above: Wye Valley, Surrey, reconstructed from James Simmons's diary; below: late nineteenth century as described in Encyclopedia Britannica, 1880. *Source* Cook and Williamson 2007, p. 114 (originals: Dr Kathy Stearne); released under a Creative Commons Attribution-NonCommercial-NoDerivatives 4.0 International License

In the functioning and layout of water meadows, we can discern three main types. Our typology is based on technical workings, which seems most relevant for a landscape approach. Leibundgut presented a different approach, based on natural habitat, technology, and social background. On these bases, he discerned four types of irrigation systems: namely, Alpine, Alpine Piedmont, Low Mountain Range, and Lowland, each of which is characterized by specific designs and cooperative arrangements (Leibundgut 2004).

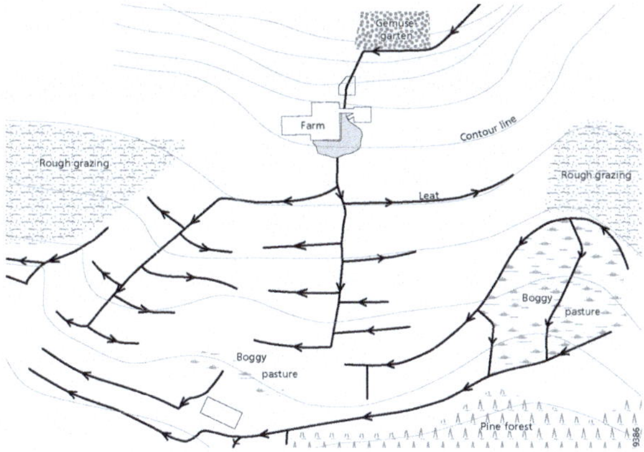

Fig. 4 Catchment work in the Black Forest, after Böhm (1990)

First in our typology are the most primitive installations or dam systems. A small dam with a sluice in a stream was sufficient for inundating adjoining meadows; these installations may be medieval or even earlier. In some cases, temporary structures, for example, wooden fences in a stream, were used—among them, the so-called wild inundations noted by Brinckmann (2015, p. 312). A related type of dam system is the deliberate flooding of land with water from rivers, meant not only to bring water but also silt to the land. Most such small-scale water meadows were developed by individual farmers; they left almost no traces in written records and did not attract much interest from early agronomists.

Catchworks, or catch meadows, are the second type of meadow and are located on slopes (shown in Fig. 4). Water flows through a leat or canal in the upper part of a meadow and then flows in a kind of film over the land. In more elaborate types, a number of embankments cause the water to flow in different steps. In a variation in Romania (Botzan et al. 2004, pp. 598–605), the water is allowed to overflow or to pass through holes dug in the embankment. In the Netherlands, such water meadows have been found in stream valleys, showing that even small differences in height can be used to effect in these systems.

Leats, the often short but sometimes very long canals that branch off streams and run almost horizontally to them, are the main indicators of catchworks. They can be spectacular: In Switzerland, they cross valleys by way of aqueducts (Konold 1994; Leibundgut and Vonderstrass 2016). They can also be much smaller; it can therefore be difficult to distinguish between smaller canals and mill leats, as irrigation and water mills use the same water. Though they seem to be in competition, this distinction is perhaps not too important. Most irrigated water collects at the lower end of a meadow and is then available to mills (Hallmann and Peters 1995). Moreover, irrigation is not a year round activity. Archival sources show that farmers would accord with mill owners on alternating use of the water. In 1663, we find that the owner of a paper mill in the Dutch village of Loenen, in Veluwe, the Netherlands, developed an agreement

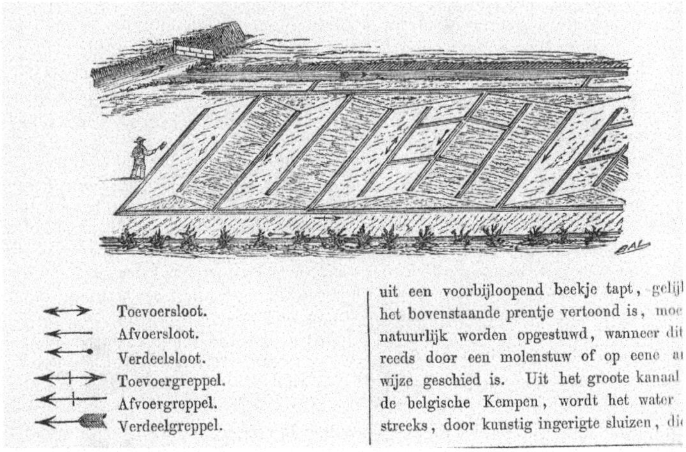

Fig. 5 Bedworks in a nineteenth-century agricultural handbook. *Source* Staring 1868, pp. 903–904; released under a Creative Commons Attribution-NonCommercial-NoDerivatives 4.0 International License

with the farming community where the miller would clean the stream in the month of May and would after this go on to water the meadows and marshes for two weeks (Kobussen 1997, p. 80). Elsewhere in Europe, many farmers profited from irrigation of their lands by mill leats (Leibundgut and Vonderstrass 2016, p. 2, 265).

The third type of meadow irrigation system, bedworks, is the most developed and most labor-intensive method; hence, it is the most expensive (shown in Fig. 5). However, it is also the type that is most easily recognizable in the landscape. The high operational costs apply to initial layout, which transforms the meadow's surface; this initial work also calls for employing a specialist to manage and maintain the meadow—who, in England, is known as a drowner. It also seems to be the most recently developed type. The oldest examples date from the seventeenth century and they are characteristic of the Early Modern Period (Cook et al. 2003). In the Netherlands, these systems principally date from the second half of the nineteenth century. The use of water meadows has been widespread in Europe. Our hypothesis is that they are most abundant in regions where meadows (hence, animal husbandry) are an important part of the agrarian system and, secondarily, where there is a reasonable population density. The third factor is that physical circumstances must be favorable (where water, e.g., contains calcium).

Historical Development

It is often difficult to date traces of farmers' activities in the landscape. That is certainly true in the case of water meadows and especially so for dam systems and catchworks. Little knowledge can be gleaned from written sources, as medieval sources are not always easy to interpret and the toponymical sources, which are important here, are particularly tricky. The toponym Brühl, found in written sources in different parts of Germany from the eighth century onwards, is one such case, which has variously been explained as designating a hunting park and, more recently, as an index of irrigated meadows (Konold 2004, p. 20). Nevertheless, many such systems have been dated to the medieval period and possibly even to the Roman period (Konold 2004, p. 19).

Medieval irrigation, primarily of meadows, was known in different parts of Europe, including areas outside the Mediterranean such as Lower Austria and the Allgäu in the twelfth century (Cate et al. 2004, p. 217) and around tributaries of the Danube in southwestern Germany during the fourteenth century. In some cases, these installations consisted of weirs (Konold and Popp 2004, pp. 23–25). Perhaps the most fascinating among them are the catchworks built by Norse colonists on Greenland, the remains of whose agricultural activities could only have been medieval, as the Norse colony on Greenland was founded during the tenth century and perished during the fifteenth century (Krogh 1982; Arneborg 2005; Adderley and Simpson 2006). Recent research has confirmed the existence of these water meadows, although aspects of the watercourses and ridges are of a natural origin (Edwards and Schofield 2013). The earliest irrigation techniques used may be connected to a bishop from Norway and field reorganization occurring around 1126 (Panagiotakopulu and Buckland 2012; Panagiotakopulu et al. 2012). This evidence suggests that catchworks were common practice in Western Norway at this time.

The High Middle Ages were a period of high population pressure and, consequently, intensive agriculture. This fact has been used to argue that a peak in the use of irrigation systems occurred in Europe during the thirteenth and fourteenth centuries (Leibundgut and Vonderstrass 2016, p. 1–26). Insufficient data causes this position to remain hypothetical. In the Netherlands' sandy landscapes, hay meadows were divided and became private property during the Late Middle Ages (Vervloet 2010), a fact which suggests intensification of agriculture during a period that is generally described as a period of crisis.

Many old irrigation systems were abandoned in the course of time (Konold 2004, p. 20) until interest in flooding revived, particularly among eighteenth- and nineteenth-century estate owners. Agronomists became newly interested, and agricultural experiments and improvements were fashionable and brought prestige to the estates. The more elaborate water meadow type appeared in handbooks and was propagated in agronomic journals (Konold and Popp 2004, p. 25). Figure 5 shows bedworks by the Dutch agronomist Staring. Staring was an exceptional case, as he also showed an interest in the simple, small-scale, older works that most scientific writers did not find worth mentioning.

The nineteenth century is often described as the last peak in water meadows. In southwestern Germany, new water meadows were laid out during the middle

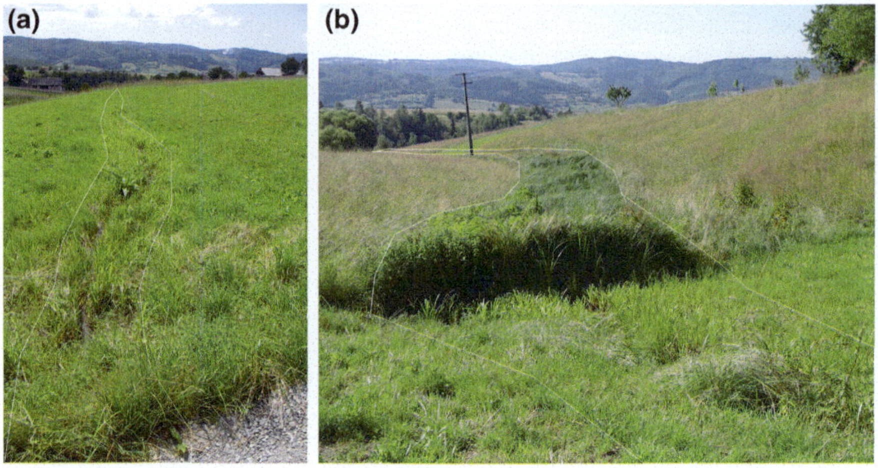

Fig. 6 Water meadows in the Hriňová-district, Slovakia: actively used (**a**) and abandoned (**b**). *Source* Dr Martina Slámová

decades of the century, although proponents of meadows were often disappointed by the slow pace of the work. Many of these survived into the twentieth century. This new interest led to the introduction of meadow irrigation into regions where it had been unknown, including Norfolk, England (Wade-Martins and Williamson 1994), the Czech Republic (Dvořák et al. 2004, p. 312), Slovakia (Fig. 6), Hungary (Fejér 2004, p. 400), and some of the poor heathland regions of the Low Countries.

Regional differences in the development of agriculture led to regionally distinct developments and chronologies in the use of water meadows. In Scania, Sweden, the growing amount of arable land outpaced the supply of manure in the middle of the nineteenth century and the following decades showed a growth of water meadows until the arrival of artificial fertilizers ended the practice (Emanuelsson and Möller 1990). In mid-nineteenth-century Switzerland, on the other hand, growing imports of grain, facilitated by a growing railway network, diminished home production of grain and triggered a shift to dairy farming. The combination of less grain and more cattle, in its turn, led to a shortage of straw. Many meadows were converted into *streuwiesen* (straw meadows) and investment in their productivity came to include irrigation (Ewald and Klaus 2010, p. 121).

Around 1950, a substantial number of water meadows were still used and maintained; most have since been abandoned. Nature reserves sometimes preserved their traces; however, industrial agriculture, collectivization in Eastern Europe, land consolidation in Western Europe, and ancillary works erased them, along with other field structures, from agricultural land.

Case Study One: The Sandy Landscapes of Northwestern Europe

The zone of sandy regions that runs from Flanders through the Netherlands and northwest Germany to western Denmark is characterized by soils with low natural productivity. The problem that occupied the agricultural sector most during the nineteenth century was how to turn the extensive heathlands in these regions into decent agricultural land.

For centuries, the heathlands had functioned within a system of mixed farming, in which small open fields and individual enclosures were used for crops, particularly for rye. Cattle and sheep were kept for meat and other products, but primarily they were held for their manure, which fertilized the arable land. Heathland occupied more than half of these sandy regions, and large-scale reclamations were impossible because of a lack of manure. Old water meadow systems seem to have existed in this region, even where a substantial part of the surface water came from heathlands and was acid as well as poor in minerals. In northwest Germany, such meadows were often organized as cooperatives (*Genossenschaften*) (Hoppe 2001). In Germany's Lower Saxony, systems of small canals, ditches, and ponds (*Weiher*) collected water from houses, streets, and stables and brought it into the meadows to increase their productivity. These systems were probably most in use during the nineteenth century.

In Belgium, new experiments were started in the 1840s. In 1849, for example, the Belgian Government began one of the largest irrigation projects to date, the reclamation of heathland near Lommel (Jansen 2015) (Fig. 7). These experiments received immediate attention in the Netherlands; experiments were also begun there soon after. From the 1840s to the 1860s, regional farming organizations and the Dutch Government encouraged farmers to build water meadows. They also invested in a new national network of canals, whose side effect was to open up some of the sandy region's more isolated parts. The canals' water level was often higher than the adjoining low-lying heathlands, which made them perfectly suited for irrigation (Fig. 8).

In the final decade of the nineteenth century, the interest in water meadows revived once again. This time, the *Nederlandsche Heidemaatschappij*, the Dutch Heathland Company, had a primary role; it had been founded in 1888 on the model of the Danish Heathland Company and also active in the construction of water meadows (Emanuelsson 2009, p. 285). Behind the society were large landowners from the eastern provinces; particularly prominent was a group of large textile manufacturers that had recently started to establish themselves as a new landed gentry. Members of this class also sat on the State Commission on Irrigation, formed in 1893, which recommended the system four years later. In the next few years, it was mainly the Dutch Heathland Company that realized a number of new water meadows (Thissen and Meijer 1991).

Fig. 7 Bedworks at Lommel (Belgium). Onroerend Erfgoed Vlaanderen (Hilde Verboven), nr DSCN0037; released under a Creative Commons Attribution-NonCommercial-NoDerivatives 4.0 International License

Fig. 8 Irrigated meadows near Weert (Limburg, NL), laid out by the Maastricht industrialist J. Chainaye during the 1850s, using water from the Zuid-Willemsvaart. With 110 ha, it was the largest complex of water meadows in the Netherlands. The system remained in use for many decades and is still recognizable in the field. Left: topographical map, scale 1:25.000, sheets 725 (1901) and 726 (1896), ©Kadaster, Apeldoorn (NL). Right: photograph J. Renes; released under a Creative Commons Attribution-NonCommercial-NoDerivatives 4.0 International License

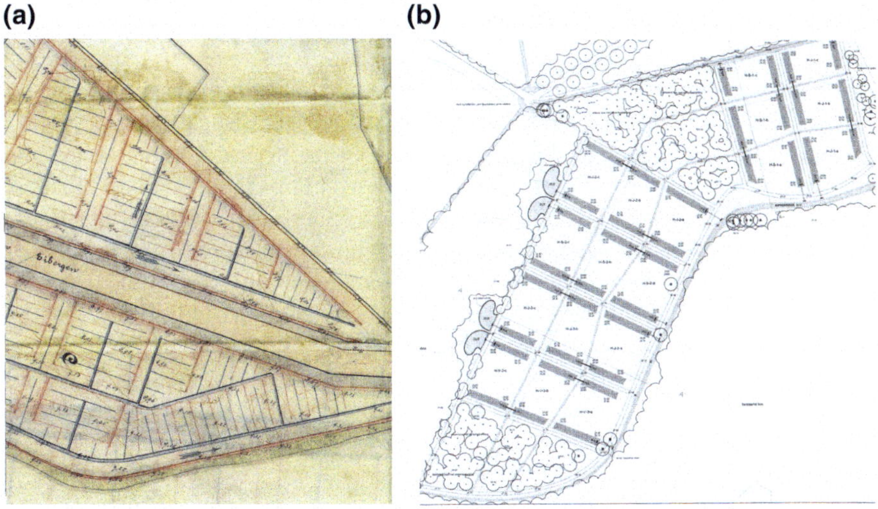

Fig. 9 Water meadows at Lankheet (Overijssel, NL). Left: two fragments of the original plan for the Lankheet. Upper right: the restoration plan by landscape architect Berno Strootman. *Source* Baaijens et al. 2011; released under a Creative Commons Attribution-NonCommercial-NoDerivatives 4.0 International License

Two Dutch historical geographers who conducted research on nineteenth-century reclamations and found evidence of water meadows started the most recent rediscovery of water meadows in a 1991 article (Thissen and Meijer 1991). They focused on water meadows as a nineteenth-century agricultural innovation. Belgian ecologist Joël Burny (1999) had already begun to interview farmers about old agricultural practices in northern Flanders's sandy landscapes. One of his 1999 book's most spectacular parts was the detailed information on innovations to water meadows made in this region (for a recent update, see Jansen 2015). Shortly after, Dutch ecologist Gert-Jan Baaijens (Baaijens et al. 2001; Baaijens et al. 2011) observed structures in the landscape that he interpreted as traces of former irrigation. Although his findings initially met with skepticism (from the author of this case study among others), his basic idea has been confirmed by archival research and is now generally accepted (Brinckmann 2015). Still more historical research is badly needed.

Some water meadows have been restored in recent years. Among the most impressive is the Lankheet landed estate (shown in Fig. 9). Here, the former water meadows are used for water storage, featuring reed swamps to purify water (by taking out phosphates and nitrates) for irrigation and nature development as well as to supply biomass for energy (Baaijens et al. 2011). The estate's restorations are an excellent example of adaptive reuse of a historic water management system.

Case Study Two: The Hriňová District, Slovakia

The Hriňová region is situated in one of five regions in Slovakia that government has classified as a traditional landscape with scattered settlements (Petrovič 2006), indicating that it continues to show many traces of the agricultural landscape and settlement patterns practiced before the collectivization of the late 1940s. Settlers from Wallachia colonized this area from the sixteenth to eighteenth centuries (Ira et al. 2008). The Riecka catchment of this region has an area of eight hundred twenty-nine hectares; more than ninety percent consists of this "traditional agricultural landscape"; remaining parts are comprised of forests and a handful of residential areas (Slámová et al. 2015). The area is near the Polana Protected Landscape Area, which became a UNESCO Biosphere Reserve (BR) in 1990. Biosphere Reserves are divided into three zones: a core zone, a buffer zone, and a transitional zone; designated "traditional agricultural landscapes," such as the catchworks, are elements of the transitional zone (Fabriciusova et al. 2015) (Fig. 10).

No literature on historical irrigated meadows in Slovakia exists. However, our recent fieldwork in the Hriňová area of Central Slovakia, which included interviews with local inhabitants, has resulted in new insights on catchworks. Using a Global Navigation Satellite System (GNSS) Leica GS05, 3736.68 m of catchworks was mapped in 2013 in the Riecka catchment (Slámová et al. 2015). It was observed that 42.82% of these catchworks were functional, 2.28% was partially functional, and 53.59% was non-functional.

Today, catchworks are mainly placed on slopes with meadows and pastures and correspond to sites and habitats of European significance (as designated by

Fig. 10 Water meadows in the Hriňová-district, Slovakia. *Source* Dr Martina Slámová; released under a Creative Commons Attribution-NonCommercial-NoDerivatives 4.0 International License

NATURA 2000) or national significance (Slámová et al. 2015). To understand the history of land use, we looked at those sites on three series of historical maps: military mapping (1764–1787), the military map (1810–1869), and topographical maps (1957–1971). The topographical maps show the sites of today's catchwork as agricultural, whereas the nineteenth-century maps show one-third of that land as still forested; the eighteenth-century maps show the majority as forested. These facts suggest that most catchworks date from the eighteenth and nineteenth centuries.

Recently, the administrative body of the Protected Landscape Area and the Biosphere Reserve Polana initiated a project on water meadows in the area which focuses on identifying landscape features and disseminating scientific knowledge among stakeholders and the community. The Action Plan for the Polana Biosphere Reserve for 2014–2018 is an innovative management model. It includes a coordination board, or advisory body that operates in the territory, in which the management of the Reserve as well as other stakeholders are represented. The Action Plan also encompasses a strategy for enhancing regional identity and landscape awareness among local inhabitants. It covers all forms of cultural, historical, and natural heritage (Fabriciusova et al. 2015).

Catchworks are a sophisticated historical irrigation system that is part of the landscape's heritage. A catchwork system retains water and supports the diversity of habitats and land covers. It is not yet clear if the spatial distribution of habitats positively correlates with those of catchworks. Although catchworks provide adequate soil conditions (that is, moisture) for the presence of habitats that include hydrophilous tall herb fringe communities found in the plains and montane to alpine

Fig. 11 Part of historical irrigation canal at Måbø farm, inner Hardangerfjord region, W-Norway (*Photograph* Sebastian Eiter/NIBIO); released under a Creative Commons Attribution-NonCommercial-NoDerivatives 4.0 International License

levels as well as moist or wet eutrophic and mesotrophic grasslands, these habitats were only observed locally around springs or streams. Permanent mesotrophic pastures, aftermath-grazed meadows, and lowland hay meadows are maintained by mowing, fertilizing, and traditional management (Slámová et al. 2015, 2016) (shown in Fig. 11).

Case Study Three: Norway

There is no specific term for water meadows in the Norwegian language. A hay meadow in general is called an *eng* or, when it refers specifically mowing, a *slåtteeng*. Most parts of Norway have always had enough precipitation for cultivation of farmland, including meadows, without recourse to irrigation. Exceptions are the farming settlements in the northern part of the Gudbrandsdalen Valley and the innermost fjord districts of Western Norway (Aamo 1972; Christensen 1997). Northern Gudbrandsdalen is located in the most extreme rain shadow, east of the main Scandes mountain ridge, and receives as little as about three hundred millimeters of annual precipitation. The inner fjord districts of Sogn and Hardanger have a somewhat higher amount of precipitation, but they feature steep slopes and coarse soil with a low capacity for water retention (Fig. 12).

Fig. 12 Half-open wooden pipe to lead irrigation water at Søre Neset farm, Vågå, E-Norway (*Photograph* Hans Joramo/De Sandvigske Samlinger Maihaugen); released under a Creative Commons Attribution-NonCommercial-NoDerivatives 4.0 International License

Historically, in this region, systems of small canals were built, which were sometimes combined with half-open wooden pipes to bring water from the mountains to farmland (Fig. 12). These canals, generally called *vassveiter*, occur with local variations. Local terms for a half-open pipe include *tro* (and its plural *trø*), *dœle,* or *lekja* (Ile 1958). These irrigation systems are up to several hundred years old and several kilometers long (Christensen 1997). They not only irrigated farmland but also sent water to farmsteads for household use (Dokken et al. 1999). Maintenance of these systems consumed substantial efforts and had to be carried out annually; maintenance was possibly also carried out more frequently on demand (Kjøk 1979; Dokken et al. 1999) and was performed with care: Even wind shields were built to prevent snow from blowing away from farmland before it melted (Sømme 1954). However, relatively high precipitation, which could be accompanied by a melting of snow right before the vegetation season, in some years and places could reduce or even eliminate the need for irrigation (Sømme 1954). Since the beginning of the twentieth century, it has become more common to use sprinkler irrigation: Motorized pumps bring river water from the valley bottom through tubes and then spray it on the fields. Farmers who could not afford to make larger investments lost their water supply for some time during the transition from the old to the new system (Kjøk 1979). Some farmers still to some degree use mountain water directly (Sømme 1954; Christensen 1997), or, indirectly, when they generate hydroelectricity to run the pumps.

Politicians quickly recognized the importance of irrigation to the productivity of meadows and other farmland. Subsidy schemes for irrigation were introduced from the early twentieth century and applied regularly around the Second World War (Borgedal 1966). In 1939 about seventy-six hectares of farmland were irrigated nation-wide; thirty-five hectares of this was grassland (Borgedal 1966). By 1959, irrigation capacity in Norway had increased to one hundred eighty hectares, with an expected further increase (Borgedal 1966).

Some irrigation canals were still maintained and functioning as late as the 1990s (Christensen 1997). The once high importance of these historical irrigation systems to the community and farming economy is reflected well in their rather strong local appreciation as cultural landscape heritage. Their descriptions can be found in local historical literature, while maintenance and restoration is performed as part of cultural history hiking trails or outdoor museums. National awareness of historical grassland irrigation, however, is clearly lower.

Cultural heritage in Norway from 1537 or earlier is automatically protected by law. Newer heritage sites must be protected individually. We do not expect many farmland watering systems to be old enough for automatic protection, nor do we know of any instances of individual protection in law or as part of a protected area. There should be ways to fund restoration and maintenance of water meadows, nevertheless, perhaps within a scheme of environmental measures in agriculture under the auspices of local authority districts (*special miljøtiltak i jordbruket*), the national cultural heritage fund (*Kulturminnefondet*), or as part of valuable agricultural landscapes (*utvalgte kulturlandskap i jordbruket*).

The Rediscovery of Water Meadows in the Twentieth Century

Following the nineteenth-century success of water meadows, new developments in agriculture led to their gradual disappearance. The main factor in their decline was the introduction of cheap artificial fertilizers that ended the problem of scarcity of manure. Although new water meadows were established during the first half of the twentieth century in some localities (Dvořák et al. 2004, p. 314), they disappeared elsewhere quickly. Most water meadows were abandoned during the twentieth century; in many cases, their very traces disappeared. The most prominent type, bedworks, was most threatened. Many were leveled since they were inflexible in use; less suited for pasture, as cattle could destroy elements of beds and ditches; and both capital and labor-intensive. A national survey of ancient meadows and pasture in Sweden (conducted from 1985–1997) produced only negligible figures for the artificial flooded meadow and artificial watered meadow categories: One hundred seventy-three and fifty-eight hectares, respectively, were found in the entire country (Ihse and Lindahl 2000, p.). In Switzerland, water meadows are still in use solely in the small region of Langenthal. Even here, they had almost disappeared (diminishing over time from seven hundred hectares around 1900, eighty hectares in 1984, to very few fields today) when local initiatives were started to revive the old irrigation systems (Ewald and Klaus 2010, p. 118; Leibundgut and Vonderstrass 2016, p. 2–297).

In recent decades, interest in water meadows has been growing again, primarily among ecologists and landscape historians (Cook et al. 2003; Baaijens et al. 2011). Several regions have restored water meadow as has been noted and listed by Leibundgut and Kohn (2014b). A landmark publication by Leibundgut and Vonderstrass (2016) describes a number of water meadows that have been restored and reused. In the eastern part of the Czech Republic, the Josefovské Grasslands Ornithological Park, a complex of irrigated grasslands in the floodplain of the Metuje River, has been in the process of restoration since 2008. In the Netherlands, recent interest from ecologists has brought the first restorations and reconstructions, notably those at Lankheet. In Germany, several nature restoration projects for safeguarding wetlands or for retaining biodiversity on alp meadows where water meadows are maintained or have been reconstructed are underway. In Western France, the region of Basse-Normandie and the European Agricultural Fund for Rural Development (Feader 2010) subsidizes farmers to preserve water meadows in the Cotentin and Bessin wetland of the Nature 2000 framework (Feader 2010). Other such projects can be found in England and Sweden (Emanuelsson 2009, p. 287).

What interests ecologists about water meadows are the specific flora and fauna that thrive there and arise due in part to the wet environment and in part to specific management practices (Cummings and Cutting 2007). Water meadows also conform to the European programs and subsidies for maintaining high nature value grasslands (Müller 2017; Keenleyside and Oppermann 2009).

Conclusion

The main reason to preserve and restore water meadows is heritage. They are remnants of past agriculture, showing the value and scarcity of meadows in a period in which the existence of farmers was always threatened by natural and economic fluctuations. Historic landscapes tell the story of past struggles for survival and show how the efforts to improve agriculture led to local solutions and specializations and, hence, to characteristic landscapes.

Western Europe in particular has a long tradition of describing and protecting the landscapes and individual relics of past agrarian practices for their cultural, aesthetic, and ecological values (Niemeier 1961; Antrop 2005; Council of Europe 2000). Officials and citizens alike see these landscapes as a building block for local and regional identity and as a major resource for rural tourism, ecological restoration, and multifunctional agriculture. There are plans to prepare a proposal for inscribing a selection of European water meadows as World Heritage Sites in UNESCO's World Heritage Program (Leibundgut and Vonderstrass 2016, p. 1–234). In recent years, the protection of historic agrarian landscapes has been extended past Europe, again through the World Heritage Program (Von Droste et al. 1995).

Although the traces of historic water meadows are subtle and will not get much attention from the majority of tourists, they add a story to the landscape interesting to small-scale ecotourists, particularly walkers. In parts of the Alps, the irrigation canals are a normal feature in tourist leaflets and footpath descriptions. Examples can be found in Tirol, where the leats are known as Waale, and in Valais (Switzerland; Crook and Jones 1999). A few places present water meadows as ecomuseums (Leibundgut and Vonderstrass 2016, p. 2–110), as, for example, the Harnham complex once painted by Constable (Fig. 13).

A final argument for protecting water meadows is water management. Their specific structures, designed for both irrigation and drainage, makes them useful for temporary storage of water. This possibility has become more visible in recent years, as water management is increasingly combined with ecological aims, stimulated by new legislation such as the EU Water Framework Directive. After centuries of focusing on maximum drainage on behalf of agriculture, water managers now seek the efficient use of local water, which means, for example, upstream water retention. Reactivating water meadows can be part of such measures that may become ever more relevant as climate change brings stronger fluctuations in precipitation.

Fig. 13 Harnham water meadows nowadays. In the background Salisbury cathedral is seen, making this one of the most iconic water meadows landscapes. *Photograph* Lorraine Blakey; released under a Creative Commons Attribution-NonCommercial-NoDerivatives 4.0 International License

Water meadows are probably not very well known to the general public, but they are an interesting part of European water heritage. In the past, they were widespread and in all their variety they are demonstrations of past creativity and innovation by farmers. Most water meadows lost their function over the twentieth century, and many systems have disappeared without a trace. However, some systems have been used until recently or even into the present, and traces of abandoned systems have been rediscovered in recent years. Their future survival depends mainly on new functions: ecology, tourism, or water retention in the face of climate change.

Acknowledgements The authors are cooperating in the Institute for Research on European Agricultural Landscapes e.V. (www.eucalandnetwork.net/), a network that discussed the subject on a workshop at Utrecht University (The Netherlands), from April 24, 2013 to April 26, 2013. The case studies are written by Hans Renes (case study 1), Martina Slámová (case study 2), and Sebastian Eiter and Oskar Puschmann (case study 3). The authors thank Ton Markus (Faculty of Geosciences Utrecht University) for drawing Figs. 3 and 4.

References

Aamo ES (1972) Ungdom og landbruk. Landbruksforlaget, (2nd ed), Oslo

Achtnich W (1980) Bewässerungslandbau. Ulmer, Stuttgart

Adderley WP, Simpson IA (2006) Soils and palaeo-climate based evidence for irrigation requirements in Norse Greenland. J Archaeol Sci 33:1666–1679

Antrop M (2005) Why landscapes of the past are important for the future. Landscape Urban Plan 70:21–34

Arneborg J (2005) Greenland irrigation systems on a West Nordic background; an overview of the evidence of irrigation systems in Norse Greenland c. 980-1450 AD'. In: J. Klápšte (ed) Water Management in Medieval Rural Economy (Pamatky Archeologicke Supplement 17, Ruralia V), pp 137–145

Baaijens GJ, Evers FH, Grootjans AP (2001) Traditionele bevloeiing van grasland; een studie naar vroegere bevloeiing van reservaten in pleistoceen Nederland, alsmede enkele boezemlanden. Expertisecentrum LNV, Wageningen (OBN-rapport 18)

Baaijens GJ, Brinckmann E, Dauvellier PL, van der Molen PC (2011) Stromend landschap; vloeiweidenstelsels in Nederland. Zeist, KNNV Uitgeverij

Böhm H (1990) Die Wiesenbewässerungen in Mitteleuropa 1937; Anmerkungen zu einer Karte von C. Troll; zur Erinnerung an den 90. Geburtstag von Carl Troll (24.12.1899-21-7-1975). Erdkunde 44 (1): 1–10

Borgedal P (1966) Norges jordbruk i nyere tid, Bind 1 Planteproduksjonen. Bøndenes forlag, Oslo

Botzan M, Haret C, Mihnea I (2004) History of irrigation, drainage and flood control in Romania. In: Csekö G, Hayde L (eds) Danube Valley; history of irrigation, drainage and flood control. ICID-CIID, New Delhi, pp 531–677

Brinckmann E (2015) Stromen en overstromen. Bevloeien van grasland: traditie met toekomst? In: Spek T, Elerie H, Bakker JP, Noordhoff I (eds) Landschapsbiografie van de Drentsche Aa. Van Gorcum, Assen, pp 312–317

Brown G (2005) Irrigation of water meadows in England. In: J. Klapste (ed) Water Management in Medieval Rural Economy (Pamatky archeologicke Supplement 17, Ruralia V), pp 84–92

Burny J (1999) Bijdrage tot de historische ecologie van de Limburgse Kempen (1910–1950); tweehonderd gesprekken samengevat. Natuurhistorisch Genootschap in Limburg, Maastricht (Reeks 42, afl. 1)

Cate FM, Leibnitz F, Mottl W (2004) History of irrigation, drainage and flood control in Austria; general review. In: Csekö G, Hayde L (eds) Danube Valley; history of irrigation, drainage and flood control. ICID-CIID, New Delhi, pp 219–234

Christensen AL (1997) Vassdragene—blodårer i landskapet. In: Larsen T (ed) Kulturminnestafetten -97: om kulturminner langs kyst og vannveier. Forbundet KYSTEN, Oslo, pp 25–28

Cook H (2008) Evolution of a floodplain landscape: a case study of the Harnham Water Meadows at Salisbury, England. Landscapes 9–1:50–73

Cook H, Cutting R (2008) The Harnham Water Meadows Trust: complex management for a heritage landscape by the Voluntary sector. J Farm Manag 13(5):1–8

Cook H, Stearne K, Williamson T (2003) The origins of water meadows in England. Agricultural History Review 51(2):155–162

Cook H, Williamson T (eds) (1999) Water management in the English landscape. Edinburgh UP, Edinburgh

Cook H, Williamson T (eds) (2007) Water meadows; history, ecology and conservation. Windgather, Macclesfield

Council of Europe (2000). *The European Landscape Convention*. Strasbourg

Crook DS, Jones AM (1999) Traditional irrigation and its importance to the tourist landscape of Valais, Switzerland. Landscape Res 24(1):49–65

Cummings I, Cutting RL (2007) The effects of floating on plant communities. In: Cook H, Williamson T (eds) Water meadows; history, ecology and conservation. Windgather, Macclesfield, pp 82–93
Dokken IS, Dahl E, Simenstad JA (1999) Ymse gamalt frå Nord-Sel. In: Bakke P, Bakke K, Bergum M, Berntsen PA, Fredriksen B (eds) Gåmålt og nytt frå Sel. Sel historielag, Otta, pp 24–33
Dvořák P, Šálek J, Janský L (2004) History of irrigation, drainage and flood control in Czechoslovakia in the Danube Basin. In: Csekö G, Hayde L (eds) Danube Valley; history of irrigation, drainage and flood control. ICID-CIID, New Delhi, pp 311–322
Edwards KJ, Schofield JE (2013) Investigation of proposed Norse irrigation channels and dams at Garđar/Igaliku, Greenland. Water Hist 5:71–92
Emanuelsson U (2009) The rural landscapes of Europe; how man has shaped European nature. Formas, Stockholm
Emanuelsson U, Möller J (1990) Flooding in Scania; a method to overcome the deficiency of nutrients in agriculture during the nineteenth century. Agric Hist Rev 38(2):127–148
Ewald KC, Klaus G (2010) Die ausgewechselte Landschaft; vom Umgang der Schweiz mit ihrer wichtigsten natürlichen Ressource. Haupt, Bern/Stuttgart/Wien (3nd ed; 1st ed 2009)
Fabriciusova V, Slamova M, Jancura P (2015) Protection of landscape values in an innovative management approach of the polana biosphere reserve. UNISCAPE En-Route 1(1):27–32
Feader (2010) Des prairies humides préservées. http://agriculture.gouv.fr/sites/minagri/files/documents/pdf/2-214I2-BN-prairieshumides.pdf. Accessed 18–1–2018
Fejér L (2004) History of irrigation, drainage and flood control in Hungary. In: Csekö G, Hayde L (eds) Danube Valley; history of irrigation, drainage and flood control. ICID-CIID, New Delhi, pp 331–462
Hallmann HW, Peters J (1995) Waale in Südtirol. Garten + Landschaft 105: 31–33
Hoppe A (2001) Verbreitung und Vegetation der Bewässerungswiesen Nordwestdeutschlands. Berichte der Reinhold-Tüxen-Gesellschaft 13: 247–250
Ihse M, Lindahl C (2000) A holistic model for landscape ecology in practice: the Swedish survey and management of ancient meadows and pastures. Landscape Urban Plann 50:59–84
Ile T (1958) Bygdabok for Øyer: Natur og kultur—innsyn og utsyn, 1. Øyer bygdeboknemnd, [Øyer]
Ira V, Huba M, Podolák P (2008) Quality of life of population living in a protected area and its sustainable development (examples of the protected landscape area Polana). Geographia Slovaca 25:97–122 (in Slovak)
Jansen A (no year [2015]) Pratum Marcidum; historische, geografische, sociale en ecologische achtergronden van de Kempense wateringen. Uitgeverij Het Punt, Dendermonde/Opglabbeek
Keenleyside C, Oppermann R (2009) A farmer-centred outlook to managing and to maintaining High Nature Value grasslands in Europe. In: Veen P, Jefferson R, de Smidt J, van der Straaten J (eds) Grasslands in Europe of high nature value. KNNV Publishing, Zeist, pp 293–303
Kerridge E (1954) The sheepfold in Wiltshire and the floating of the watermeadows. Agric Hist Rev 6(3):282–289
Kjøk E (1979) Garmo sokn. in: Bjørgen S, Kjøk E, Mo E, Rustad A, Venås K (eds) Garmo kyrkje og sokn, pp. 19–30. Garmo sokneråd, [Garmo]
Kobussen M (1997) Loenense Molenbeek; oude levensader van een Veluws dorp. Commissie Floralia/Dorpsverfraaiing, Loenen
Konold W (ed) (1994) Historische Wasserwirtschaft im Alpenraum und an der Donau. Wittwer, Stuttgart
Konold W (2004) General review. In: Csekö G, Hayde L (eds) Danube Valley; history of irrigation, drainage and flood control. ICID-CIID, New Delhi, pp 17–22
Konold W, Popp S (2004) History of meadow irrigation in the Danube area in Württemberg. In: Csekö G, Hayde L (eds) Danube Valley; history of irrigation, drainage and flood control. ICID-CIID, New Delhi, pp 23–37

Krogh KJ (1982) Erik den Rødes Grønland. Nationalmuseet, [Copenhagen]
Leibundgut C (2004) Historical meadow irrigation in Europe—a basis for agricultural development. In: The Basis of Civilization—Water Science? (Proceedings of the UNESCO/IAI-1S/IWHA symposium held in Rome, December 2003). IAI IS Publ. 286:77–87
Leibundgut C, Kohn I (2014a) European traditional irrigation in transition. Part 1: Irrigation in times past—a historic land use practice across Europe. Irrig Drainage 63(3):273–293
Leibundgut C, Kohn I (2014b) European traditional irrigation in transition. Part 2: Traditional irrigation in our time—decline, rediscovery and restoration perspectives. Irrig Drainage 63(3):294–314
Leibundgut C, Vonderstrass I (2016) Traditionelle Bewässerung—ein Kulturerbe Europas. Merkur, Langenthal (2 Vols)
Müller IB (2017) The influence of traditional flood irrigation on biodiversity, plant functional composition and plant nutrient availability in Central European grassland. Universität Koblenz-Landau, Diss
Niemeier G (1961) Agrarlandschaftliche Reliktgebiete und die Morphogenese von Kulturlandschaften im atlantischen Europa. Geografiska Ann 43:229–235
Panagiotakopulu E, Buckland PC (2012) Irrigation at Garđar, SW Greenland and its North European context. Water History 4:197–211
Panagiotakopulu E, Greenwoord MT, Buckland PC (2012) Insect fossils and irrigation in medieval Greenland. Geografiska Ann A 94:531–548
Petrovič F (2006) The changes of the landscape with dispersed settlement. Ekologia Bratislava 25:65–89
Slámová M, Beláček B, Jančura P, Prídavková Z (2015) Relevance of the historical catchwork system for sustainability of the traditional agricultural landscape in the southern Podpolanie region. Agric Agric Sci Procedia 4:10–19. https://doi.org/10.1016/j.aaspro.2015.03.003
Slámová M, Jančura P, Fabriciusová V, Beláček B, Zrníková K, Prídavková Z (2016) Traditional agricultural practices, land cover diversity and biodiversity in the southern podpolanie region. In: Agnoletti M, Emanueli F (ed) Biocultural Diversity in Europe. Springer International Publishing Switzerland (Environmental History 5), pp 247–268
Sømme A (1954) Jordbrukets geografi i Norge—Geography of Norwegian agriculture. A. Tekstbind. J.W.Eides forlag, Bergen
Staring WCH (1868) Huisboek voor den landman in Nederland. Funke, Amsterdam
Thissen PHM, Meijer M (1991) Rug- en hangbouwbevloeiing in Nederland; een negentiende-eeuwse cultuurtechnische innovatie die niet doorzette. Landinrichting 31(4):16–22
Vervloet JAJ (2010) Zandlandschap. In: Barends S et al. (red) Het Nederlandse landschap: een historisch-geografische benadering. Matrijs, Utrecht (10e edn), pp 132–161
Von Droste B, Plachter H, Rössler M (eds) (1995) Cultural landscapes of universal value. Fischer, Jena/Stuttgart/New York, pp 272–288
Wade-Martins S, Williamson T (1994) Floated water-meadows in Norfolk: a misplaced innovation. Agricultural History Review 42(1):20–37
Williamson T, Cook H (2007) Introducing water meadows. In: Cook H, Williamson T (eds) Water meadows; history, ecology and conservation. Windgather, Macclesfield, pp 1–7

Hans Renes is Historical Geographer at the Department of Geosciences at Utrecht University and Professor of heritage studies at Vrije Universiteit Amsterdam, the Netherlands. He publishes on many different aspects of landscape history of the Netherlands and Europe as well as on the relation between landscape heritage and planning.

Csaba Centeri is Associate Professor and Head of the Institute of Nature Conservation and Landscape Management at Szent István University, Gödöllö, Hungary. His main research interest is soil: in nature conservation, landscape management, landscape quality, agricultural landscapes, wildlife management, forestry, ecosystem services, renewable energy systems, and botany in general.

Sebastian Eiter is a geographer and landscape ecologist. He is Research Scientist in the Department of Landscape Monitoring at NIBIO, The Norwegian Institute of Bioeconomy Research. His recent research topics include causes and consequences of agricultural landscape change, cultural heritage, biodiversity, public participation, renewable energy and landscape quality, and urban agriculture.

Bénédicte Gaillard is a political scientist in heritage studies. Her research interests include the management of cultural landscapes and World Heritage Sites, focusing on potential conflicts resulting from development pressure. In her doctoral thesis, she investigated the conflicted process of delisting the Dresden Elbe Valley from the World Heritage List.

Alexandra Kruse studied agriculture, landscape development, nature protection, and geology in Kiel, Germany. She is Director of insitu World Heritage consulting and Secretary General of the Institute for Research on European Agricultural Landscapes (EUCALAND), a German nonprofit association which is compiling a database of European agricultural landscapes (https://www.feal-future.org/eatlas/en). Her projects also address UNESCO world heritage and the revitalization of rural areas.

Zdeněk Kučera is Assistant Professor in the Department of Social Geography and Regional Development at Charles University, Prague, Czecholslovakia. His research interests include historical, cultural, and regional geography, encompassing the study of heritage, regional identities, transformations of landscapes and settlements, and historical and cultural aspects of the development of borderland, mountain, and rural areas.

Oskar Puschmann is a landscape geographer and photographer. He is Senior Scientific Advisor in the Department of Landscape Monitoring at NIBIO, The Norwegian Institute of Bioeconomy Research. His interests include Norwegian national landscape character assessment, and agricultural landscape history and landscape change, and his recent landscape photography project is *Tilbakeblikk—Norwegian Landscapes in Retrospect.*

Michael Roth is Professor in landscape planning and landscape information systems at the Faculty of Landscape Architecture, Environment, and Urban Planning at Nürtingen-Geislingen Polytechnic, Germany. He has written extensively on GIS and on modern landscape developments and has taken part in many European research projects on landscape classification, landscape heritage, and energy landscapes.

Martina Slámová is Assistant Professor in the Department of Landscape Planning and Design, Faculty of Ecology and Environmental Sciences, Technical University in Zvolen, Slovakia. Her current research explores cultural landscapes, historical landforms, landscape visualizations, and territorial planning. She is Leader of the international ERASMUS KA2 + project FEAL on multifunctional and sustainable agriculture in European agricultural landscapes.

Dutch influence is evident in Holler colonies all over Europe, both in the construction of the land and in the built heritage, Alexandra Kruse, released under a Creative Commons Attribution 3.0 Unported License

Chapter 7
Holler Colonies and the Altes Land: A Vivid Example of the Importance of European Intangible and Tangible Heritage

Alexandra Kruse and Bernd Paulowitz

Tell me what you pay attention to and I will tell you who you are.
—José Ortega y Gasset

Abstract The Holler colonies are settlements in European marsh and dyke landscapes created starting in the High Middle Ages through land reclamation by Dutch water engineers who had been hired by local leaders. Some Holler colonies were later abandoned; others remained. Surviving colonies, some of which remain and are largely intact, was often changed by later land reclamation processes (Renes and Piastra 2011: 24). Today, the remainders of these landscapes are visible proof of the intangible and tangible heritage of European economic and social history. These created landscapes—Holler landscapes—are easily recognized by their linear landscape structure and small parceling. Other characteristic features are hydraulic engineering structures, such as dykes, drainage systems, receiving waters, and the pumping stations used to clear the marshes of water and strip parceling, linear settlements, and infrastructure, houses and farms in a row, dykes, roads and byways, ditches, and channels. Holler colonies are a European landscape typology in several respects. First, they were the consequence of Dutch water engineers' large-scale knowledge transfer within Europe, are always built at the invitation of local leaders, and always encompass the diffusion of Dutch laws, traditions, production, and models of society. Second, they may be considered a European landscape type that defined European wetland areas along large rivers from the Middle Ages onwards. Their uniqueness is based on the techniques applied, the social processes involved, and their long duration.

Keywords Landscape history · Cultural heritage · Rural development · Landscape resilience · People's identity · Landscape heritage

---

A. Kruse (✉)
insitu World Heritage Consulting, Buc, France
e-mail: akruse@whconsult.eu

B. Paulowitz
Hamburg, Germany

C. Hein (ed.), *Adaptive Strategies for Water Heritage*,
https://doi.org/10.1007/978-3-030-00268-8_7

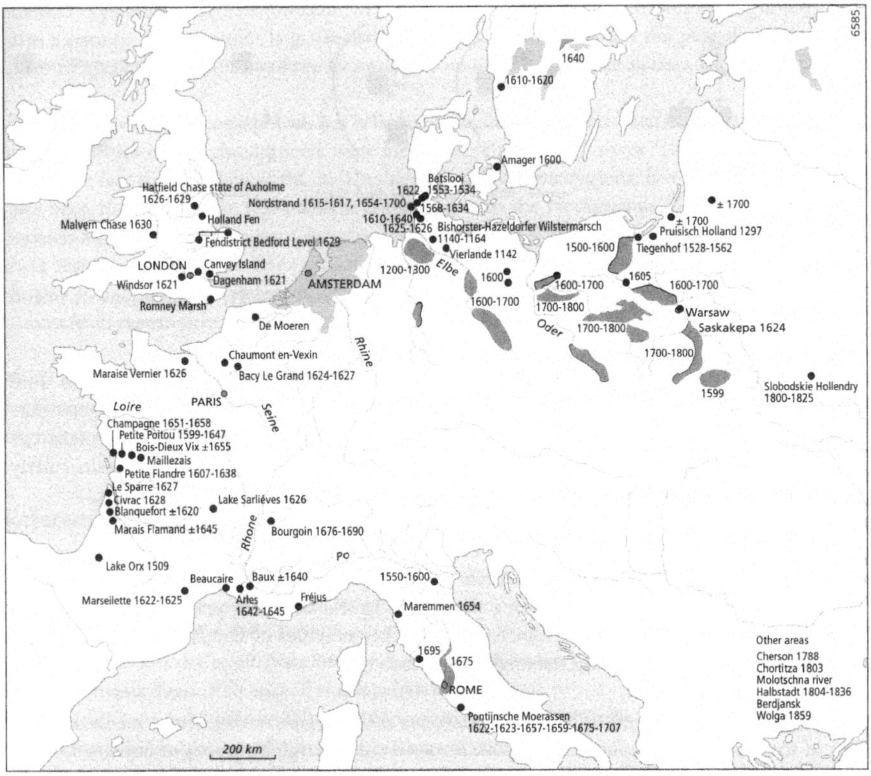

Fig. 1 Land reclamation and improvement measures carried out by the Dutch. *Source* Renes (2005: p. 27), based on Van Veen 1955: p. 52; released under a Creative Commons Attribution-NonCommercial-NoDerivatives 4.0 International License

Holler colonies were early European settlements made possible by a distinctive style of land reclamation. (Renes 2005: 27f, Fig. 1) Techniques for cultivating the marsh and peat regions of the Dutch Lowlands (van der Linden 2000: 292) were first developed in the eleventh century. Dutch engineers spread these techniques to many other European countries (Hofmeister 2009; Renes 2005: 27ff) in a process of *inner colonization or inner expansion*, centrally steered migrations that revitalized abandoned and less populated areas, often closely linked to new parceling (Bartlet 1993). The name Holler colonies points to the people from the Low Countries, often called *Holländers*, who reclaimed the land; it designates both this type of settlement and its surrounding landscape. This chapter details the characteristics of Holler Colonies and why they are distinctive in Europe.

This transfer of technology is the defining feature of Holler Colonies. Marsh and dyke land reclamation was planned and not random, not exactly the same everywhere but everywhere followed the same principles. In particular, this marsh and dyke reclamation exemplifies *Cope cultivation*. The term's origin in the Dutch and Low

Fig. 2 Aerial view of the Altes Land shows the linear structure, very much related to the so-called mother landscapes in the Netherlands, e.g., Lopikerwaard, Wowbrugge, Teckop, which can be found again in other Holler Colonies, e.g., Malborg in Poland. *Source* With kind permission of Martin Elsen; released under a Creative Commons Attribution-NonCommercial-NoDerivatives 4.0 International License

German term "kopen", meaning "to buy," conveys that the Dutch settlers bought the land which they were asked to reclaim. This dynamic made their imported culture sustainable and distinguished the Holler colonies from other land reclamation areas.

The created landscapes—Holler landscapes—are easily recognized by their linear landscape structure and small parceling (Fig. 2). Other characteristic features are hydraulic engineering structures, such as dykes, drainage systems, receiving waters, and the pumping stations used to clear the marshes of water and strip parceling, linear settlements, and infrastructure, houses and farms in a row, dykes, roads and byways, ditches, and channels. Their linear structure is still visible all over Europe in the layout of fields, infrastructure (roads, paths, and ditches) and built heritage. The ensemble of features forms a unique and clearly identifiable landscape, still found in Germany (Kaup 2005), Poland (Chodyla 2005), Denmark (Stenak 2005), France (Toussaint 2005), and England (Williamson 2005). According to Lewandowski and Szewczyk (2008: 7), Holler settlements in Poland are particularly under pressure today and could well be inscribed on a list of endangered landscapes akin to the International Union for the Conservation of Nature's Red Book for Threatened Species (http://www.iucnredlist.org/).

It may be useful to group or cluster these European landscapes as *daughter* landscapes of a Dutch *mother* landscape (Verein 2009: 109). Here, we concentrate on one of the best preserved examples of a daughter landscape, the Altes Land, literally "the Old Country," a region south of the Elbe River in Northern Germany, near the city of Hamburg. In examining the region, we are able to gain a short overview of the land-shaping process and demonstrate how a specific historic sociocultural development plays an essential part in our identity today. We also show how water, water infrastructure, and heritage have defined the identities and ways of life in these regions.

The Low Countries as a Center of Innovation in Water Management

The period of population growth of the tenth to the early fourteenth century was a demographic shift which triggered dramatic changes in land use; land reclamation and clearance in most parts of Europe accommodated the new populations (Renes 2010). This population increase was especially problematic for the Low Countries, where land for settlement and agriculture was scarce (Bartlett 1993) where large uninhabited fenlands, in most cases too wet for agricultural production, dominated the landscape. In response to these circumstances, the Dutch invented techniques to reclaim land from the marshes (Renes 2005: 13ff). With the invention of dykes in the tenth century, new areas became habitable and usable (Lebecq 1979:145). Soon, the Dutch developed large-scale drainage systems, including main channels and secondary ditches, to supplement the dykes. Another important new measure: a sustainable system for maintaining the dykes included a civic organization which was responsible for watching over and, if necessary, repairing them day or night at any time of year. These responsibilities were set out in contracts.

According to current knowledge, planned drainage commenced in the thirteenth century (Renes 2005: 27ff; Verein 2009: 27f). Almost immediately after its onset, Dutch technical know-how was in high demand by land owners and authorities elsewhere across Europe, who soon invited the Dutch engineers to come to help drain their lowlands. A later period of population growth, in the long sixteenth century (from 1450 to 1650), triggered another phase of reclamations. Especially in the Low Countries, this activity was characterized by an intensive use of existing arable and pasture lands (Renes 2010) and created less clearly visible and regionally distinct landscapes.

The Dutch hydraulic engineer Johan van Veen provides an overview of the history of Dutch reclamation activities throughout Europe in Fig. 1; although it dates from 1955, this is still the most complete map of its kind. Toussaint (2005: 121ff) has found that Dutch water engineers traveled to different areas, spreading knowledge of the embankment and its value throughout Europe, thus creating a unique continuity. The Center of Research on Holler Colonies in the Altes Land (founded in 2012) has started a deep investigation into the rationale for Dutch settlers' drainage measures and land reclamation technologies and of their societal consequences.

The Dutch engineers and settlers brought other types of activities from Holland to the rest of Europe: They enriched communities with their often advanced knowledge and expertise in labor, new forms of social and political organization, commercial and economic networks, in addition to disseminating knowledge of how to drain and to maintain the wet landscapes which had been more or less unproductive (Lewandowski and Szewczyk 2008: 62f; Hofmeister 2009; Verein 2009: 15). Further contributions included legal traditions, capital investment, and, particularly in the Altes Land, place names. The material interventions were, from the time of origin, always closely linked to immaterial heritage. Water and heritage have now become a unity in these landscapes.

Land Reclamation Process Under Dutch Influence and Introduction of Dutch Political and Social Models in the Altes Land

The Altes Land, at 170 km^2, approximately 30 km long, and, at its widest, 10 km wide, is limited by natural borders. The Elbe River lies in its northeast, a dense peatland of 0.5–3 km thickness in the southwest, and the Schwinge River in the northwest. The whole area of medieval land reclamation in the Altes Land (= 240 hides ~ 8000 ha) has a distinctive legal history, leading not only to a recognizable landscape but also an independent mentality.

In 1113, Archbishop Frederick of Hamburg-Bremen invited a group of Dutch people to reclaim and cultivate the area along the Weser River. The Charta of 1113 ('*Cope Contract*'), a contract between the Archbishop of Bremen and the settlers that confirmed and specified the terms of the arrangement, shows that the Low Countries were reputed for their expertise in the reclamation of marshland. In it, settlers were guaranteed their "individual freedom and the ownership of the land reclaimed by them." (Hofmeister 1979). In other words, they were allowed to enjoy the same rights as in their home country (Hofmeister 1979). The Charta also granted them legal liberty, which led to a strong and distinctive cultural identity and tradition (Verein 2009: 15f). It was Priest Heinrich (cf. Fig. 3), born in Rijnsaterswoude in the Netherlands, working in Steinkirchen, who subsequently brought the Dutch water engineers to the area (Hofmeister 1979) of the Altes Land. The tithe of the Altes Land belonged at that time to the Archbishop Adalbero of Bremen and Hamburg (1123–1148), who was a key actor in shaping the region: He set up the governing structure in three districts (each called *Meile*, which is the German term for mile) that formed a legal, social, and economic unit until the first half of the nineteenth century. In 1140, Adalbero handed the tithe in Huttfleth (1st Meile) to the Hamburger cathedral chapter, and the one in Steinkirchen to the cloister Harsefeld. In 1148, he also added to Hamburg the tithe of Thitgerscoph (Hollern, 1st Meile). Adalbero's successor, the Archbishop Hartwig (1148–1168), who was also the duke of Stade, gave in 1149 to settlers in his territory the same rights as at the Weser and in Stade defining the legal

Fig. 3 Priest Heinrich symbolized the colonization process in the Altes Land from 1106. This 1992 bronze sculpture by Carsten Eggers stands in front of the church in Steinkirchen. *Photograph* Kruse (2011); released under a Creative Commons Attribution-NonCommercial-NoDerivatives 4.0 International License

framework for the colonists. While authority over the area was still often divided between the local dominant family and the Archbishopric of Bremen and Hamburg, the main colonization of the Altes Land ended in 1236, when Archbishop Gerhard of Lippe (1219–1258) gained full sovereignty over the second and third districts.

As early as 1361, the territory had a specific seal, which was inscribed *Sigillum Communitatis Veteris Terrae* (i.e., the Seal of the Municipality of the Altes Land). That this occurred is important: it shows that the Altes Land was autonomous and therefore different from many other regions in central Europe at a very early stage. It was only after two events, the 1932 disbanding of the political district of Jork and the 1937 establishment of Greater Hamburg, that the unity of the Altes Land— which had held for many centuries—was suspended.

Thus, the Dutch model of society offered a great deal of freedom: political representation, sovereignty over local districts, land ownership, and local maintenance of the landscape. One result was a strong sense of citizenship. Such freedom and responsibility had been relatively unknown to local farmers in the High Middle Ages (Hofmeister 2009; Verein 2009: 15f) and were perhaps as a big a change for local political norms as reclamation was for the landscape.

The Altes Land: Tangible and Intangible History and Heritage

The transfer of technology and knowledge over the centuries was also critical to local identity. Fruit trees dominate the landscape and are, after landscape history, the second most important feature of local identity. Inhabitants began to cultivate fruit here (Fig. 4) very early: the first records are dated 1320, the year when Paul, son of Basilius Curia, sold to Johannes Tymbeke of Asse a yearly pension financed through his orchard (Verein 2009: 34). By roughly 1600, fruit had achieved a notable value. When the Archbishop of Bremen banned Hamburg beer from the Altes Land, the council of the city of Hamburg retaliated by banning cherries and other fruits from the Altes Land in the so-called cherry war. According to the rectification protocol (a legal provision on taxes) of 1657, the Altes Land had 743 fruit farms with 202 ha (Verein 2009). Fruit production gained even more importance after 1870, reaching its greatest importance in the 1960s.

Skilled crafts and trades developed in the seventeenth century in the Altes Land independently of the rights belonging to tradesperson offices in the neighboring cities of Stade and Buxtehude.

The design of private houses is of a particular high quality and refinement. It is clear that a great deal of money was spent on arranging their interiors (cf. Fig. 5). Pieces of regionally distinctive Altländer furniture are often still found in private houses and are on display in the Museum in Jork (Verein 2009). These include suitcase chests, settles, chairs, and the Hamburg Schapp, a distinctive baroque hall closet. These furniture pieces illustrate the durability of the Altländer society (Verein 2009).

The technical and cultural achievements of drainage, irrigation, and land reclamation, a highlight in European internal colonization processes, are still visible in the landscape, while the intangible heritage has often been lost. As one would expect,

Fig. 4 Fruit plantations still follow the medieval plots. *Source* Kruse (2011); released under a Creative Commons Attribution-NonCommercial-NoDerivatives 4.0 International License

each of the Holler landscapes, whether in France, Poland or elsewhere, now look somewhat different from each other. Some of them, in particular in regions with a high self-administration, like Germany, retain many of the original cultural, social, and technical features, while others, where a strong national administration dominated politics, like France, retain only the distinctive landscape structure. In most places, later practices of land forming (like combining fields) overwrote the medieval patterns, either due to new societal needs (infrastructure) and or because of the modernization of land use and therefore the creation of larger plot units. We find as well the opposite phenomena, which mean that land was abandoned due to a lack of performance and/or productivity. The latter fact applies especially for Holler Colonies in Poland.

The very visible linear structures, the canal, ditch, and dyke system, as well as the social organization in the Altes Land have survived mainly because natural conditions made a different, more profitable use impossible. The Altes Land stands out today as best preserved example of the historic diffusion process originating in the "Low Countries," more so than in any other European area and even more recognizable than in the current-day Netherlands. As a result, we find a stunning integrity in this regional structure on a fairly large scale.

Finally, yet importantly, another feature of cultural and historic significance survives today: Continuous historical records and a well-kept archive. This too differentiates the Altes Land from the other landscapes where people from the Low Countries settled. Indeed, much of the research on Holler colonies in Europe originates here.

Fig. 5 Precious and colorful worked wooden furniture and interior decoration of the houses as well as symbols and figures at the roof tops are material proof of wealth and proudness of the *Altländer* people. All photographs taken by Alexandra Kruse; released under a Creative Commons Attribution-NonCommercial-NoDerivatives 4.0 International License

Landscape, Identity, and Development in the Holler Colonies—Based on the Findings in the Altes Land

In the Holler landscapes, landscape identity is not only geographic, but also very strongly social. The concept of landscape identity can be explained by the help of the so-called Swiss landscapes. These are wide and hilly landscapes appreciated highly since romanticism. Swiss landscapes are a kind of master landscape (idealtypische Landschaft), and therefore the terminology is used for similar landscapes in foreign countries, for example, Saxonian Swiss or Holsteinisch Switzerland in Germany. One may speak of an early marketing trick. However, the Swiss landscape model is primarily based on geographical features.

Holler colonies owe their distinctive character as ongoing cultural landscapes to the interactions between humans and nature, with a strong awareness and appreciation of traditional costumes and customs, buildings and architecture, dances, and dialect. Recently we find, for example, the costumes of the "blossoms queen (Blütenkönigin)"

Fig. 6 Every year the blossom festival celebrates the fruit production; it is a magnet for thousands of tourists. The photograph shows the blossom queen with her maid in traditional costumes. *Photograph* with kind permission of the Association for the Recognition of the Altes Land as UNESCO World Heritage e.V.; released under a Creative Commons Attribution-NonCommercial-NoDerivatives 4.0 International License

(cf. Fig. 6) and traditional wedding dress. Female tourist guides wear special outfits based on traditional dresses but newly designed and reintroduced.

Agricultural cultivation and other land use have fostered an anthropogenic designed landscape. The orchards were created on narrow marsh hides, so-called beds, delimited by parallel dykes (cf. Fig. 4). Other artificial elements of paramount importance to the cultural landscape include (see Fig. 7):

- the Altländer farms and courtyards
- the linear-shaped settlements
- medieval roads and trails
- the long strip parceling
- the number of bettors and straight drainage ditches, which underscore the overall linear structure.

The Altes Land has a long history of the peaceful settlement of foreigners of different origins, whose cultural, human, and societal impact is felt to this day. Findings from the Stone and Bronze Ages prove early settlement; from 200 AD Saxons came and settled only in the higher, dry areas (Hofmeister 1979). No fights or conflicts

Fig. 7 Linear structures and elements form the landscape of a Holler colony, as here in the Altes Land. (From upper left to lower right: Estebrügge, Zesterfleth in Borstel, Deichhufendorf Mittelnkirchen Ort; bed-ditch structure in the orchards, Neuenschleusener Wettern, back dyke (Hinterdeich) at the Landwettern in Moorende. *Photographs* Silvia Hotopp-Prigge and Andrea Rachow, provided by the Association for the recognition of the Altes Land as UNESCO World Heritage e.V.); released under a Creative Commons Attribution-NonCommercial-NoDerivatives 4.0 International License

between the settlers originating from Saxony and the new ones from the Low Countries are recorded. The exchange and communication with the Dutch originating areas is still apparent and lived. This exchange was and is not only limited to landscape aspects but also is particularly rich and to be found in ideas and traditions. The strong link between the originating and colonized landscape was in earlier times organized through the church (church records of the seventeenth-century bear witness) and has been taken up since by cultural associations (Raap 2010).

Dutch Society Models as Driving Force for Development

As we have seen, one of the outstanding features of this cultural landscape was and is the freedom and self-determination of the citizens, an imported Dutch model. This idea of freedom led to many of the other characteristic features of the Altes Land. One is that women always held a very strong position here. For example, neither party to a jointly composed will could unilaterally change the text to the detriment of the other (Verein 2009). Women also had the right of succession. This is one reason why we find farms that are in the possession of the same family—undivided—since the fifteenth century, leading to a strong historical consciousness and social and regional cohesion (Verein 2009: 28f).

The social system established in the fourteenth century—with independent self-determination, regional auto-administration, and representatives—also formed the preconditions for a flourishing economy.

Popular traditions and a distinctive dialect, still taught to children today, are sound evidence of a strong traditional cultural identity. This traditional identification made it possible for people to self-organize in cooperatives in order to carry out the necessary works and construction of the dykes (Hofmeister 2009). Citizens rather than politicians took responsibility for that work.

On a different scale, the development of larger rural municipalities as sustainable local cooperatives in parishes or peasantries with regional councils, a representative system, goes back to the eleventh and twelfth century; it was first observed in the Frisian populated coastal region and in Dithmarschen (Schleswig-Holstein). This political and social model, including its terminology, was soon to be found as well in the Dutch settled areas along the River Weser. One has to particularly highlight the separation of jurisdiction and administration that developed with the establishment of administrative sectors (*Hauptmannschaften*) in the fourteenth century; it preceded similar systems in central Europe of the nineteenth century. This autonomous administrative and judicial organization remained more or less intact in the Altes Land until the late nineteenth century. In short, as Verein (2009: 28f) wrote, "Since the fifteenth century, courts have remained the undivided property of one family for generations. Ownership structures, topography, rule-based conditions, contractual arrangements, and legal relations shaped the daily lives of people in the Altes Land in a special way over the past centuries."

Current Status of the Landscape and Challenges to European Heritage

Today, a variety of changes threaten the Altes Land. Due to its geographical proximity to Hamburg, the second largest town in Germany, there is a pressure from urban development, including the demand for industrial land use by the city and port of Hamburg. The shipping industry and other means of transportation have affected land use along the river Elbe, as well as the ground-water level. There is a proposal to further deepen the Elbe. The effects of that change on the marshlands and cultural landscape are yet unclear, but it will certainly change the soil conditions and therefore microclimates and agriculture. Ditches and canals risk drying out, which will not only affect biodiversity but also will make water-regulating features like locks and bridges obsolete—possibly even the former water bodies themselves. Further construction is a new physical, social, and cultural problem, constantly reducing the open area of the Altes Land, especially the farmland. Recent construction has included a new highway along the border between the Altes Land and higher elevations (*Geest*), and new industrial, commercial, and housing developments.

The Altes Land now belongs to two different Federal States, or Länder. The difference in development pressure is visible: The part belonging to the Federal State of Lower Saxony still retains the historically evolved landscape and historic features while the part belonging to the Federal State of Hamburg is undergoing constant change. In particular, the construction of the deep sea harbor and the resulting artificial hills of the dugout material altered the topography and therefore the visible integrity of the land. Similarly, the huge industrial development by the airplane builder Airbus destroyed parts of the historic landscape.

Another threat comes from outside. As mentioned earlier, the local agriculture is based on fruit production, mainly apples (in 2007 88.1%), followed by cherries, pears, and plums (Verein 2009: 40). This system is under pressure from large-scale production all over the world that does not face the same environmental and topographic restrictions as we find here. It is also under pressure also from constantly renewed EU-regulations, which often do not take into account small-scale production when targeting production methods.

Several new environmental regulations restrict possibilities in regionally and topographically specific contexts. In the past, the farmers of the Altes Land always had to find a way to adapt their production methods to the traditional water management of dykes and ditches, and to fit their fields into the historical parceling. The traditional average distance between two ditches in the Altes Land measures 16–20 m. But the European Water Framework Regulation demands 5-m distance from a water body (in each direction) when using chemicals (European Commission 2000a), putting half the arable land out of production and leaving only a strip of 6–10 m for production—too small to be profitable. Not so long ago, many farmers addressed this problem by sealing the ditches, a lasting negative change not only in the established water management, but also, and this is even more important for our discussion, in the cultural landscape. An internal transformation process is underway, favoring high-quality and organic-based production, thus actually fostering preservation of both immaterial and material heritage. Organic agriculture would be a way for farmers in the Altes Land to follow regulations while keeping farming as a main income source, maintaining and reinforcing the historic cultural landscape, preserving traditions, and making the most of historic water management. Finding this balance might be the most important challenge facing the Altes Land.

Today, 4.1% of the farmers in the administration district of Stade, in which the Altes Land is located, produce under ecological farming labels (Kompetenzzentrum Ökolandbau 2017). The European Commission EIP-AGRI, a "Special area Altes Land" Operational Group financed by the EU rural development policy, not only looked specifically at organic farming but also found that some conventional processes and machinery are transferable to organic production. It will adapt a number of technologies so that all fruit farmers can apply them, bringing together organic and non-organic farmers as well as a number of research and advisory centers.

In both organic and conventional farming, avoiding spray drift when applying plant protection products helps to reduce adverse effects on the environment. The

Operational Group is testing tunnel-application spraying systems. Vertical anti-drift nets are also being tested: Farmers set up these dense, 4-m-high insect nets around the orchards to catch a substantial part of spray drift (European Commission 2018). These programs might just the stimuli that Altländer agricultural sector needs to gain economic competitiveness while preserving its cultural heritage.

Heritage, Region-Building, and the Future of the Altes Land

A local group of citizens, the Association for the Recognition of the Altes Land as UNESCO World Heritage e.V." (Verein 2009), has been working since the beginning of this millennium to protect the historic landscape. At the beginning, its goal was to nominate the Holler Colony to UNESCO as a candidate for the World Heritage List. The group met local resistance, in particular, from farmers who feared the status would bring new restrictions. They pointed out their bad experience with environmental protection regulations, especially Natura 2000, a Europe-wide conservation network. (European Commission 2000b) The citizens' association had to become much more professional and enduring, capitalizing on growing awareness of UNESCO World Heritage and its status as an income source for many municipalities and state governments. The people from the Altes Land appropriated the idea that sustainable economic progress can also foster sustainable protection of their heritage. The association became expert in many planning issues and political debates, even as the concept of World Heritage itself evolved into a vehicle for regional development.

Today, the World Heritage nomination goal plays an important role in the local discussion of how to develop the Altes Land into a living area marked by high quality and prospects. Therefore, the goal is part of a new landscape plan (Samtgemeinde Lühe 2017). The farmers are correct that protection status would impose some limitations or even obligations on them, for example, requiring them to keep historic ditches intact. On the other hand, they could realize that it could help safeguard the Altes Lands as a high-quality agricultural production region and as a tourist destination (tourism is the second largest income sector in the region).

The local reflection and participation process that the Association has fostered over the last 15 years fits very well with the whole concept of the World Heritage Convention. In some ways, this road to World Heritage is more important than the status itself. Indeed, the citizens' initiative uses the World Heritage goal to work for a common future of the region.

The association members have organized many campaigns to raise local, regional, and global awareness of the history of the region and the unique historic cultural landscape that is a Holler colony. It organized several national and international symposia and exhibitions for locals and visitors as well as the national and international scientific community. Although not without encountering difficulties and negotiating continuous discussions, the association succeeded in getting farmers, municipalities, tourist organizations, and the general public behind the common goal.

Fig. 8 Traditional Altländer houses possess a high degree of authenticity. *Photograph* Kruse (2008); released under a Creative Commons Attribution-NonCommercial-NoDerivatives 4.0 International License

In addition, and even more importantly, the Association has become a focal point for local development and protection planning in general. They were a driving force and competence center in developing the *Altländer Charta* (Samtgemeinde Lühe et al. 2011a), which sets out a central guiding principle (Leitbild, Samtgemeinde Lühe et al. 2011b) for tourism, planning, and economic development for the historical Holler colony heritage of the Altes Land. It sets out guidelines for local people and decision makers on how to engage with the region; it also describes the history, characteristics, and current status of the landscape. Similarly, it published a building guide (Baufibel), focusing on the protection of built heritage (Figs. 8 and 9). Most largely, it sets up a participatory process in which citizens and policymakers came up with a regional development plan, the first for the region. The Centre for Research on Holler colonies, founded in 2012, coordinates further participatory processes, enabling locals to base their positions and arguments based on sound knowledge.

The nomination to World Heritage remains in the air—it was not included in the renewed 2014 German Tentative List but rescheduled with the demand for further specifications. But the Association has achieved many successes. Cooperation with Holler colonies in other European countries is ongoing, and signs are everywhere that awareness of the fact of being a Holler Colony is stimulating regional development.

Fig. 9 Typical Altländer Farm, with residence in front, outbuildings and later storage facilities behind. In the foreground is the Brautpforte (bride's door) which the family traditionally only used for either weddings or funerals. The building structure follows the long stripe land clusters. *Photograph* Kruse (2008); released under a Creative Commons Attribution-NonCommercial-NoDerivatives 4.0 International License

In 2016, the Association was recognized at the international level: it was nominated for the European Union Prize for Cultural Heritage/Europa Nostra Awards 2017.

Conclusion

The Holler colonization produced clear, regional distinctive cultural landscapes across Europe, dominated by the presence of water as a land-shaping factor; in turn, water must be constantly managed in them in order to maintain agricultural space. In the Altes Land, the large-scale container ships that sail down the Elbe to the sea loom in impressive contrast with the small parceled landscape dominated by apple trees. Both, the immaterial heritage of the shaping as well as their ongoing maintenance, make this landscapes a recognizable and unique heritage found all over Europe. Together, they help the people who live in the region to maintain their specific history and identity, and construct a sustainable future.

These elements are of course under pressure from modern-day developments, including the need to keep the rivers navigable and national as well as European

Union regulation of agriculture. Modern European norms often defy the European past and historical identity, which we need in order to define ourselves. Defining the Holler colonies as cultural heritage may be the only way to safeguard the important ongoing dynamics that created and maintain these unique cultural landscapes.

The Holler colonization is living evidence that migration and cultural exchange can be driving forces in social and economic development. It also shows the resilience of traditional landscape concepts to changes imposed from outside (Janssen et al. 2017). In this respect, the colonies may well be a source of best practices for addressing climate change.

Today, the historical consciousness of the Holler Colonies landscape is growing. More and more locals in Altes Land, including farmers, understand the significance and also the market value of keeping the historic ensembles intact. This awareness does not stop at the maintenance of houses and farms but includes also the fields and ditches as well as traditions and techniques. In line with its European past, this common heritage will be made accessible more widely as part of the Hollerweg, a European cycling route that will start in the Altes Land and connect different Holler Colonies in Poland and the Netherlands. The Hollerweg itself will be part of the Cultural Route program of the Council of Europe (Council of Europe 1987). Awareness of the historic importance of the Altes Land offers the Europeans the opportunity to not only build on heritage and to find new responses to development pressures but also to secure legal protection for the cultural and physical landscape (Hofmeister 2009).

References

Bartlett R (1993) The making of Europe: conquest, colonization and cultural change 950–1350. Princeton University Press, Princeton

Chodyla Z (2005) The Dutch contribution to the development of the hydraulic system and cultural landscape in Poland, 1547–1945. In: Danner HS et al (eds) Polder pioneers. Netherlands Geographical Studies, pp 33–55

Council of Europe (1987) Cultural routes. https://www.coe.int/en/web/cultural-routes

European Commission (2000a) Water framework directive—framework for Community action in the field of water policy. Directive 2000/60/EC

European Commission (2000b) Natura 2000. http://ec.europa.eu/environment/nature/natura2000/index_en.htm

European Commission (2018) EIP-AGRI, Inspirational ideas: organic and "conventional" farmers joining forces for innovation. https://ec.europa.eu/eip/agriculture/en/news/inspirational-ideas-organic-and-%E2%80%9Cconventional%E2%80%9D. Date of access: 2018/03/18

Hofmeister AE (1979, 1981) Settlement and constitution of the Stade Elb Marsh in the Middle Ages". Volume I: The Stade Elb Marsh before the colonisation in the 12th century. Publication of the institute for Historical Land Research University Göttingen, Hildesheim 1979 [in German], ISBN 3-7848-3642-9, Volume II: The holler colonisation and the land community Kehdingen and Altes Land, Publication of the institute for Historical Land Research University Göttingen, Hildesheim 1981, ISBN 3-7848-3644-5 [in German]

Hofmeister AE (2009) Holländersiedlungen im Alten Land – Gedanken zu den Hollerkolonien. Speech on the occasion of the inauguration of the Competence centre Hollerkolonies 2009/06/25, Jork (in German)

Janssen J, Luiten E, Renes H, Stegmeijer E (2017) Heritage as sector, factor and vector: conceptualizing the shifting relationship between heritage management and spatial planning. Eur Plan Stud. https://doi.org/10.1080/09654313.2017.1329410

Kaup M (2005) Niederländer in Norddeutschland, vor allem Brandenburg. In: Danner HS et al (eds) Polder pioneers. Netherlands Geographical Studies, pp 57–76

Kompetenzzentrum Ökolandbau Niedersachsen GmbH (2017) Öko-Flächenanteile in den niedersächsischen Landkreisen. http://www.oeko-komp.de/wp/angebot-statistiken-analysen-marktberichte/. Visited: 24/10/2017

Lebecq S (1979) De la protohistoire au Haut Moyen Âge : le paysage des "terpen" le long des côtes de la mer du Nord, spécialement dans l'ancienne Frise. In: Actes des congrès de la Société des historiens médiévistes de l'enseignement supérieur public. 10e congrès, Lille, 1979, pp 125–151. https://doi.org/10.3406/shmes.1979.1331

Lewndowski W, Szewczyk R (2008) Poland—disappearing landscapes. Sport I Turysyka, Muza SA, Warsaw

Raap E (2010) Oproep aan inwoners van Woubrugge en omgeving! Article on "Gemeentepagina Kaag en Braasem" of the local paper. http://witteweekbladkaagenbraassem.nl/ [archive 2010, not available anymore]

Renes H, Piastra S (2011) Polders and politics: new agricultural landscapes in Italian and Dutch Wetlands, 1920s to 1950s. In: Landscapes I, pp 24–41

Renes H (2010) European agricultural landscape history. In: Pungetti G, Kruse A (eds) European culture expressed in agricultural landscapes—the Eucaland project book. Palombi Editori, Roma, 224 p

Renes H (2005) Water management and cultural landscape in The Netherlands. In: Danner HS et al (eds) Polder pioneers. Netherlands Geographical Studies, pp 13–32

Samtgemeinde Lühe (2017) Landscape plan—continuation, 1. Draft, Nov 2017. Lühe (in German)

Samtgemeinde Lühe et al (2011a) Building code (Baufibel) Altes Land. Hamburg (in German)

Samtgemeinde Lühe et al (2011b) Altländer Charta: Gemeinsames Leitbild für die Kulturlandschaft Altes Land. Lühe (in German)

Stenak M (2005) Dutch influence on Danish Water management—with particular respect to Lammefjorden. In: Danner HS et al (eds) Polder pioneers. Netherlands Geographical Studies, pp 77–101

Toussaint B (2005) The Dutch-Flemish role in reclamation projects in France. In: Danner HS et al (eds) Polder pioneers. Netherlands Geographical Studies, pp 121–149

Van der Linden, Hendrik (2000) Die Königsrute. Eine Revision der „Cope-Untersuchung" bezüglich der mittelalterlichen Kultivierungssystematik. Siedlungsforschung 18(S):269–296.

Van Veen J (1955): Dredge—drain—reclaim; The art of a nation, 1st ed (1948). Nijhoff, The Hague

Verein zur Anerkennung des Alten Landes zum Welterbe der UNESCO e.V., Jork (2009) Das Alte Land – einzig, nicht artig. Jork (in German). http://welterbe-altes-land.de/en/ (in English) date of access: 2018/03/18

Williamson T (2005) Dutch engineers and the draining of the Fens in Eastern England. In: Danner HS et al (eds) Polder pioneers. Netherlands Geographical Studies, pp 103–119

Links

http://holland.org.pl/art.php?kat=obiekt&id=486&lang=en. Catalogue of monuments of Dutch colonization in Poland https://de.wikipedia.org/wiki/Kujawien [16/06/2017]. Wikipedia entry about the region Kujawien in Poland

Alexandra Kruse studied agriculture, landscape development, nature protection and geology in Kiel, Germany. She is the director of insitu World Heritage consulting and secretary general of the Institute for Research on European Agricultural Landscapes e.V. (EUCALAND) (https://eucaland.net), a German nonprofit association which is compiling a database of European agricultural landscapes (https://www.feal-future.org/eatlas/en). Her projects also address UNESCO World Heritage and the revitalization of rural areas.

Bernd Paulowitz a historian and political scientist, is World Heritage Coordinator at the Department for Heritage Preservation in Hamburg and World Heritage Manager of the WH-Site "Speicherstadt and Kontorhaus District with Chilehaus". He has been a consultant in cultural heritage for UNESCO and CEO of insitu World Heritage consulting.

Statue of Tsuto Nitobe in Taiso Jinja in Towada city with his son Jujiro Nitobe's in the back. *Source* Photographed by Yoshiyuki Kudo, published in Towadashi Shiseki Bunkazai map, Towada city Cultural Heritage Preservation Committee, June 2016

Chapter 8
Archaic Water: The Role of a Legend in Constructing the Water Management Heritage of Sanbonkihara, Japan

Izumi Kuroishi

Abstract Until the end of the nineteenth century, agriculture formed the basis of cultural identity in the many parts of Japan where land and water were locally maintained, managed, and sustained (Wigen 1995; Toyama 1993). This chapter expands the Japanese idea of heritage beyond the question of beauty to include agricultural social systems and water management, exploring the long-term interactive relationship between water resources, the riverine landscape, and local people (Soja 2003). It does so by unraveling the formation and transformation of the legend of the irrigation pioneers. Tsutō Nitobe, a samurai of the Nanbu clan, is legendary for his pioneering water resource management and land reclamation in Japan's Sanbonkihara region in the nineteenth century (Ministry of Agriculture, Forestry and Fisheries 2018). After his death, his irrigation project was the pride of the region; part of the legend is that subsequent irrigation projects inherited the spirit of his work (Northeast Agricultural Administration Bureau 2018). The legend changed over time: the Meiji government and postwar governmental officials developed the nation in Nitobe's name; and today Land Improvement Districts (LID), or water user associations, and agricultural bureaus exalt Nitobe as the spiritual symbol of agricultural society and its cultural landscapes. However, in on-site research, we discovered that many untold people put together these projects. This chapter shows that the tangible heritage of the irrigation project is inseparable from the area's historical identity or intangible heritage.

Keywords Inaoi River · Irrigation · Land reclamation · Legend · Tangible heritage · Japanese modernization

I. Kuroishi (✉)
Aoyama Gakuin University, Tokyo, Japan
e-mail: izumi-k@sccs.aoyama.ac.jp

C. Hein (ed.), *Adaptive Strategies for Water Heritage*,
https://doi.org/10.1007/978-3-030-00268-8_8

Fig. 1 The photo of Inaoi River landscape at the beginning of Meiji period showing Nitobe's huge scale in old agricultural landscape. *Source* History of Land Improvement Project of Inaoi River, 2010; released under a Creative Commons Attribution-Non Commercial-No Derivatives 4.0 International License

Introduction

Legends circulating throughout Asia describe water as sacred and highlight its importance in sustaining life. Although this chapter will not examine mythological understandings of water in Japan, we hold that it is nonetheless important to recognize the close relationship between mythology, historical legend, and tangible water heritage (Bruner 2003) and examine how legends and personal narratives about water have been kept alive and allowed to thrive in the face of political and economic restructuring. In particular, we unravel the formation and transformation of the legend of Tsutō Nitobe. A samurai of the Nanbu clan, Nitobe is legendary for his pioneering water resource management and land reclamation in Sanbonkihara in the nineteenth century . He created the first intake from the Oirase River to create Inaoi River, which later connected the Oirase Valley and Towada Lake to the Pacific Ocean. Oirase Valley and Sanbonkihara are now listed as Japanese water landscape heritage structures. However, in on-site research, we discovered that the legend changed over time, and in parallel to the change in the story of Nitobe the significance of water in this region also shifted, from an agricultural resource to a treasured physical and symbolic element in the national landscape. Historians have consistently excluded other actors and dynamics to create the symbolic images of Nitobe (Fig. 1).

The History of Water Management Systems and Society in Japan

Before discussing the role played by Nitobe in greater detail, it is necessary to convey the history of Japanese agricultural water systems, to illustrate the ways in which the community system and legal systems of water management (particularly that of agricultural paddy field water) are intertwined and form the community's values. This work frames the memories and legendary stories of the ancestor Nitobe–an intangible heritage–which, in turn, helps to create the communal meaning of cultural identity, as it is inscribed in the tangible heritage of land reclamation (Anderson 1983).

Ingenious political systems integrated water management within the village community system from the Edo period until after World War II (Kimura 2010; Komori 1996; Yamasaki, ed. by Ishii et al. 1996). In the Edo period (1603–1867), new laws and policies shaped the relationship between the water right, ownership of land, the village community system, and taxation. The Tokugawa government formed the water management association of villages in each region to organize irrigation–in effect, to ensure the collective operation and maintenance of water facilities as well as to regulate both water rights and distribution systems in each village (Akimoto 2004). As a result, both land and water were managed and owned by all village residents and agricultural works and environmental management became an everyday matter in village governance.

The government regulated the water right much as it had earlier regulated river systems (Ishii et al. 1996). In the Meiji period, all water and land with no clear ownership belonged to the government: they were made public resources. In 1896, the Meiji government established the River Law under which the national government decided every issue pertaining to rivers. In 1898, the Meiji Civil Code approved the organized water right system of the Edo period as customary law and established a rule to prioritize the older, local water right over the newer national one (Okuda 2010; Yamasaki 1996). In 1909, the government revised regulations for the improvement of the drainage system; and in 1923, it started to improve the main drainage network and to invest heavily in land improvement projects to modernize the agricultural water right system (Miyazaki 2009).

In 1949, the Land Improvement Law extended pre-war water management systems into the postwar agricultural reform. The law formed Land Improvement Districts or water user associations. The LIDs then maintained and organized water systems and coordinated water management with individual landowners. They improved and managed agricultural water facilities that supply and drain water; improved agricultural land infrastructure, such as underground drainage; created landmarks, agricultural land, and grasslands; constructed farm roads; repaired irrigation ponds and installed disaster prevention devices to agricultural land; and managed rivers that served as resources for farming and consumption, as well as surrounding landscapes that contained sewers and roads (Shobayashi et al. 2017; Ministry of Agriculture, Forestry and Fisheries Production 2017). Later, with the advent of industrialization of water use and cities' growing demand for water, the government (in 1964) revised the River Law, such that, while they respected the customary water right, public institutions controlled the water supply. So, then, government sustained both the LID's and customary law, in order to stabilize the economic and social structure of the nation as well as to codify the water environment during this period of modernization of Japanese society (Kide 1984).

Review of Existing Studies

Intangible heritage, memories, and the history of water resource management are in fact connected to the tangible heritage of the Japanese water landscape. While agricultural scholars, ethnographers, historians, and geographers have analyzed how the social management system formed agricultural society and have examined historical figures and narratives, they have not extended their studies to landscape and heritage (Yanagita 1997; Nitobe 1898). Landscape heritage study in Japan was founded on the idea of identifying beautiful and characteristic historical landscapes for tourism (Shiga 1995). And these scholars did not move beyond these bounds to critically examine the tripartite relationship between spatiality, sociality, and historicity (Watsuji 1979; Higuchi 1975). To provide some context, until the end of the 1960s, Japanese regional and urban planning studies emphasized effective planning to modernize rural areas from an engineering viewpoint (Ishida 2004). It was in the 1980s that urban studies scholars began to study the water environment in Japan, joining architectural scholars (Kawahara 2001). And although Japan established a chapter of ICOMOS, the international NGO which protects heritage in 1972, it did not become a party of the UNESCO World Heritage Convention until 1992, as the national government could not promise to protect historical monuments and sites as public values until then.

Heritage historians like Emma Waterson and Laurajane Smith have addressed the difficulty of integrating social history into discussions of tangible heritage. They argue that definitions of heritage often contain idealized conceptions of community that perpetuate constructions of "others" which the heritage process then institutionalizes (Waterson and Smith 2010). For example, studies of historical and agricultural social life have often constructed nationalist and romantic images of rural areas, which the tourism industry then uses in its advertising campaigns. The modernization of rural Sanbonkihara and the stories of that modernization help us think about constructions of community and heritage, and explore the interrelationship of agricultural water systems with local landscapes and local histories.

General History of Sanbonkihara's Reclamation Developments

Sanbonkihara was long a barren land filled with the volcanic ash and gravel left behind by the eruption of Mt. Hakkoda a million years before and the formation of the Towada Lake caldera. The Sanbonkihara plateau is in the east of the northern part of the main Japanese island of Honshu and includes many municipalities (Towada City, Rokunohe Town, Shimoda Town, Hyakkoko Town, Misawa City, Towada Lake Town, Shichinohe Town, and Kamikita Town). The area measures twenty-seven kilometers from east to west and eight kilometers from north to south. The substantial difference in elevation—of more than thirty meters—between the Sanbonkihara plateau and the Oirase River, coupled with poor water retention capacity of the soil, made the plateau difficult to irrigate and thus unsuitable for rice cultivation. Sanbonkihara is furthermore susceptible to cold winds in the summer. Inhabitants often suffer famine. Farmers on the plateau planted field crops, like millet and soybeans, and reared horses as agricultural activity. These difficult conditions have made Sanbonkihara one of Japan's three major historical cultivation regions that have overcome natural difficulties to achieve agricultural productivity (the other two are Kawaminami cho of Miyazaki prefecture and Yabuki cho of Fukushima prefecture) (Fig. 2).

From the seventeenth century, small scale primitive hydrophobic developments utilizing natural terrain were constructed between several villages in Sanbonki area and Oirase River (Fig. 3). In 1853, Tsutō Nitobe gathered sixty-five comrades to initiate a first massive irrigation project in Sanbonkihara. They dug a six thousand meter tunnel through the Tengumori and Sakura mountains to take water up from the Oirase River through a water weir. This newly created conduit, the Inaoi River, reached approximately 10.3 km in length, boasting three hundred hectares of paddy fields cultivated in its basin in 1859. In 1861, Tsutō's son Jujiro took over the work to expand the clan's territory, having studied various industries around Japan and proposed a plan to extend the Inaoi River to Ogawara Lake close to the Pacific Ocean. Following his son's tragic death in 1867, Tsutō persisted in realizing the project and completed 1.1 km of the second conduit. Unfortunately, following the Meiji restoration in 1868, funding for new paddy irrigation—provided by the Nanbu clan—ceased and the project was shelved, relegating Sanbonkihara to ruin (Aomori Kenshi Hensan Kingendaibukai, v6 2015).

Approximately fifteen years later, in the Meiji period, the Inaoi River irrigation project was revived and received renewed support from the state. With the help of the Japanese government in 1884, Shigeaki Fujita established the Kyoritsu Reclamation Company, restored the old conduits constructed by the Nitobe family, and resumed reclamation of Sanbonkihara. In 1888, Eiichi Shibusawa, an industrialist who is widely recognized as the father of Japanese capitalism and friend to Nitobe, bought out the Kyoritsu Reclamation Company, opening the seventeen-hundred and one-hectare Shibusawa farm in Sanbonkihara (Sanbonkihara Kaihatsu Kenkyukai 1967; Ogasawara 1994; Mizuno 1961). The Inaoi River had been lengthened tremendously, to 39 km, reaching the Pacific Ocean by 1897. In 1905, Kyoritsu Reclamation Company became a stock-issuing company named the Sanbonki Reclamation Corporation, an event which signaled a marked growth in the scale of the irrigation project. The agricultural bureaucrat and scholar Saburo Mizoguchi (1948), reflecting on these events from his post-World War II perspective, explained these later Meiji period projects as the successors to Nitobe's endeavors (Fig. 4).

In 1920, Noriyoshi Mizuno, both an agricultural engineer and a pupil of Shibusawa, took office as the manager of the Shibusawa farm and worked to transfer its water management to the government (Ogasawara 1996). Following Mizuno's proposals to the government, in 1937, the Tohoku Promotion Research Council selected the Sanbonkihara district as a development area. It then set into motion a large-scale Sanbonkihara national reclamation project that lasted from 1938 to 1944, with several administrative bodies regulating the movement of water from Lake Towada to

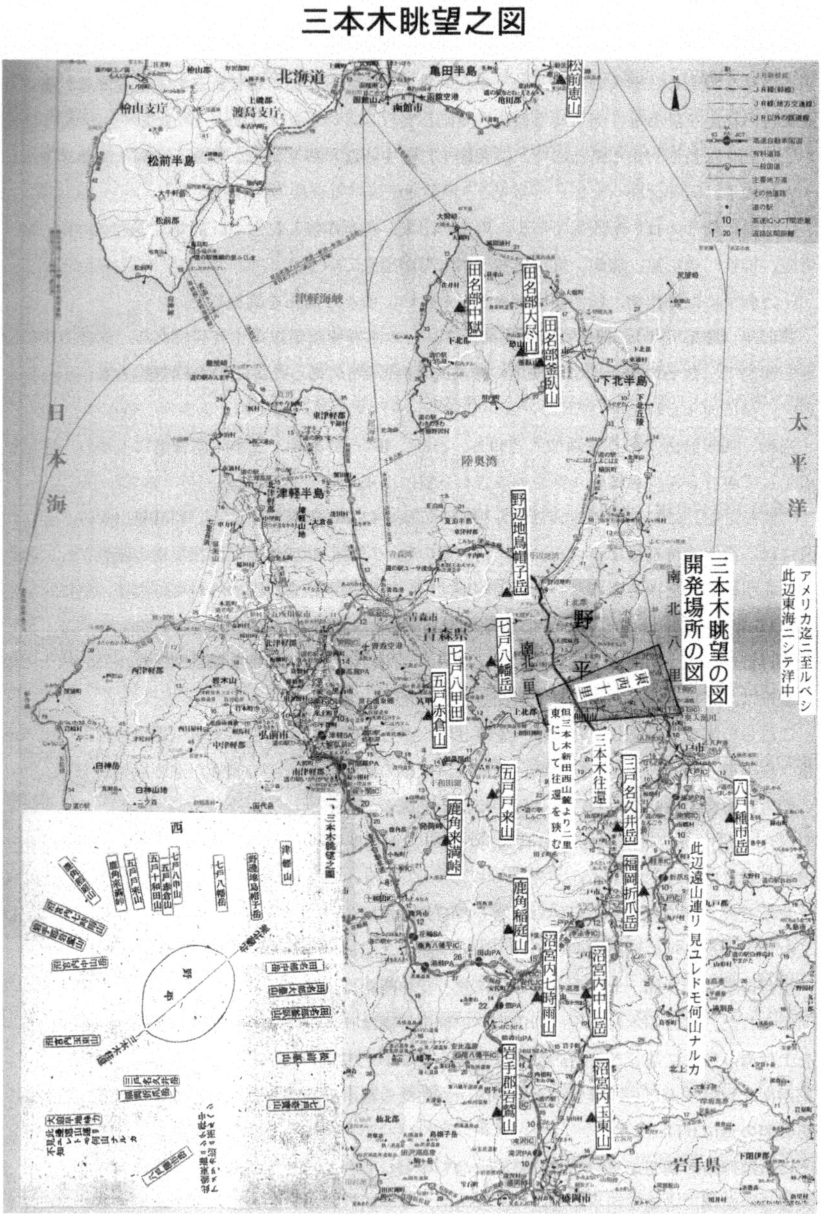

Fig. 2 Map of Sanbonkihara in the northern part of Japan with Tsutō Nitobe's proposal diagram for future development. *Source* Aomori Kenshi Hensan Kingendaibukai 2007, *Aomori Kenshi Shiryohen Kingendai (Aomori Prefecture History, Appendices Modern) 1*, Aomori prefecture; released under a Creative Commons Attribution-NonCommercial-NoDerivatives 4.0 International License

Fig. 3 Old weir utizing natural terrain connecting Oirase River to Fujisaka village. *Source* Photo taken by the author 2010; released under a Creative Commons Attribution-NonCommercial-NoDerivatives 4.0 International License

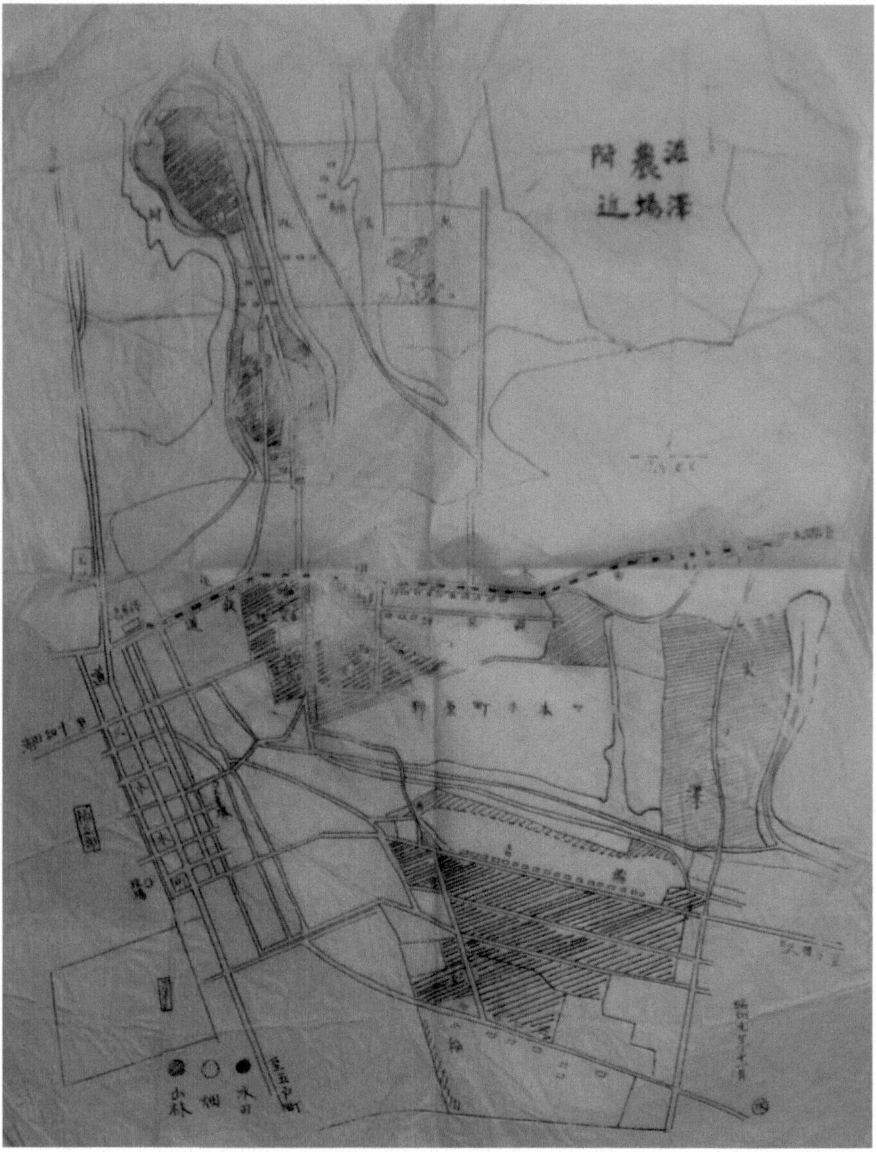

Fig. 4 The purple area indicates Shibusawa farm territories showing how they extended Nitobe's irrigation project to cover the whole area of Inaoi River in the 1930s. *Source* Shibusawa Bunko archive in Towada City; released under a Creative Commons Attribution-NonCommercial-NoDerivatives 4.0 International License

the Oirase River. The project achieved a dual goal: it secured water for the landscape of Lake Towada and the Oirase Valley while providing water for cultivation and hydroelectric power generation in Sanbonkihara. The project constructed an additional Sanbonki national trunk conduit to uptake water from the Oirase River, allowing the cultivation of large-scale paddy fields. The impact of these works was demonstrable: In 1944, the water supplied by the Oirase River irrigated a record 2500 ha and generated a record 27,000 kW of electricity.

The second period of intense governmental development in Sanbonkihara followed World War II; indeed, the entire Towada area was eventually slated for national development (Northeast Agricultural Administration Bureau 2016, 2017) to address a postwar crisis in housing and feeding war refugees and veterans. In this effort, two thousand veterans and soldiers were settled on the 4445 ha former Sanbonkihara military site and made farming and state-controlled cultivation their livelihood. In 1948, the government began the Ogawara pond irrigation near the Pacific Ocean as an extension of the Sanbonkihara National Reclamation Project. When this project was completed in 1966, 3376 ha of paddy fields and 5947 ha of crops had been developed, and the waterway was finally extended to seventy-one kilometers to the Pacific Ocean. This massive infrastructure became the foundation of the regional economy. Since then, the maintenance and management of rivers, water supply, and dams have been conducted under prefectural administration in collaboration with LIDs (Fig. 5).

Challenges to the Historiography of the Legend as Heritage in Sanbonkihara

When we conducted on-site research on water heritage and compared local literature to local history, we discovered that this historical framework did not tell the whole story. It is true that some local and institutional history textbooks, such as *Watashitachino Furusato Towadashi (Our homeland Towada City)*, published by the educational committee of Towada City, described Nitobe's project as a pioneering development. But we discovered many structures from as early as the Edo period, suggesting that Sanbonkihara had already been developed before Nitobe started his work. We also found historical records in private museums and in the museums of neighboring towns that showed that many migrants to the area contributed to local water reclamation on smaller scales over the years. The Tomabechi family from Osaka developed Sanbonkihara's first paddy field in 1673 in an early reclamation project (Aomori Kenritsu Kyodokan 1994; Aomori Kenshi Hensan Kingendaibukai 2007; Inaoi Gawa Tochi Kairyoku 2003). Immigrants from the Uesugi clan introduced fortress construction technology which was then used in irrigation projects by migrants and low class samurais. In 1845, Morita Kiuemon of Shichinohe developed more paddy fields in Sanbonkihara. The Fujita family of Gonohe, originally from Shiga Prefecture, also played a part in developing Sanbonkihara, as well as other villages (Gonohe, Asamizu, Ichikawa, and Osaka). *Ema* boards or wish plaques with

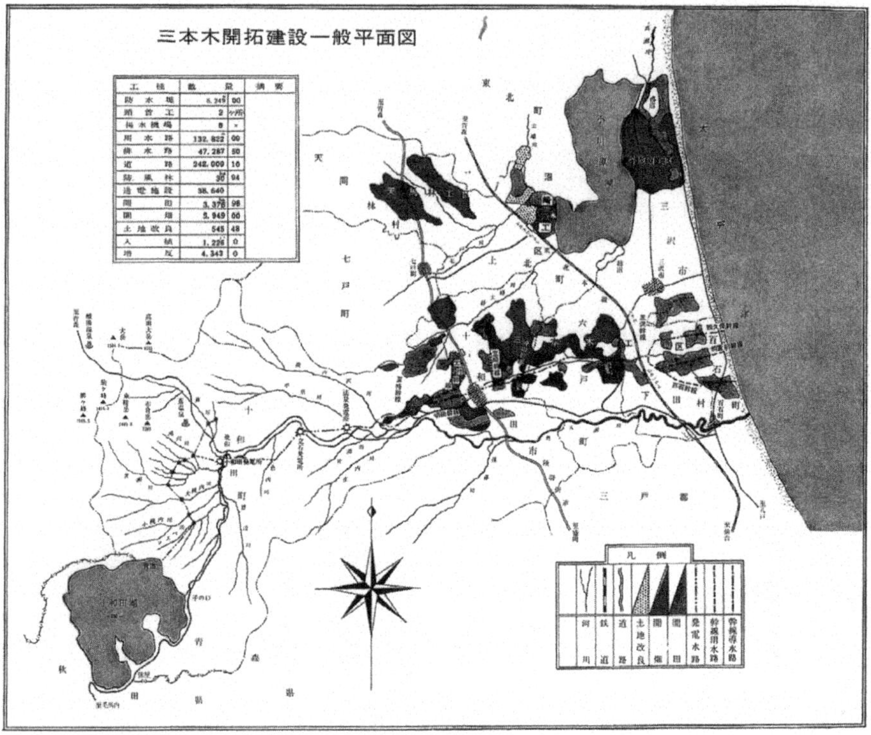

Fig. 5 Development of Sanbonkihara reclamation reaching from Towada Lake to the Pacific Ocean. *Source* History of Land Improvement Project of Inaoi River, 2003; released under a Creative Commons Attribution-NonCommercial-NoDerivatives 4.0 International License

pictures in local shrines, tell the tragic and heroic stories of those who committed suicide to calm the river god's anger and to protect the area from floods in Edo period (Towada shi kyoiku iinkai 1982; Ministry of Literature 1939; Towada shi kyoiku kenshu senta 1988).

The absence of these peoples' works from local and institutional historical treatments raises questions. Why did authors keep writing that Nitobe was the pioneer and stressing that all later developers were spiritual followers? and why was Sanbonkihara a symbol and a priority of national development until after World War II? (Kahoku Shinpo Newspaper, October 20th, 2017). To answer these questions, we draw on Eric Hobsbawm's idea of historiography, which posits that some historical narratives signal the dominant values of those in power; he argues that such narratives show the range of those who can belong to the nation (Hobsbawm 1990). We found three dynamics essential to our reframing of the legend of Nitobe: first that not only Nitobe but a range of actors developed the reclamation project; second, that national political affairs affected policies on water supply and constructions of national community; and third and last, that cultural events and urban planning bridged tangible and intangible heritage.

Beyond Nitobe: Other Actors and the Creation and Maintenance of Water Resources

Although legend attributes the Sanbonikihara reclamation project to Nitobe, in fact, a range of individuals and associations realized it. These actors harbored multiple agendas that evolved over time, held substantially distinct economic ambitions, and worked at different territorial scales which varied according to time. The irrigation project of the Edo period created paddy fields for rice cultivation. In the Meiji period, the Shibusawa farm and military camps introduced new agricultural technology which led to profit from forestry, stock farming, and ranching. Following this success, in the pre-war period, the government established hydroelectric power plants to stabilize water intake with strong pumps and to accelerate the industrialization of Sanbonkihara. Finally, in the postwar period, Yukio Sugimoto, a pupil of Shibusawa, made his ground-breaking economic contributions by establishing a tourism industry in Sanbonkihara (Sugimoto 1981; Inagaki 1998; Sasamoto and Ogasawara 1996; Towada Kanko Newspaper 1957–1968).

In its several different incarnations, the Sanbonkihara reclamation project required a range of niche skills and techniques as well as a very large investment of new technology, capital and labor. Many skilled workers and engineers from outside Towada were involved. Before the Nitobe family, the Uesugi clan and migrants from Shiga had begun to develop Sanbonkihara. After Nitobe, Eiichi Shibusawa and his disciples Mizuno and Sugimoto continued the reclamation work and regional industrial development. Many outsiders came during political upheavals. During the Meiji Restoration, over 17,000 people in the Tonami clan fled from Fukushima to Sanbonkihara as political refugees. Historical textbooks in Sanbonkihara describe how, under the orders of the Meiji Government, Tsutō Nitobe took care of some of them at his farm in 1871, and how they had difficulties getting along with other local residents of the Nanbu clan. Shigeaki Fujita was a descendant of this Tonami clan who later resumed the irrigation project after sixteen years' abandonment. After World War II, many former military personnel and displaced Japanese settled in Sanbonkihara to complete the national irrigation project.

Meanwhile, water resource management frequently switched hands. The irrigation project transitioned from being under a local clan, to a private company and to a government entity. Initially, it was under the Nanbu clan, then the Kyoritsu Reclamation Company, and the Sanbonkihara Reclamation Cooperation. After the Sanbonkihara Reclamation Cooperation dissolved in 1921, the Inaoi River Irrigation Association took over water resource management, and after the war, the Inaoi River Land Improvement Organization assumed leadership. Over time, local people managed and steered these companies, sometimes working in their own interests.

These historical analyses clarify that the actual *process* of the developing reclamation, the core of intangible heritage, has frequently changed direction, with multiple agents shaping and negotiating changes in the meaning of water (Spirn 1998).

The Politics of Nitobe's Legend in the Management of Water and Community

Another way of telling this story is to point out that neither the local community nor the community working on the reclamation project was homogenous. But Sanbonkihara needed a symbolic leader in order to integrate outsiders into Sanbonkihara society, to control tensions in sharing water and land use, and to manage the difficulties of political and industrial modernization overall. Nitobe stood for a social ethos shared by a range of community members that could bridge the government, capitalists, immigrants, and local individuals. These political dynamics shape the story of Nitobe as the legendary founder of Sanbonkihara in books by the Nitobe Memorial archive, local children's textbooks, ministry reports, and the LIDs' books; these narratives exclude other characters for the same political reasons.

Tsutō and Jujiro Nitobe carried out their projects in the transition from the Edo to the Meiji periods. Although political red tape hindered many projects, the Nitobes were highly regarded in the national government, and they played a pivotal role in guaranteeing the legitimacy of the Meiji government in northern Japan. On two occasions in 1877 and 1881, Emperor Meiji visited the Nitobe family's house and honored Tsutō for his accomplishments but for his own interests: to solidify his own historical and ethnic authority, to associate himself with living gods among the people, and to advance his status as a demigod (Osaka 1998). At the same time, this association strengthened the authority of the Nitobes and the power of the local administrative agencies in the region.

In 1884, Emperor Meiji nationalized a large-scale budget for the Sanbonkihara area to support the Kyoritsu Reclamation Company in Nitobe's name. When the Emperor ascended to the throne in 1867, his government had formulated policies to resolve conflicts with the clans he defeated, including state-supported agricultural land development projects (Osaka, Ibid.) Sanbonkihara was the cornerstone of this overall project in the 1880s to make industry the new foundation of the nation: other big projects included Tazawa Lake in Akita and Asaka Hydrophobic in Fukushima (Miyamoto et al. 2002; Northeast Agricultural Administration Bureau 2018). Later, from 1937, as war brought an influx of external laborers for state agricultural cultivation, the legend of Nitobe helped again to inculcate a regional ethics and cultivate a state that diminished discrimination toward immigrants. Thus, as intangible heritage, more than just a local historical story, Nitobe's legend had a lasting and far-reaching impact on the identity of Sanbonkihara.

Cultural Events and Urban Planning Bridging Tangible and Intangible Heritage

Festivals, statues, and educational activities also all celebrated Nitobe's works as legendary. These cultural events bridged intangible legend and tangible structures.

In Towada city, Taiso (Archaic) festival is held annually (from 3rd to 5th May) in Taiso Shrine during the season of cherry blossoms, commemorating what is known as Water Flowing Day. People place bronze statues and other memorials to the Nitobe family in the garden of Nitobe archive, tangible heritage for the area. Elementary schools bring children to the Shrine to teach them the legend.

The aura of legend of around the accomplishments of Tsutō Nitobe extends to the urban planning of his son, Jujiro Nitobe. Local historical textbooks celebrate Jujiro's application of Kyoto's urban planning method to Towada in 1860 as unique at the time, particularly his use of the grid system of roads dividing the town into blocks of 1.3 × 1.3 km; they claim it as the beginning of modern Japanese urban planning (Taiso Kensho kai 1998). They note that Jujiro established the Inari Shrine, Clear Moon Temple, and Rinen Temple out of concern for the people's psychological and spiritual well-being. He designed a canal to bring water from the Inaoi River not only for agricultural purposes but also for domestic use and fire protection in town and designed the city around a canal. He planned a new commercial district in the new town of Inaoi-machi along the Oshu Kaido Road, with two-story shops lining an arcade, and demarcated new districts for different industries (dishware, silk, leather) and even a horse market. At several locations, forests were to be planted to shield against strong seasonal winds.

However, Towada City never actually built the water canal from Inaoi River in its center, and it is unclear how Jujiro's ideas could have contributed to the city's later urban planning. In 1885, well after Jujiro had died in 1867, the Army Battalion Bureau opened a new training center and expanded Jujiro's original grid system to form the current city area of about 4 km^2, sandwiched by Inaoi River and Oirase River. In the updated grid, Taiso Shrine (1965, much later than the planning described here) is clearly separated from the surrounding town by a torii gate with a designated approach on a strong spatial axis, which suggests its intentional alienation even from the Army's revision of Jujiro's original idea. Despite the legend, then, Towada City was not developed according to Jujiro's planning. It was largely influenced by the pre and post WWII planning, which followed the planning of Sapporo city (Towada shi kyoiku kenshu senta 1981) (Fig. 6).

Transformation of the Territory and Values of Heritage

The gaps between legend of Jujiro Nitobe and the actual landscape of Towada city also echo the shifts in the historical relationship between the legend of the elder Nitobe and the landscape. The clearest evidence of this shift is that the area named Sanbonkihara itself changed through history. After the formation of Aomori and Iwate prefecture in 1871, the city territory changed from Sanbonki-cho to Sanbonki city and finally Towada city, with various city mergers extending Nitobe's legendary status geographically. What "Sanbonkihara" means in the legend has also expanded from the actual project site: adding Tonami clan's new settlement after 1869 in the nearby Ninohe and Sannohe area, Shibusawa farm's wide ranch area along Inaoi

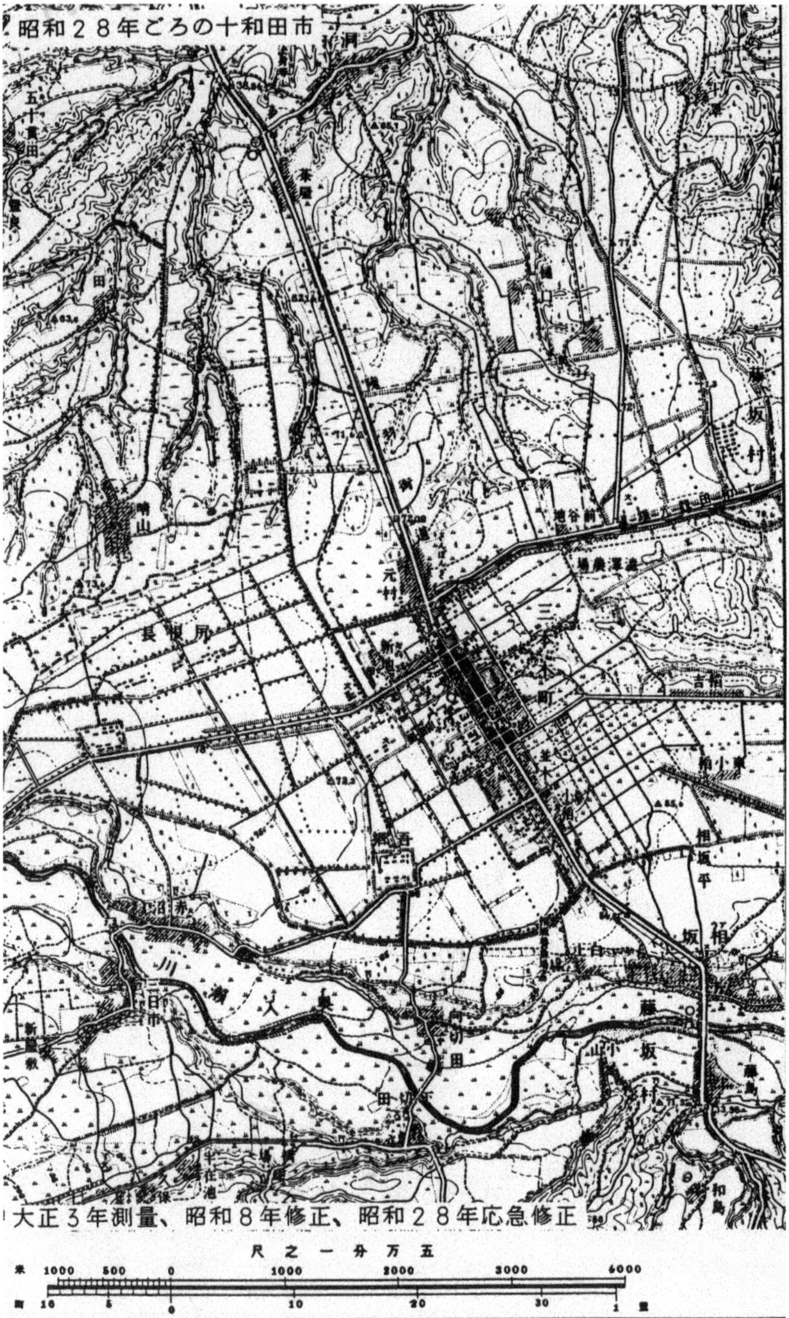

Fig. 6 Whole area map of Towada city downtown area showing post WWII grid roads extentions in the former Army Battalion area. *Source* Towada shi kyoiku kenshu senta; released under a Creative Commons Attribution-NonCommercial-NoDerivatives 4.0 International License

River, the national irrigation projects connected to Towada and Ogawara lakes, and Sugimoto's tourism from Towada to the Misawa area (Aomori Kenshi Hensan Kingendaibukai 2007). In each stage, the image of the landscape of Sanbonkihara in Nitobe's legend is significantly different; paddy field, ranch field, broader farming land reaching to the Pacific Ocean, and natural valley with deep woods and Towada lake (Fig. 7).

Another significant shift is apparent in the aesthetic value of water in each stage of the story. The early texts of the legend of Nitobe did not mention the beauty of the landscape. Then, in the Edo period, Japan celebrated the beauty of the west side of Towada Lake (in Akita prefecture) as a kind of heritage, and 1930s state projects and 1950s works by Sugimoto recognized the beauty and cultural values of the length and landscape of the rivers from Towada Lake to Sanbonkihara (Okubo et al. 1993). In building the Towada Science Museum in 1953, Ogawara Lake Ethnic Museum in 1958, and Saigyodo park in 1980, Sugimoto combined local folklore with the landscape, bringing together intangible and tangible heritage. Claiming Nitobe and Shibusawa as his spiritual predecessors, Sugimoto rode the waves of capitalist success and began a nationwide tourism project between the Oirase River and Tokyo in 1967 (Nakazono 2012).

Conclusion

This chapter reframes the relationship between water and people in the Sanbonkihara region and the contribution of Nitobe's legend to local society to include the effects of the ever-evolving interactions between people, space, and society in and after Nitobe's time. The value of water and the landscape was not a simple outcome of Nitobe's project but developed in the historical process of agricultural expansion, river conservation, and redevelopment, and in the historical process of spatializing and formalizing the legend. Political, economic, and social restructuring kept creating a new sense of community and new interrelationships between tangible and intangible heritage (Tuan 1993).

Though the legend of Nitobe faded with time, its spirit remains in the natural and everyday spaces of Sanbonkihara, continuing to evolve with the people who visit or live there (Kahoku Shinpo Newspaper 2017). In this period of declining agriculture and traditional social ethos in Japan, it is necessary to re-establish water not merely as a natural resource or as an asset of agricultural landowners but as the whole community's historical, cultural and social inheritance; heritage, with which diverse people have to actively engage and cultivate for their and future generations' survival.

Fig. 7 Expansions of Sugimoto's tourism industry from Sanbonkihara to Tokyo. *Source* newspaper Towada Kanko, published in 1966; released under a Creative Commons Attribution-NonCommercial-NoDerivatives 4.0 International License

Acknowledgements This study was made possible by the material and support of Towada city, Shibusawa Bunko, Aomori Prefecture office, Inaoi Tochi Kairyoku, Towada Kanko, and Nitobe Kinenkan. The original research started with funding from the Toyota Foundation and later developed through support from the Ministry of Literature. The project evolved to its current theme thanks to the advices by Professor Masamitsu Ogiwara and Mr. Mamoru Okubo, and to the conference on Water and Heritage for the Future organized by ICOMOS, the Center for Global Heritage and Development (CGHD), and Delft University of Technology. I appreciate the support of Professor Carola Hein for inviting me to the conference and helping edit my text.

References

Anderson B (1983) Imagined communities: reflections on the origin and spread of nationalism. Verso, London

Aomori Kenritsu Kyodokan (1994) Inaoi gawa to Dobuchi seki. Aomori, Aomori ken

Aomori Kenshi Hensan Kingendaibukai (2007) Aomori Kenshi Kingendai Shiryohen (Aomori Prefecture History, Modern, Appendices) 1, Aomori Prefecture

Aomori Kenshi Hensan Kingendaibukai (2015) Aomori kenshi shiryohen, Kinsei (Aomori Prefecture History, Modern, Appendices) 6, Aomori Prefecture

Bruner JS (2003) Making stories: law, literature, life. Harvard University Press, Cambridge

Komori H (1996) Nogyo suiri to nihon gata shakai (agricultural water system and Japanese society). In: Shokei Ronso, no 45, pp 1–19

Higuchi T (1975) Keikan no Kozo: Landscape to shiteno Nihon no Kukan (Structure of landscape: Japanese space as landscape). Gihodo Shuppan, Tokyo

Hobsbawm E (1990) Nations and nationalism since 1780: programme, myth, reality. Cambridge University Press, Cambridge

Inagaki M (1998) Sugimoto Yukio Monogatari. Ryoko Yomiuri Shuppann, Tokyo

Inaoi Gawa Tochi Kairyoku (2003) Inaoi Gawa Tochi kairyokushi (History of Inaoi River LIDS). Inaoi Gawa Tochi Kairyoku, Sanbonkihara

Ishida Y (2004) Nihon Kingendai Toshikeikaku notenkai (Transformation of the modern and contemporary Japanese Urban Planning)1868–2003. Jichitai Kenkyusha, Tokyo

Ishii M, Nagaoka A, Harada T (1996) Kokudo Riyo no henyo to chiiki shakai (Transformation of the use of National Land and local society). Daimyodo Hakko, Tokyo

Akimoto H (2004) Wagakuni Kindai niokeru Nogyo suiri chitsujo no Saikento (Reexamination of the agricultural water right order in the modernization of Japan). Keizai Ronshu 29(2):1–11

Kahoku Shinpo Newspaper (2017) Oct. 20th, Kahoku Shinpo, Sendai

Kahoku Shinpo Newspaper (2017) Nitobe memorial archive was ordered to close its building for seismic problem by Towada city and is appealing to a court. http://www.kahoku.co.jp/tohokunews/201710/20171021_21036.html. Accessed 2 Jan 2018

Kawahara I (2001) Chikyu Kankyo to Tokyo (Global environment and Tokyo). Chikuma shobo, Tokyo

Kide I (ed) (1984) Suiri no Shakai kozo(Social structure of water management). Kokusai rengo daigaku press, Tokyo

Kimura S (2010) Nihon Nogyo shi. Yoshikawa Kobundo, Tokyo, pp 256–320

Ministry of Agriculture, Forestry and Fisheries production (2017) Tochikairyo ku, Tochikairyo Jigyodantai rengokai ni tsuite (LIDs and association of LIDs). file:///Users/Izumi/Desktop/農林水産省:土地改良区・土地改良事業団体連合会について.html. Accessed 10 Dec 2017

Ministry of Agriculture, Forestry and Fisheries. Sanbonkihara Daichi wo Kaitakushita Nitobe Tsutō. http://www.maff.go.jp/j/nousin/sekkei/museum/m_izin/aomori/. Accessed 2 Jan 2018

Ministry of Literature (1939) Tohoku Dokuhon (Reading about the Northern part of Japan), Gekan, Towada city, 1–3, 42–45

Miyazaki J (2009) Kanko suiriken no ruikei to sono koryoku(typology of customary water right law and its effectivity). Mizu shigen, kankyo kenkyu 22:1–12

Miyamoto T et al (2002) Aomoriken niokeru Shizoku jusan to tsugaru ai sangyoka henokokoromi (The Industrialization and education of Samurai class and the creation of dying industry in Aomori). In: Review of the researches in educational department of Hirosaki University, vol 87, pp 89–98

Mizoguchi S (1948) Uton, Suiri Shidan (Rainy village, history of water management). Yukei sha, Tokyo

Mizuno N (1961) Taishi wo tsuide. Sanbonkihara Kaihatsu Kenkyukai, Towada city

Nakazono H (2012) Totestu ga hashitta jidai. In: Toou nippou, 2.14-25

Nitobe I (1898) Nogyo Honron (Principles of agriculture). Mokabo, Tokyo

Northeast Agricultural Administration Bureau. Aomoriken, Towadakoshuhen, Sanbonkihara Daichimonogatari (Aomori prefecture, Area of Towada lake, Sanbonkihara Plateau story). http://www.maff.go.jp/tohoku/nouson/seibi/rekisi/6story01.html. Accessed 2 Jan 2018

Northeast Agricultural Administration Bureau (2016) Kokuei Kangai Haisui Jigyo: Ousakagawa sagan chiku (National Irrigation and wast-water project: Ousaka river left bank area

Northeast Agricultural Administration Bureau (2017) Suido no Ishizue (Foundation of water and earth). file:///Users/Izumi/Desktop/陸奥の先進地%20—福島県%20国営会津北部農業水利事業%20—水土の礎.html. Accessed 20 August 2017

Northeast Agricultural Administration Bureau (2018) The histories of these three projects. http://www.maff.go.jp/tohoku/nouson/seibi/rekisi/6story01.html. Accessed 2 Jan 2018

Ogasawara K (1996) Shibusawa noujo to sanbonkihara no yoake. Seikoen shuppann, Towada city

Ogasawara Y (1994) Shibusawa Nojo no Kaihatsu to Keiei, Towada city

Okuda S (2010) Nogyo Suirichitsujo to Suiriken no arikata (Order of water distribution and the way of water right control). In: Aoyama Homu Kenkyu ronshu (Aoyama Legal research collection), Aoyama Gakuin publication, no 3, vol 1, pp 107–131

Okubo H et al. (1993) Towadako, Oirasegawa no Fuchihozennto kaihatsu tono chosei (Study on the Planning Methods for Coordination between the keeping scenic beauty place and the development of water resources in Towada Lake and Oirase River Basin). In: Nodo shi, no 61, vol 10, pp 927–932

Osaka H (1998) Daijo kanki Chiho Junko no Kisoteki kenkyu (Basic examination of the rural tour by Meiji Emperor in his early ruling period). In: Hosei Shigaku. Tokyo, Hosei Shigakkai, no 50. March, pp 62–72

Sanbonkihara Kaihatsu Kenkyukai (1967) Kaitaku no Oni, Towada city

Sasamoto K, Ogasawara K (1996) Chosen, Tokyo, Jitsugyo no Nihon sha. Towada Kanko Newspaper, Towada shi, 1957–1968

Shiga S (1995) Nihon Fukeiron. Iwanami Bunko, Iwanami Shoten, Tokyo

Shobayashi M, Okajima M (2017) Kikan suiri sisetuno jizokutekina koshinnotameno aratana seidoteki wakugumi (New Legal framework to sustainable renewal of the main irrigation facility). In: Nogyo Noson Kogakkaishi (Agriculture and agricultural area engineering academy research review), vol 85, no 9, pp 21–27

Soja E (2003) Writing the city spatially. In: City: analysis of urban trends, culture, theory, policy, action, vol 7, no 3, Routledge, London, pp 269–281 (13), November 2003

Spirn AW (1998) The literature of landscape and survival and imagination: reading and telling the meaning of landscape. In: The language of landscape. Yale University Press, New Haven, pp 27–46, 47–81

Sugimoto Y (1981) Shibusawa Keizo sensei ni tsukaete
Taiso Kensho kai (1998) Towadashi, Sanbonkihara Kaitaku to Nitobe sandai no Rekishi gaidobukku (Guide book about history of the development of Sanbonkihara in Towada city and the three generation of Nitobe). Nitobe Kinenkan, Towada city
Towada shi kyoiku kenshu senta (1981) Chugakko shakaika fukudokuhon Wagakyodo Towada shi, 78, Aomori
Towada shi kyoiku iinkai (1982) Watashitachino Furusato Towadashi (Our home land Towada city), Towada city
Towada shi kyoiku kenshu senta (1988) Kyoudogakushuu shiryou, zokuzoku medemiru towadashi no daichi (Material for the program of learning homeland and additional material of visual survey of the area of Towada city). In: Sanbonkihara wo hiraku(Reclaim Sanbonkihara plateau), Towada city
Towada Kanko News paper (1957) 8.1-1968, 10.15, Towada Kanko Dentetsu
Toyama K (1993) Nihon no Kome: Kanko to Bunka wa kakutsukurareta (Japanese Rice: this is the way how tourism and culture have been developed). Tokyo, Chuko Shinsho, pp 28, 122, 154–159
Tuan Y-F (1993) Space and place, Chikuma Gakugei Bunko, tr. hiroshi yamagata, pp 265–282
Waterson E, Smith L (2010) The recognition and misrecognition of community heritage. Int J Heritage Stud 16:4–15
Watsuji T (1979) Fudo: Ningengakuteki kosatsu (Climate: Examination as humanistic study). Iwanami Bunko (Originally it was published in 1938), Tokyo
Wigen K (1995) The making of a Japanese periphery, 1750–1920. University of California Press, pp 1–21, 70–79
Yamasaki K (1996) Kanto chiho no mizu shigenno hippakuto sonohaikei(Study of the critical shortage of the water resource in Kanto region and its background). In: Ishii M, Nagaoka A, Harada T (eds) Kokudoriyouno henyo to chiikishakai (Transformation of the Use of National Land and Local Society), 26. Daimyodo Hakko, Tokyo
Yanagita K (1997) Jidai to Nosei (Contemporary Society and Agricultural policies), vol 2, Anthology of Yanagita's Works, Tokyo, Chikuma shobo (originally published in 1911)

Izumi Kuroishi is professor of the School of Cultural and Creative Studies of Aoyama Gakuin University, Japan. Her recent publications include *Constructing the Colonized Land* (Ashgate 2014) and "Urban Survey and Planning in Twentieth-Century Japan" (JUH, 2016). She currently works on disaster relocation and social resilience after the 2011 tsunami and on domestic space and house planning in wartime Japan.

Shihmen Dam in Taoyuan County, Hua Lin, released under a Creative Commons Attribution-NonCommercial-ShareAlike 2.0 Generic license

Chapter 9
How Citizens Reshaped a Plan for an Aerotropolis and Preserved the Water Heritage System of the Taoyuan Tableland

Sinite Yu, Chung-Hsi Lin, Hsiaoen Wu, Wenyao Hsu and Yu-Chuan Chang

Abstract Heritage preservation in Taiwan was for many years limited to a small group of art experts, who focused on masterpieces in museums or on magnificent architecture, disconnected from people's daily lives and current society. Others made efforts to enhance local cultures and encourage grassroot participation in heritage preservation, but, as heritage sites continued to be treated as single, unrelated objects, these efforts remained only loosely tied to local communities. A recent campaign for the preservation of the pond–canal system heritage in the Taoyuan tableland suggests a change of direction and indicates new opportunities for increased attention to and participation in heritage. In 2008, a plan for developing an airport-based metropolitan region, a so-called Aerotropolis, threatened the local water heritage infrastructure, including its distribution network and the local sustaining eco- and social systems. Opposition arose among a wide range of environmental and social groups, which weighed in on the importance of the water system. Together their stories wove a fine and unbreakable "web of narratives," which became a shield that successfully protected the water heritage. Water heritage, as a system of multiple sites that is by nature more relevant to people's daily lives, can benefit from such participatory preservation. Moreover, it is possible, too, that the conservation of heritage of all categories could find inspiration in this strategy.

Keywords Water heritage · Taoyuan tableland · Pond–canal system · Aerotropolis · Public participation

---

S. Yu (✉) · H. Wu · W. Hsu
Taiwan International Institute for Water Education, Taiwan, Republic of China
e-mail: dryu@tiiwe.org.tw

C.-H. Lin
National Yunlin University of Science and Technology, Taiwan, Republic of China

Y.-C. Chang
HsingWu University, Taiwan, Republic of China

C. Hein (ed.), *Adaptive Strategies for Water Heritage*,
https://doi.org/10.1007/978-3-030-00268-8_9

Introduction: Theoretical and Methodological Implications of Participatory Heritage Preservation

The heritage debate of Taiwan has focused since the mid-twentieth century on constructing a narrative in line with the government's political objectives. After World War II, the Chinese nationalist government took Taiwan over from Japan and four years later retreated there from mainland China. With the hope of reconquering the mainland, the postwar government defined "cultural heritage" as the treasures brought from mainland China, including Chinese antiques and art masterpieces. This emphasis on the ties between culture and national identity was further enhanced when the Chinese Cultural Revolution began in 1966: the nationalists propagandized Taiwan as both the lighthouse of democracy and the guardian of Chinese culture. The most precious heritage, as taught in school, was the Chinese treasures in the Palace Museum in Taipei, a place designed in the style of a Chinese palace. Many famous buildings of this period, like the Grand Hotel, were built in similar style on the site of former Japanese temples, to proclaim the triumph of the Chinese over the Japanese in the war as well as the blooming of classic Chinese culture in Taiwan.

Ideas about cultural heritage in Taiwan were limited to single and mostly tangible objects for decades, quite disconnected from people's daily lives. Even after political ideologies became less important and influential in the 1980s, dominant approaches continued to view "cultural heritage" as individual buildings, constructions, and other objects that embody some important aspects of human civilization. They highlighted the splendor of single sites, but, when these places ceased to be active in their original roles, they became disconnected from the society they belong to and distant from people's lives or memories.

In recent years, more and more preservationists have adopted a new approach to enhance the value of cultural heritage: weaving a web of narratives that view multiple sites or items as a heritage network or system and inviting the people of the local community to participate in this narrative creation. This new method in many aspects echoes the concept of "cultural route" that emerged in the international community of heritage preservation in this century. But this chapter focuses on the distinctive Taiwanese context and gives credit to local devotees who, without being informed of the latest theories on cultural routes, developed something on their own. Their method has won some significant successes. This chapter discusses a recent campaign in Taiwan's Taoyuan tableland for protecting regional ponds and canals and suggests that participatory narrative weaving empowered that campaign. It further argues that this example shows how water infrastructure that is both parts of a network and close to people can boost both new heritage appreciation and participatory preservation in modern society.

Heritage preservation focusing on individual objects or sites has its benefits and setbacks. Most sites in this kind of preservation projects have some magnificence or beauty that can be easily discovered and demonstrated. Preservationists, researchers, and the government can thus invest their energy and resources on these single objects and make the most out of their investment. But this approach tends to handle heritage sites as treasures displayed in the museum window, and, as a result, often disconnect them from the context in which they were created. They are dead relics of past splendors. This focus could have some further negative consequences. First, the less a heritage site is connected to the society to which it belongs, the less important it becomes. When society changes and the heritage site ceases to function, this disconnection worsens and makes the site no longer relevant to its community. Second, disconnecting an item from its context often makes it difficult for people to apprehend the value of the context itself and the role this item plays in relation to other things in the same context or network (See: Lin 2017).

This is the problem that many heritage sites in Taiwan face. Take one example: The Lin family's mansion in Wufeng, Taichung, form the largest pre-modern residence owned by a civilian; it was built and expanded many times during the nineteenth century. Its importance lies in its architectural beauty, the Lin family's legendary wealth, and some major historic events related to the family. Lin Xian-Tang, a clan chief, actively advocated democracy during Japanese colonialism and sponsored the island's literary and art societies—many writers and artists met at the Lin mansion and garden. Though the Lin family occupies quite a few pages in Taiwan's history textbooks, and though the cultural values of the site are widely recognized by art and architectural historians, local people in Taichung stopped finding this heritage relevant to their community after the Lins ceased to be active in politics (See: Liu 2003).

Narrative weaving takes a very different approach to heritage preservation from what is described above (See: Lin 2017). It views single treasures as essential components of a larger network or system incorporating tangible and intangible elements. This more holistic view takes into account the context that produced, nourished, benefitted from, or expressed itself with the treasures. As the scope of inquiry enlarges, it is possible to link the system of treasures to more aspects of human history and society and appeal to some universal values. In this way, the heritage network becomes relevant to the experiences or memory of more people, and more important to the local community also increased. In addition, narrative weaving is a participatory project that invites all people related to the heritage site or system, members of local communities in particular, to take part by contributing their stories and opinions as well as discovering new meanings of the heritage site in the modern world.

Participatory narrative weaving arouses communal awareness of this heritage and stimulates greater participation in the definition and preservation of the heritage. The more these stories are told, the more important the heritage becomes in the life, history, and memory of the local community. This creates an interwoven Web, enriching the heritage's relevance and binding it with people's daily life and shared memory, and makes the heritage an indispensable part of the community. The significance and influence of the heritage may hence expand beyond the very specific context in which it was born and embody more universal values. Moreover, unlike the heritage preservation which was studied, defined and decided by a small group of experts and government officials, participatory narrative weaving encourages the local community to be very active, and participants are able to create new meanings and functions for the heritage and prolong its significance in today' fast-changing society. Finally, although participatory narrative weaving appears to need more resources than traditional preservation, paying attention to multiple sites, and including more participants, it is a more efficient use of resources in the long run: committed community members take heritage protection as their interests or duties and willingly donate their time and efforts. Today, the efficiencies are even greater as people can use Internet and social media to spreads news and information more easily and to solicit wider support.

Water heritage, particularly sites and systems still in operation, benefits in particular from this approach. As water penetrates every part of human life and human society, it is all too easy for people to overlook its importance. At the same time, participatory narrative weaving invites people to look at the contributions of the infrastructures and institutions that manage water resources for them, by bringing together stories of local water's close ties to the community's existence, industry, culture, etc. (See: Yu 2015). Any new story about the site or system, or any new participant in the heritage conservation project, empowers a virtuous cycle for the heritage's prolonged well-being.

Historical Ponds and the Taoyuan Main Canal System

Historical records of irrigation systems for agriculture in Taiwan can be traced back to the seventeenth century. After it took over southern Taiwan in 1624, the Dutch East India Company (VOC) brought in Chinese farmers to grow tropical crops and sugarcane and installed wells and ponds (See: Chang 2016). The commonly used concept and technology of water-retaining works (see Fig. 1a) in Taiwan can be attributed to the presence of the Dutch VOC on the island, as many man-made ponds (or *Bei*) are called "ponds of the red-haired" in historical maps and oral traditions (See: Yu 2017). Taoyuan, which lies in north Taiwan and was not occupied by the VOC during the seventeenth century, saw its first man-made pond on record in 1741. Documentations of later pond construction abound as Chinese immigrants brought the privatization of land ownership and the use of written contracts with them (See: Chen 2003a). A land lease of 1843 (see Fig. 1b) shows how a community settled land and water rights involving ponds and canals. The "Maps of Taiwan Fortresses", a survey published by the Japanese colonial government in 1904, were the earliest topographic maps of Taiwan (See: Government of Taiwan 1996). The survey (with 1/20,000 scale) detailed more than ten thousand man-made ponds (see Fig. 2a), varied in size and location, dotting the 900 km^2 Taoyuan tableland like a starry sky, all built before the twentieth century. Google Earth's 2016 updated version shows that many of the ponds are still in existence today (see Fig. 2b). The comparison between the topographic evidence explicitly indicates the existence of many ponds on Taoyuan tableland landscape over centuries (See: Yu 2017).

Before the twentieth century, farmers in Taoyuan would build ditches or culverts to transport water from ponds to individual farmlands (Chen 2003a). But these networks of water management remained dispersed and on a small scale. After the Qing Empire

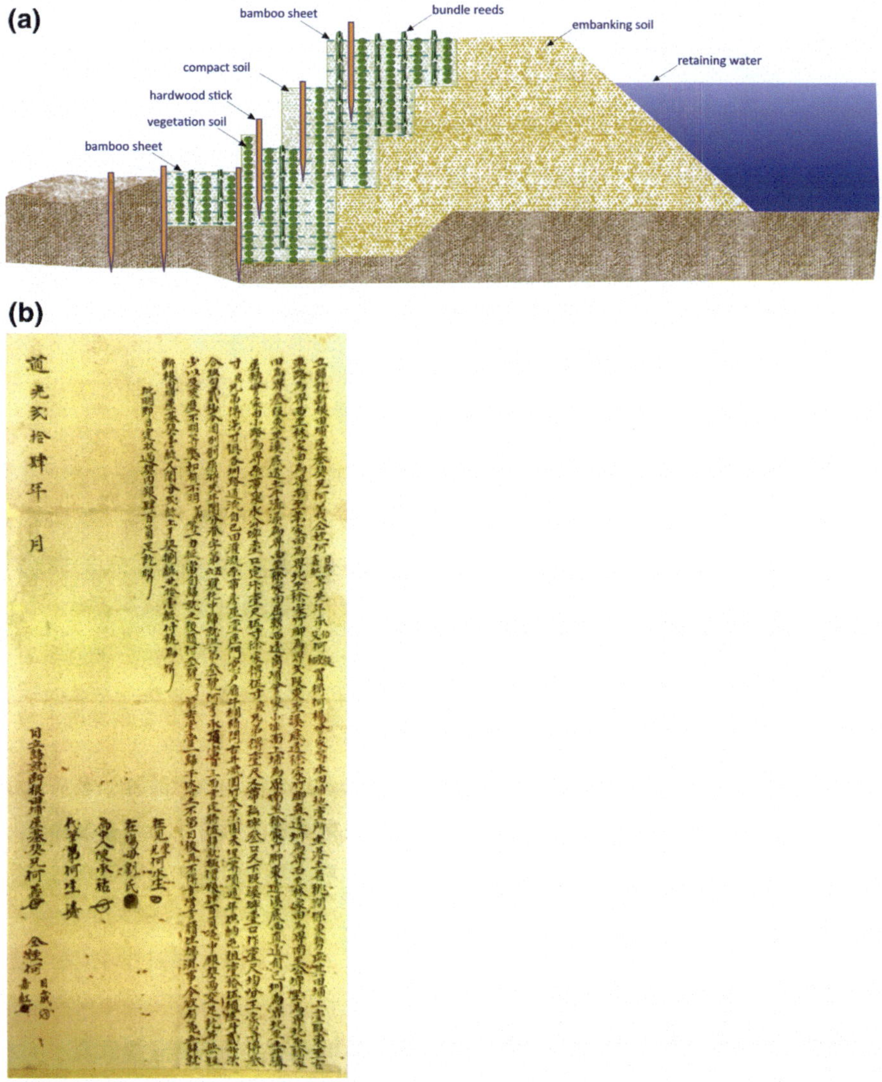

Fig. 1 Historic evidences of pond construction: **a** Schematic diagram of Dutch water retaining work in Taiwan (left), **b** Land lease copy in 1844 (right). *Source* https://blog.xuite.net; released under a Creative Commons Attribution-NonCommercial-NoDerivatives 4.0 International License

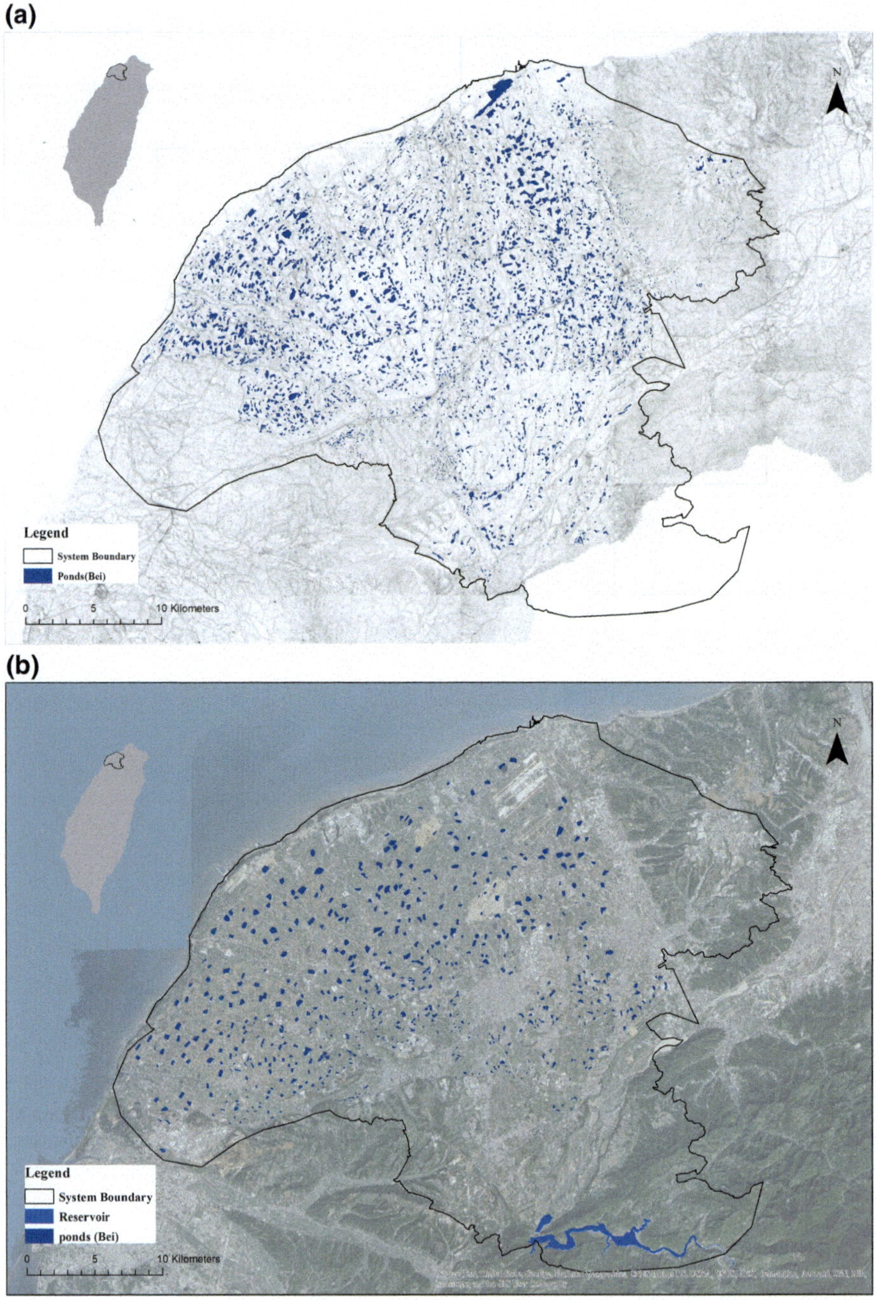

Fig. 2 Ponds at Taoyuan tableland: **a** On "Maps of Taiwan Fortress" in 1904 (above), **b** More than a hundred years on current Google earth (below); released under a Creative Commons Attribution-NonCommercial-NoDerivatives 4.0 International License

of China handed Taiwan to Japan in 1895, the new colonial government issued a bill in 1901 transforming the private management into government-run public facilities. In 1913, a severe drought occurred in the Taoyuan area, prompting the colonial authority to initiate the construction of the Taoyuan Main Canal, a water network linking all existing ponds with canals (TMC, see Fig. 3a), with an irrigation altitude of less than 110 m; this was according to the canal inlet location for water to transport by gravity to achieve in the Taoyuan tableland terrain (See: Chen 2003a). The plan also aimed to bring in water from sources over mountains. In short, it managed the water resources of the region as an integrated system. It also reorganized, rearranged, merged and modified the historic ponds into fewer than three thousand modern ponds and by this expanded the whole volume of water supply. TMC Construction began in 1916 and was completed in 1924 (See: Chen 2003a). At its inauguration, a memorial tower was set up for the people who had died during construction. Taoyuan Irrigation Association (TIA) has operated this canal system thereafter and increased the irrigated farmland from twelve thousand to twenty-five thousand hectares (See: Chen 2003b).

The Shihmen Reservoir and the Shihmen Main Canal System

A significant portion of the Taoyuan tableland area above the elevation of 110 m was not included in the TMC irrigation network. Residents there continued to suffer from the unstable water supply during dry seasons, especially in the drought of 1953–1954. As part of official efforts to improve Taiwan's postwar agricultural infrastructure and economic self-sufficiency, the Shihmen Development Commission was founded in 1955 to oversee the construction of the Shihmen Reservoir (Shihmen Reservoir Construction Committee 1966). This reservoir was the fruit of collaboration with American funds and technology, and its initial design went through some major changes after the collapse of the French Malpasset Dam in 1959. During the construction period, in 1961, TIA began to operate a large-scale rotary irrigation system in its irrigation area, to conserve water usage for more farmland (See: Ku and Lee 2009). This water conservation system distributed water to different service areas by turns, not incessantly, and thus provided each unit of farmland just enough amount of water needed. With this more efficient use of water, the TMC system created nearly 4700 ha of new irrigation areas. After the reservoir was completed in 1964, the water stored could often reach 200 m elevation. The Shihmen Main Canal (SMC, see Fig. 3b) was built in 1974 to irrigate the areas at the elevation of 110–195 m; a new local water management organization, Shihmen Irrigation Association (SIA), was founded to be in charge of it (See: Shihmen Reservoir Construction Committee 1966). Thereafter, the new reservoir served as a pivot for both TMC and SMC systems, extending the irrigated area to up to more than forty-two thousand hectares in the tableland region.

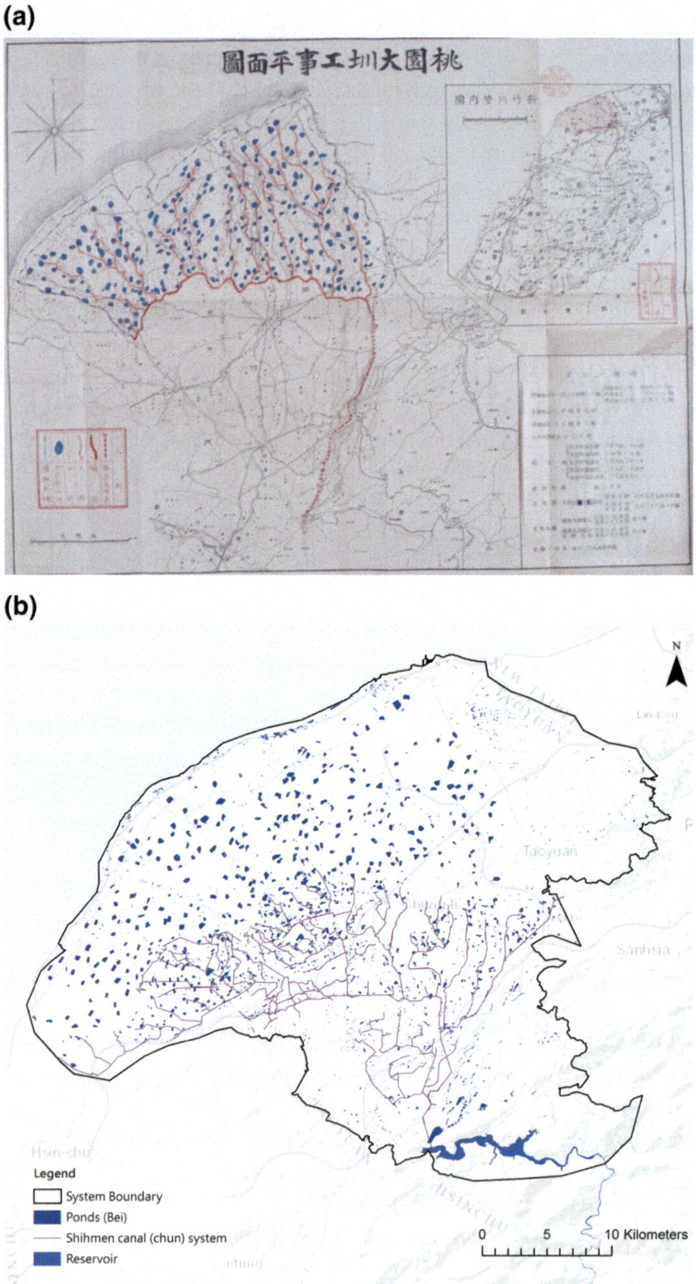

Fig. 3 Canal systems and Reservoir on Taoyuan tableland: **a** Taoyuan Main Canal fortification plan in the 1920's (above), **b** Shihmen Reservoir and Shihmen Main Canal system in 2016; released under a Creative Commons Attribution-NonCommercial-NoDerivatives 4.0 International License

The Cultural Heritage of the Taoyuan Tableland Water System

People in the Taoyuan tableland say "two droughts in three years" to describe the uneven distribution of rainfall (See: Yu 2017). Natural water sources are scarce, especially in the drought season from September to April. In the pre-modern period, conflicts between families or clans over water rights occurred frequently. Artificial ponds were their solution to this severe water shortage, saving rainwater for dry seasons and allocating water to daily life and agricultural uses. Ingeniously, early engineers used no windmills or pumps in this water system; rather, they sited these water facilities according to the terrain, so that gravity would help transfer water. The ponds partially ceased the feuds over water and strengthened social stability. Land cultivation and migrant communities flourished with steadier water supply.

However, the new water could not completely meet the demand, and in dry years, clashes between clans unavoidably continued. When population and agriculture expanded, the need for water grew too, leading to more social upheavals. Hence, the constructions of the water system with ponds, Taoyuan Main Canal, Shihmen Reservoir, and Shihmen Main Canal helped settle these conflicts. Since water was so precious, ponds and canals became an important part of daily life and rural landscape in the region. Many places were named after their local ponds. For example, "Lun-Bei (崙埤)" means "pond on the hill"; "Bei-Liao (埤寮)", literally "cabin by the pond", indicates the existence of a cabin for the pond guard in the past (See: Yu 2017).

Building a pond or a canal was not an easy task. It required the efforts of an entire family or clan. Thus, a pond or a canal was often the property of an extended family, closely bound to its prosperity. And people located most ancestral temples near ponds or canals. Thus, the tangible and intangible elements of Taoyuan's water landscape are crucial to the region's culture.

This water system was a hybrid of technologies from the Dutch VOC, Taiwanese indigenous people, Chinese immigrants, Japanese colonists, and American engineers. It is an excellent demonstration of how driving forces from different corners of the world have contributed to Taiwan's history. It also exemplifies the intensified interactions between different regions—globalization—since the era of Columbus. In recognition of the culture and history embodied in this distinctive water landscape, in 2009 the Taoyuan tableland water system was listed as a potential world heritage site in Taiwan.

Threats to the Water Heritage, Ecosystem, and Society on the Taoyuan Tableland

Facing challenges from land development and climate change, this water management system has become a battlefield of growing demands for lands and water.

Postwar Taiwan witnessed an economic and demographic boom. Taoyuan tableland's population grew from less than 500,000 in 1960 to 730,000 in 1970, crossed over one million in 1980, and reached two million in 2010. This explosive growth led to the destruction of historic irrigation ponds, reducing their number by more than 70% in the past decades: many ponds and canals have been filled up for residential blocks, schools, government facilities, industrial plants, public roads, and the island's largest international airport. The consequence: The water supply capacity of Taoyuan's irrigation system dropped from 34,568 ha of farmland in 1970 to 28,241 ha in 1984, 26,235 ha in 1989 and 25,967 ha in 1999. This drop in capacity also reflected the shrinking of agricultural land area as a whole over this period. Before the 1970s, agricultural sectors took 95% of all water supplied by Taoyuan's water management system. Now the agricultural portion has been reduced to 50%, with domestic and industrial demands increased to 50%.

With the reduction of farmland, many water rights for agriculture gradually shifted to non-agricultural sectors. Today, Taoyuan's water management system supports agriculture as well as the domestic water usage of the whole Taoyuan tableland and half of the Taipei basin, the island's largest metropolitan area, or daily water for five million people daily water and agricultural irrigation of thirty-six thousand hectares'.

It is standard practice in Taiwanese water right administration to put domestic and agricultural uses before industrial demands. However, when a drought warning is announced, the government has prioritized industrial water use to prevent significant economic losses and has compensated the farmers affected. In the long run, this adjustment damages agricultural food production. With the threat of global climate

Fig. 4 Aerial view of Taoyuan Aerotropolis region in 2014 (*Photograph Credit* Jnfakimo CC BY SA 2.0)

change becoming acute, the allocation of water in drought seasons has become a crucial issue, especially with the aforementioned shift of water rights away from agriculture.

In 2008, a Taoyuan Aerotropolis development plan, which had been under discussion since the turn of the century, was officially announced, forecasting a new and tremendous change for the environs of the Taiwan Taoyuan International Airport in the Taoyuan City (See: Construction and Planning Agency 2014). Supported by the city and the Ministry of Transportation and Communications (MOTC), the Aerotropolis plan aimed to expand the international airport and make the surrounding 6800 ha into a new city. Within the proposed area were some 4600 ha of farmland with ponds and canals (see Fig. 4), about one-tenth of all the agricultural land on Taoyuan Tableland. The plan proposed adding 3000 more hectares, which would dislocate over 26,000 people from farmlands and villages. This has been viewed as the worst forced community demolition in Taiwan's history, hurting both human rights and social sustainability.

The acquired farmlands were to be converted into zones for business, manufacturing, exhibition, leisure, and housing (see Fig. 5a), permanently altering the landscape of the Tableland, the network for water distribution, and the tableland water management heritage. Specifically, paving farmland and backfilling water bodies (e.g. ponds and canals) would significantly reduce the capacity of the area to retain water. The loss of this capacity would then make the surrounding region of Taoyuan Aerotropolis much more vulnerable to both droughts and floods—increasingly frequent in recent years as climate change haunts Taiwan. And the disappearance of farmlands and water bodies means the loss of their temperature moderating functions. This would worsen the existing severe heat island effect of global warming. In addition, two side effects of the development plan might well imperil social stability: The reduction of farmlands could damage Taiwan's food production and threaten the island's food security during extreme weather conditions, and the massive-scale forced relocation would create discontent and conflicts.

The Aerotropolis plan roused debate among all sectors of Taiwanese society. Some people worried that the economic globalization accompanying the new Aerotropolis would worsen the competition for water, labor, and land resources between sectors. Some argued that the proposed Aerotropolis area was unnecessarily large, making the project too costly, and imposing some irreversible impacts on the ecosystem of the region and the life of local residents. Heritage conservationists objected that the construction of the Aerotropolis could destroy two important pieces of cultural heritage, an old navy airport, and the pond–canal system.

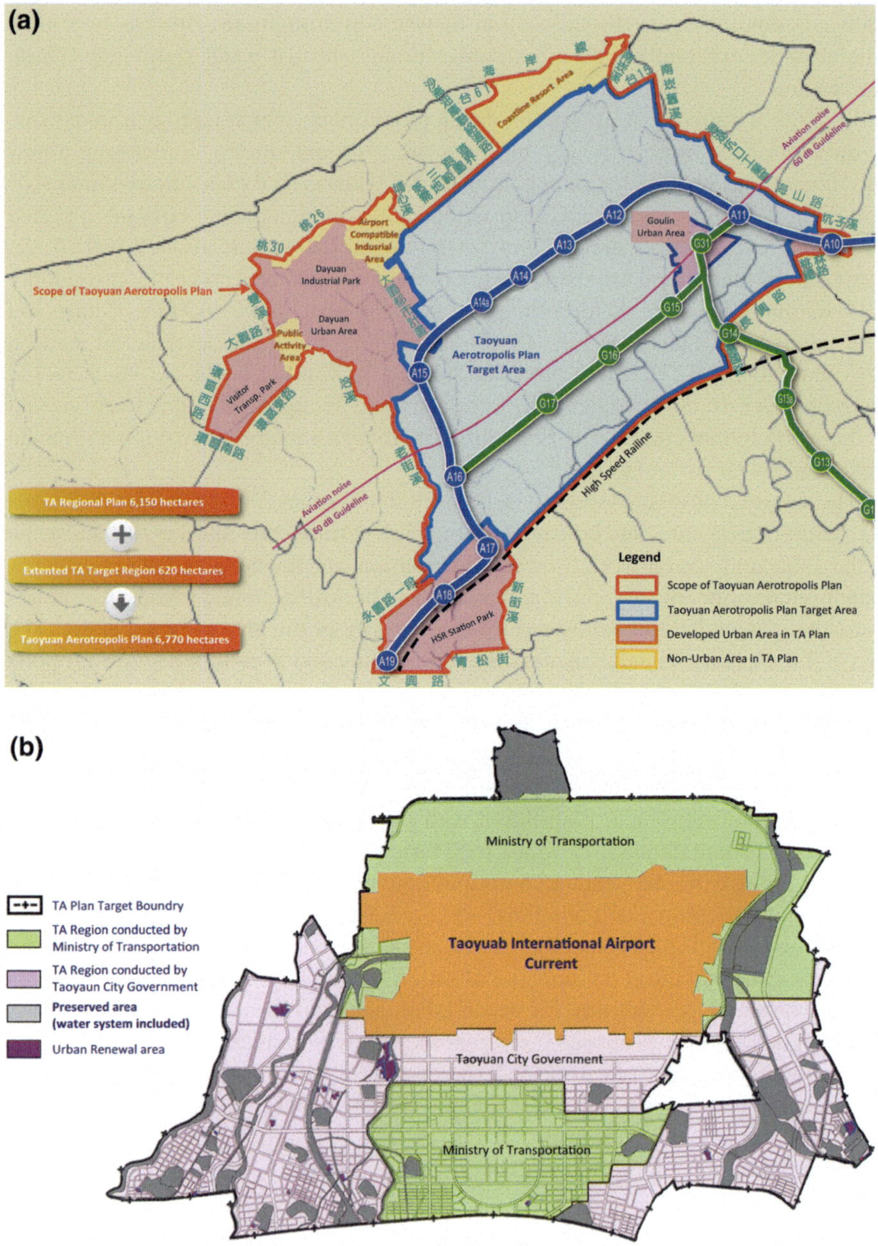

Fig. 5 Taoyuan Aerotropolis Plan layout: **a** Initial development version (above), announced in 2012; **b** Cultural water heritage preserved version (below), announced on August 11, 2017; released under a Creative Commons Attribution-NonCommercial-NoDerivatives 4.0 International License

The Legal Evolution of Cultural Heritage Preservation in Taiwan

The Cultural Heritage Preservation Act (CHP Act) of 1982 was the legal cornerstone of Taiwan's heritage preservation. It was the fruit of a growing awareness of heritage in the late 1970s, when cultural policies took a turn to embrace the lives of local and living people.

In 1999, a severe earthquake in central Taiwan killed thousands of lives and destroyed hundreds of thousands of buildings, both historic and modern. In order to recover from the devastation, heritage conservationists and cultural professionals initiated many comprehensive community development programs on local scales. When they tried to apply the CHP Act in these projects, however, the Act's defects became apparent: its targets of preservation were single objects such as architectural works and their ancillary facilities, archaeological or historical sites, and antiquities most of which were only loosely connected to the communities they belonged to. Moreover, the Act mandated that the only way to preserve a heritage site was to "freeze" it in its earliest shape and historical context, preventing people from creating new meanings and functions. Not surprisingly, communities were often against this kind of preservation of their heritage. When something costs a lot to maintain but brings no benefit—be it material, emotional, or spiritual—no community would bother to keep it. Hence, the campaign for a revision of the CHP Act that would mobilize bottom-up or community engagement in heritage preservation and reutilization.

The second version of the CHP Act (called CHP Act 2.0), enacted in 2005, provided the legislative ground for public participation in heritage management. It broadened the study, protection, and management of heritage sites or systems beyond a small group of experts (See: Lin 2005). However, it continued to view heritage preservation as many single, unconnected tasks and did not integrate heritage management with regional development plans. This weakness of the newer CHP Act has been all too visible in many recent attempts to reallocate land and water resources.

Public Participation as Conflict Resolution in Preservation and Development

Many actors—including historians, writers, artists, environmentalists, rural protectors, farmers, ecologists, and even water engineers—formed an alliance for the preservation of Taoyuan's water heritage infrastructure as heritage. They demanded that policymakers include the public in the early stages of urban and land planning projects, and deployed their expert knowledge to demonstrate that the historic ponds and canals are indispensable to today's Taoyuan society. For example, experts in water resources showed that water storage and conveyance capacity in agriculture are essential to human existence in the face of climate change. Scientists noted that biodiversity in ponds prove that a water system is a valuable repository of treasures for local ecology. Historical traces of water management reflected the transformation of an immigrant society during the seventeenth and eighteenth centuries. Poetry, photography, and other forms of art depicting Taoyuan's water landscape illuminated local people's memories of the past. Moreover, both Taoyuan and Shihmen irrigation associations had come to understand the value of rice paddies to the bionetwork and began working with local communities to preserve the memories and landscapes associated with Taoyuan tableland's irrigation ponds and canals system.

This alliance of protectionists raised widespread awareness in Taoyuan society of the area's water heritage and heated up the debate over the Aerotropolis plan. Policymakers and land developers for the project at first proposed to replace the backfilled ponds with community parks, not the more profitable residential or commercial quarters. At public hearings, the protectionists responded that the reduced water volume would still be insufficient for the area's hydrologic cycle. The developers then adjusted the plan so that there would be no net loss of water, which was rejected again as the project would still damage biodiversity. During these exchanges of opinions, one plan even proposed relocating water habitats. But the opposition's key point, which in the long run won the argument, was in a different realm altogether: The significance of Taoyuan's water heritage system lies not merely in the locations of associated monuments, its scale, its history, or its contribution to the profit of different sectors. Rather its significance is in its holistic context: That residents of Taoyuan could relate each of these issues and water heritage systems as a whole to their everyday lives and memories.

With the aid of social media and online forums, stories about the endangered ecosystems, reports on unfair compensation to local farmers, and others spread quickly and widely. In the spring of 2014, a political storm, the Sunflower Movement, in Taiwan prompted many young people to be engaged in all kinds of social activism and to demand greater transparency in policy making. Unsurprisingly, many focused on the already controversial Aerotropolis plan. Public hearings were livestreamed online by volunteers to invite greater participation in forming consensus. At the end of the year, when voters elected a new mayor, they chose the candidate who promised to thoroughly review and modify the Aerotropolis project. As more and more people expressed personal concern over the fate of the ponds and canals, it became more difficult for the developers to win public support for the project.

Then a severe drought ignited an even broader discussion on water resource management. In September 2014, the amount of rainfall on the Taoyuan tableland was only 20% of the normal seasonal average. In January 2015, a full-scale water right shift transferred all the irrigation water to the domestic sector. Cultivation was suspended on 27,000 ha of farmland; about 20,000 farmers found it difficult to survive the drought, even with higher than usual compensation. This sparked great anger among the farmers and society. Meanwhile, several farmers in the region were unaffected by the water suspension, as their respective ponds helped them maintain irrigation. The difference between farmlands with ponds and those without poignantly illustrates why the pond-canal-reservoir system should not be sacrificed for the profits of industrial economic growth. People wondered if the Aerotropolis project might cause further setbacks from which the region would not recover.

Heritage preservationists now joined with other groups to promote another amendment of the CHP Act 2.0. They suggested that public participation and narrative weaving were essential for preserving non-traditional heritage structures such as water management networks. This new concept of narrative and heritage was adopted in an amended CHP Act of 2016 (aka CHP Act 3.0). The new stricter rules significantly increased the sustainable risk for the Taoyuan Aerotropolis plan, adding to the maintenance and financial burden of the developers. Any failure to fulfill their promises—whether of no net loss of the water body, water habitat relocation, adjustments for social and environmental impacts, or other concessions—would incur further punitive payments.

Several years had passed since the 2008 announcement of the Aerotropolis project; the government and the developers had failed to soothe opposition or start major constructions. Instead, new obstacles such as the CHP Act 3.0 rose during this long process of negotiation. Eventually, the Aerotropolis plan was revised (see Fig. 5b). The Taoyuan city mayor decided to exclude all disputed areas from the project; this major change meant that the whole water heritage system and its cultural landscape could be preserved.

Conclusion

In Taiwan, it has been more difficult to draw public attention to water heritage systems than to monuments, historic buildings, or archaeological sites. Preserving them is thus a challenging mission. In past decades, devoted preservationists in Taiwan have learned that the accumulated tacit knowledge of water heritage systems may be excavated from local community's oral traditions and legends and be used to express the residents' loved memories and feelings. When professional investigations, public awakening, and social mobilization all come together to form a common ground for public participation, one result can be that people cherish a water heritage system even more; another might be that the system brings economic profit.

A distinctive characteristic of water facilities is that they always work and function in a network. That is, a water heritage structure would consist of multiple sites dispersed in different locations. Except for grand constructions such as dams or canals, most of these essential components appear mundane and without historical or cultural value. However, water facilities are so essential to daily life that, like the air people breathe all the time without always noticing its existence, they can be easily introduced into different narratives that cover all aspects of human life, society, and history. A whole network of water facilities could be represented as a heritage or a "water museum", with participatory narratives woven around the network or system.

Water heritage, particularly those systems still functioning, is more relevant to local people's daily lives than the heritage of other categories. Any change to these systems, as the case presented here shows, would affect more stakeholders and raise greater public concern. This is why, at the announcement of the Aerotropolis Plan, local farmers and residents—whose lands and existing lifestyle were put under threat—stood up with professionals to voice their concerns in the first place. The protection of the Taoyuan water system was thus initiated by a mixed troop of people from various backgrounds and with a range of issues to address. This diversity helped them reach to more sectors of Taiwanese society and gain wider public support. As the case of the Taoyuan water heritage shows, when a local community becomes interested in the waters around it, public opinion can form a protective shield for endangered heritage against new urban development plans.

References

Chang Y-C (2016) Report of strategical planning for engaging world water system heritage program. Water Resources Planning Institute, Water Resources Agency, Ministry of Economic Affairs, Taiwan (in Chinese)

Chen H-T (2003a) From manmade ponds to Taoyuan main canal: the development of irrigation system on Taoyuan Tableland. Dong Hwa J Humanit 5:183–208 (in Chinese)

Chen C-P (2003b) A survey of Taoyuan main canal and Guan-Fu canal systems. Executive Yuan, Taipei (in Chinese)

Construction and Planning Agency, Ministry of the Interior (2014) Proposed development plan for Taoyuan international airport and its surrounding areas (in Chinese)

Government of Taiwan (1996) Taiwan map of 1904. YLib, Taipei (in Chinese and Japanese)

Hsu Y-D, Li H-Y (2006) The preservation of architectural heritage in a multicultural country. J Taiwan Land Res 9–2:75–96 (in Chinese)

International Council on Monument Sites (2008) The ICOMOS charter on cultural routes

Ku Y-W, Lee T-H (2009) Interview of Kuo-Yun Fu, Director of Rotary Irrigation. Shihmen Reservoir. http://epaper.wra.gov.tw/Article_Detail.aspx?s=3CCE023F9B10350C (in Chinese)

Lin H-C (2005) https://www.boch.gov.tw/information_189_49251.html (in Chinese)

Lin Y-H (2011) A brief History on cultural heritage preservation of Taiwan. Bull Natl Museum Taiwan 64–1:75–106 (in Chinese)

Lin H-C (2014) The legalization of Taiwan's post-war heritage preservation. J Cult Resour 8:27–55 (in Chinese)

Lin C-H (2017) Cultural heritage preservation 3.0—cultural and ecological system in water system heritage. International Water Heritage Forum, Bureau of Cultural Heritage, Taichung of Taiwan (in Chinese)

Liu Y-Y (2003) The Wufeng family awaits daylight. Control Yuan, Taipei (in Chinese)

Shihmen Reservoir Construction Committee (1966) Records of the construction of the Shihmen Reservoir. Shihmen Reservoir Construction Committee, Taoyuan (in Chinese)

Yu S (2015) Report of the pilot framing for fused innovation with water and heritage of Taiwan. Bureau of Cultural Heritage, Ministry of Culture, Taiwan (in Chinese)

Yu S (2017) Report of investigation and evaluation of water system heritage in Taiwan (I). Bureau of Cultural Heritage, Ministry of Culture, Taiwan (in Chinese)

Sinite Yu is a research fellow at Taiwan International Institute for Water Education. He has taught at National Chiayi University and Huafan University. He started his professional career in soil and water resource management and water treatment; he later pioneered the research of water heritage in Taiwan. He now seeks to cultivate awareness of water heritage among different sectors of Taiwanese society.

Chung-Hsi Lin is a professor of cultural heritage conservation at National Yunlin University of Science and Technology, Taiwan. A prominent advocate in Taiwan for innovative cultural theories and community participation in heritage conservation, Lin writes on a wide range of topics to prompt public discussion on cultural policies. He now also serves as an advisor to Taiwan's Ministry of Culture.

Hsiaoen Wu is an assistant research fellow at Taiwan International Institute for Water Education. She researches the historical and cultural contexts in which heritage was created, used, and perceived. As an author of popular history, she writes stories about the remote past to help people understand connections between the past and the present.

Wenyao Hsu is an associate research fellow at the Taiwan International Institute for Water Education. Her research explores citizen participation in water resource management and urban planning. She has conducted many water-related projects and authored multiple articles, including "The Citizen Participation Plan for Promoting the Operation and Management of Zengwen Reservoir."

Yu-Chuan Chang is a professor of tourism management at Hsing Wu University, Taiwan. Trained as an agricultural engineer, he acquired extensive experience in irrigation systems during the early years of his career. He now works primarily on the sustainable use of water for irrigation and the restoration of historic irrigation canals.

Part III
Land Reclamation and Defense

The Hachirogata Region and the Hachirogata Central Polder (Photo by Yasunori Kitao)

Chapter 10
Reassessing Heritage: Contradiction and Discrepancy Between Fishery and Agriculture in Planning the Hachirogata Polder and Its Surrounding Lagoon in Mid-Twentieth Century Japan

Yasunori Kitao

Abstract The Hachirogata central polder in Akita Prefecture in Japan is an important example of a completely planned and constructed agricultural production landscape and also of an industrial heritage landscape built in the context of the myth of modernism and large-scale planning intervention. The land was reclaimed from the Hachirogata lagoon in the middle of the twentieth century with the help of Dutch engineers. Today, the polder is one of the most productive landscapes in Japan for growing rice. For centuries, the fishery was the main industry of the lagoon. But the government was interested in developing agriculture by turning the lagoon into agricultural land, and it did not initially consider fishermen's rights or the traditional lifestyle and heritage of the region in its planning. This chapter explores the planning of the land reclamation from about 1930 to the late 1950s. Drawing on articles in local newspapers and on official planning documents, it explores the conflicting interests and heated debates between the central government, the prefecture office, local municipalities, and local people regarding agriculture and fishing in local history. It contributes insights on food production, environmental issues, and local traditional industries to both heritage debates and the planning of future industrial heritage landscapes.

Keywords Polder · Regional development · Regional planning · Fishery · Agriculture · Brackish water

---

Y. Kitao (✉)
Kyoto Women's University, Kyoto, Japan
e-mail: kitaoy@kyoto-wu.ac.jp

C. Hein (ed.), *Adaptive Strategies for Water Heritage*,
https://doi.org/10.1007/978-3-030-00268-8_10

Introduction

The Hachirogata central polder, located in Akita prefecture on the northern end of the Japanese main island, Honshu, is well known for its large-scale, environmentally conscious rice production (General Affairs and Planning Section of the Ogata Mura Village Office 2011). Ogata Mura, the only rural village in the polder, is known as an advanced case of local autonomy (Miyata et al. 2012). The Hachirogata polder is also an important example of a completely planned and constructed agricultural production landscape in Japan. The construction of the polder is celebrated as a great achievement: the polder museum in Ogata Mura presents it as a success of modern large-scale agriculture in line with the most advanced international practices, and a symbol of Japan's engineering heritage; and foreign literature holds it up as an example of government intervention aimed at creating an ideal society, and the buildings and structures that serve it (Van den Heuvel et al. 2008).

These celebrations focus only on the rural development planning phase in the 1960s when city planners, architects, and other specialists collaborated to build a new village on the polder (Kitao 2008). Such a writing of the polder's heritage effectively constructs a rather limited assessment. It ignores the heritage of the water area that remains of the Hachirogata lagoon, formed at the same time as the Hachirogata central polder.

This paper assesses the remaining lake as part of the industrial heritage of the area. The polder heritage is effectively composed of two different historic practices: agriculture on the newly created land; and fishery in the remaining lake. The relation between the lake and the polder is Yin and Yang, balanced, and always together. This chapter, therefore, explores the hidden history of the water landscape in order to fully understand the development of the entire lagoon area. Then, it will be possible to evaluate the central polder as a part of the modern industrial heritage of the region.

The Hachirogata polder was reclaimed from the Hachirogata lagoon in the middle of the twentieth century with the help of Dutch engineers. Japan has a history of modern land reclamation dating to the beginning of the twentieth century, which coincided with an increased interest in modern polder projects in the Netherlands. As early as 1914, the Ministry of Agriculture and Commerce in Japan published a report, The Great Planning of Expanding Arable Lands in the Netherlands, outlining the Zuider Zee project, a large man-made system of land reclamation, dams, and dykes; it also analyzed the Dutch Government's policy of providing fishermen with funding for bigger boats so that they could fish in the North Sea rather than the Zuider Zee. Shortly thereafter, in 1916, the Fisheries Experiment Station of Akita Prefecture began a detailed investigation of the fishery resources of the Hachirogata lagoon and the agricultural implications of creating the Hachirogata polder. Despite extensive studies and discussions of the lagoon and its historical significance, the research did not result in a plan that satisfied all parties. As the records make clear, fishermen strongly opposed the impoldering project. (In this text, "fishermen" refers to collectives such as fishermen's cooperatives or fishermen's informal groups).

Fig. 1 An old map of the Hachirogata lagoon. This map was made in the nineteenth century for the national defense policy at that period. The "※" mark indicates that the lagoon is located at a longitude of 140° east and a latitude of 40° north. *Source* Akita Prefectural Library; released under a Creative Commons Attribution-NonCommercial-NoDerivatives 4.0 International License

This chapter explores the history of competing visions in the planning process for the Hachirogata lagoon, particularly between creating a landscape supporting fishing and one supporting agriculture (The Fisheries Experiment Station of Akita Prefecture 1916). (Figures 1, 2 and 3). We draw on local newspapers, project reports, and related literature to illuminate the hopes and worries of the different stakeholders, starting in the early 1930s, when the wartime regime came to power, and ending in the late 1950s, when the conflicts were resolved and the construction work was launched.

Local Fishery and Governmental Policy

The Hachirogata lagoon (called Hachiroko lake in the past: "gata" means "lagoon," and "ko" means "lake") was a brackish body of water located about 20 km north of Akita city, at a latitude of 40° north and a longitude of 140° east. Before the

Fig. 2 Hachirogata central polder. A pomp station and drainage is in the center of the photograph (Photo by Yasunori Kitao); released under a Creative Commons Attribution-NonCommercial-NoDerivatives 4.0 International License

Hachirogata polder was completed by draining some of the water, the lagoon was the second largest in Japan. The total area of the lagoon was about 22,000 ha, and measured 12 km from east to west, and 27 km from north to south. The lake shore was 90 km around, the average water depth was 3.2 m, and the maximum depth was about 4.5 m (General Affairs and Planning Section of the Ogata Mura Village Office 2011). The aim in constructing the polder was to create 17,000 ha of arable land, including coastal polders (Jansen 1954); it was the largest polder project in the modern period of Japan. It was started by the Ministry of Agriculture and Forestry (MAF) in 1952 as a research project. Construction eventually began in 1958, and the last piece of the project, the settlement on the polder, was completed in 1977. Over this period, the plan for this polder and its village changed many times, but after discussing it for around 25 years, the state finally completed the modern polder (Kitao 2008).

In the early 1930s, the Hachirogata lagoon was known as the "Lake of Death," because fishermen's livelihoods had deteriorated markedly due to the bad catch in 1932 (Akita Sakigake Shimpo 1935-5-30). Many local fishermen went to work as migrant workers in the Sea of Okhotsk fishing industry in order to earn a living, and other fishermen became farmers. Some lakeside villages began public construc-

Fig. 3 Fisherman in the Hachirogata lagoon at the beginning of twentieth century. A specific point of the fishing boat is that the boat's bottom is flat because of the shallow water of the lagoon. *Source* The Fisheries Experiment Station of Akita prefecture, "The report on the using the water area of the Hachirogata Lagoon; 八郎潟水面利用調査昱告書," Akita Prefecture, 1916; released under a Creative Commons Attribution-NonCommercial-NoDerivatives 4.0 International License

tion projects, reclaiming the lagoon to allocate the reclaimed arable lands to these fishermen/farmers (Akita Sakigake Shimpo 1935-5-31).

The fishermen thought that the reasons for the bad catch were overfishing and damage by agricultural land expansion to the shallow lake bed that fostered the young fish. They also pointed to the reclamation of the estuary of the lagoon, noting that it prevented the free movement of fish from the sea to the lagoon (Akita Sakigake Shimpo 1935-5-30). Lawmakers of the Prefectural Council and fishermen's associations asked the Akita Prefecture Office (APO) to construct a fish way at the estuary of the lagoon, to give the fish a way to get around the barrier (Akita Sakigake Shimpo 1936-9-8). The fishermen asked the APO to take measures to prevent illegal poaching, to control lake bank reclamation projects, and to promote the fishery as an industrial reinforcement policy. (Akita Sakigake Shimpo Newspaper 1935-5-30). The fishermen were seeking to preserve their way of life on the shores of the lagoon (Akita Sakigake Shimpo 1937-2-5), which depended on the brackish ecosystem. In response, the APO began studying the fostering of the fish, and indicated that they would permit only a minimum of landfill construction projects which would not affect the fishery; ultimately, it promised to promote fishery as a policy (Akita Sakigake Shimpo 1935-6-21).

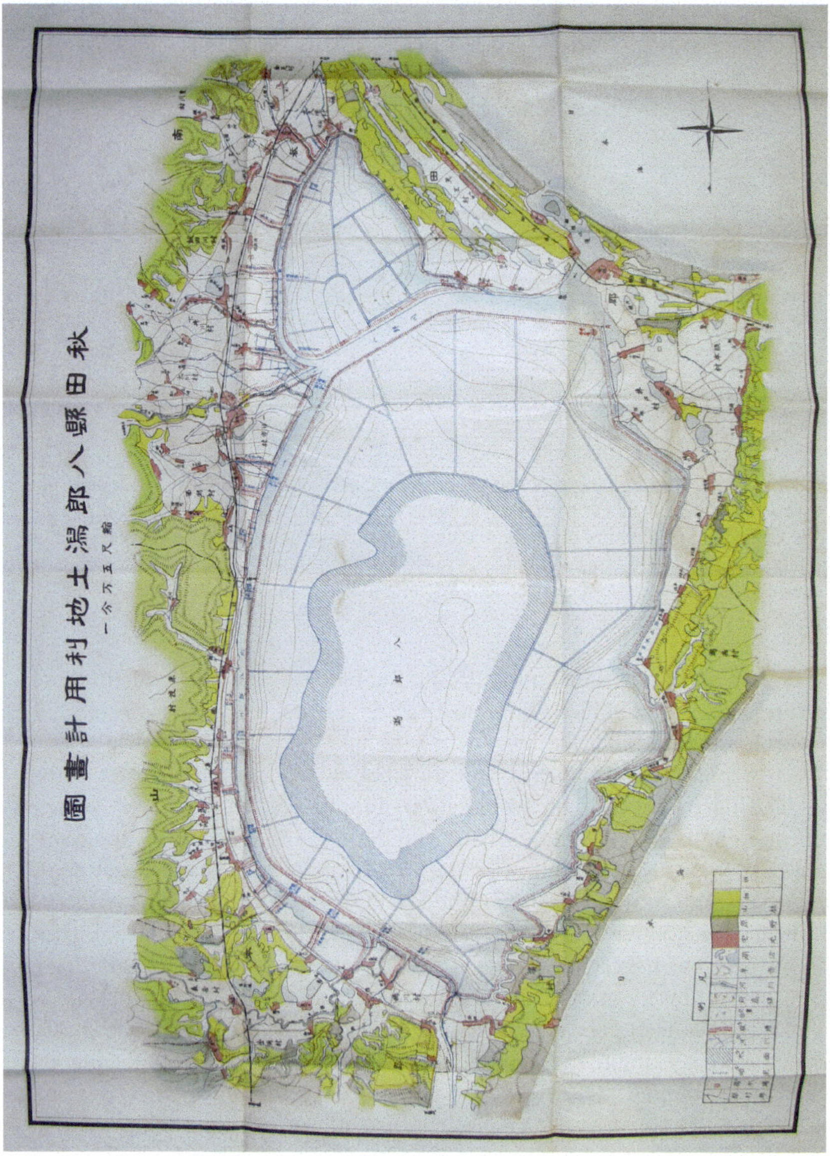

Fig. 4 MAC Plan, 1924. *Source* The Public Archive of the Akita Prefecture, Japan; released under a Creative Commons Attribution-NonCommercial-NoDerivatives 4.0 International License

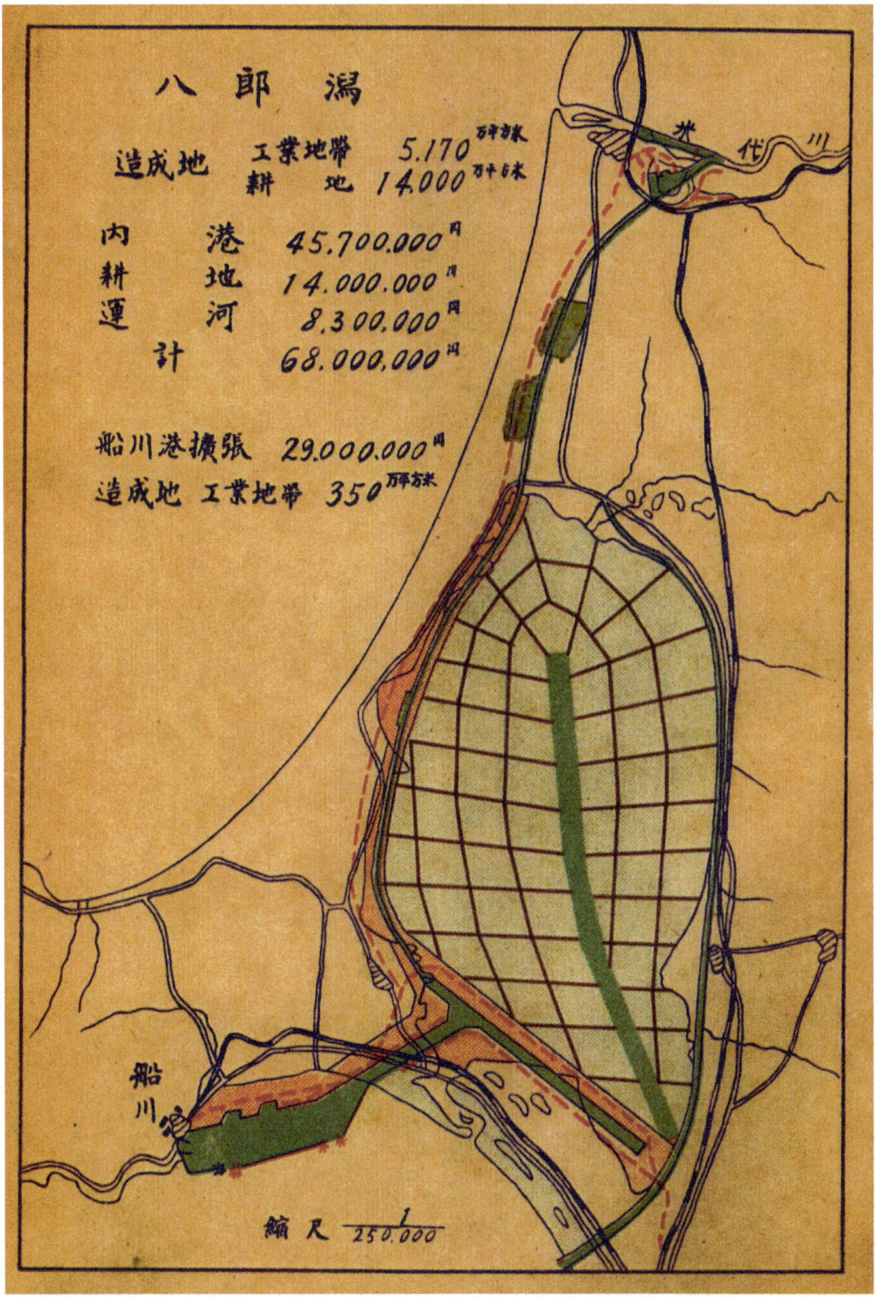

Fig. 5 MHA Wartime Plan, 1940. *Source* Akita Prefectural Library;"The Greater East Asia Co-Prosperity Sphere and Hachirogata; 大東亜建設と八郎潟";

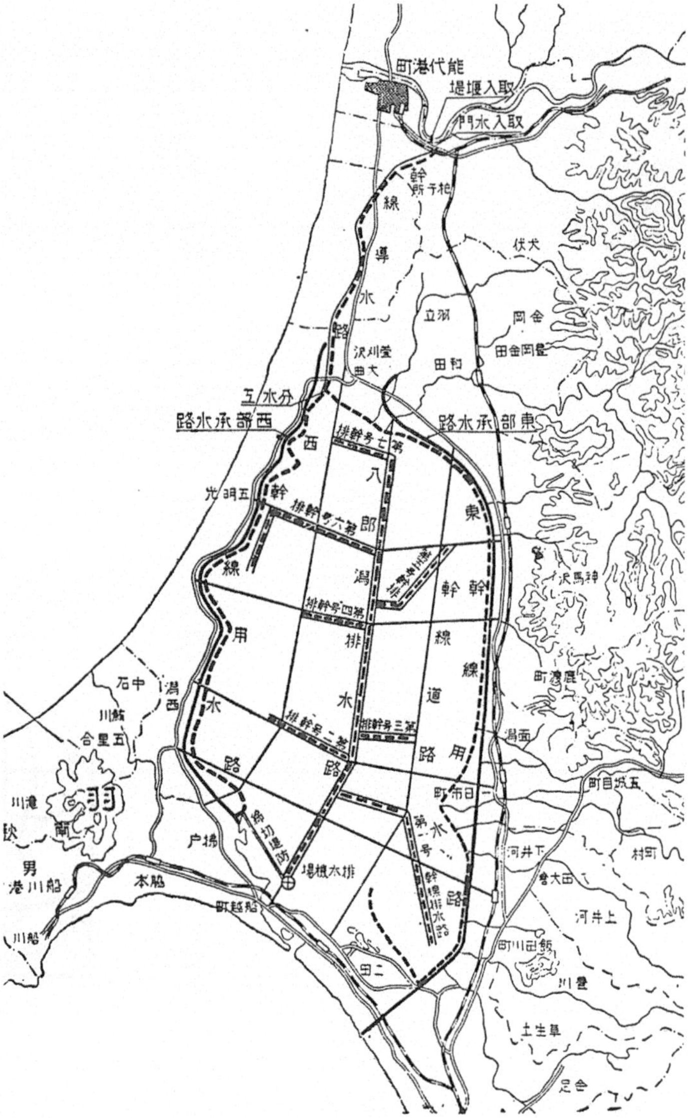

Fig. 6 MAF Wartime Plan, 1941. *Source* The Institute of National Development Projects; 国土開発調査会; "Developing the Hachirogata Lagoon;拓けゆく八郎潟", 1960, p 24; released under a Creative Commons Attribution-NonCommercial-NoDerivatives 4.0 International License

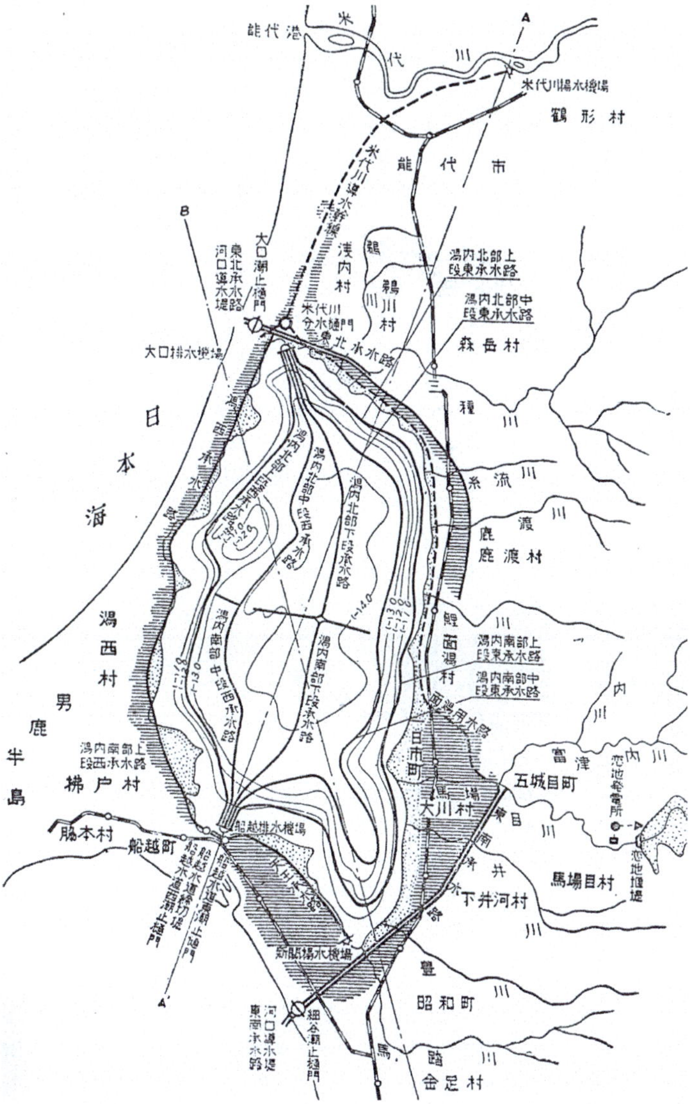

Fig. 7 MAF Post-Wartime Plan, 1948. *Source* The Institute of National Development Projects;国土開発調査会;"Developing the Hachirogata Lagoon;拓けゆく八郎潟", 1960, p 25); released under a Creative Commons Attribution-NonCommercial-NoDerivatives 4.0 International License

Aggressive government intervention in lagoon planning began under the wartime regime in the early 1930s. The Ministry of Agriculture and Forestry (MAF) questioned the effect of opening a fishway in the estuary, but the Ministry of Home Affairs (MHA), like the fishermen, was interested in creating one (Akita Sakigake Shimpo 1938-9-23). In 1939, the APO and MAF began considering creating 18,000 ha of agricultural land by dividing the land reclamation into three phases (Akita Sakigake Shimpo 1939-4-23). The MAF conducted a survey, but they did not take the fishery resources into consideration (Akita Sakigake Shimpo 1939-4-23); this time, the MHA did not express an interest in opening a fish way either. Instead, it proposed plans for developing a canal, factory areas, and port facilities along with the full reclamation of the land (Kanamori 1940a, b). This plan became a national plan for an industrial area, and also a national project for increasing food production during wartime. The MHA and MAF both proposed full reclamation, turning the entire lagoon into agricultural land. But they had different motivations: the draft plan of the MHA aimed to promote world trade, contrasting with the MAF plan for activating the economy of the region (Figs. 4, 5, 6 and 7).

After the outbreak of the Pacific War, the MAF conducted a survey for the project, but did not actually start construction due to a lack of materials and manpower (Morooka 1960a, b). The government emphasized the importance of the fishing resources of the lagoon to the war effort (Akita Sakigake Shimpo 1941-12-10). A Navy officer acknowledged the necessity of building port facilities in the Bay of Akita, but he said that land reclamation was unnecessary (Akita Sakigake Shimpo 1942-4-18). The following issues related to rivers were also being discussed under the wartime regime: building flood countermeasures (Sawa 1937); reducing the civil engineering budgets of local municipalities (The Japan River Association 1939); effectively using water resources (Mizutani 1940); and handling fishery rights (Tokusaki 1938). The various parties debated priorities. Considering the maintenance costs such as the revetment of rivers, one opinion was that building flood countermeasures was more important than increasing food production (Horikiri 1942). Another was that it was better to play a role in the war by using the water resources of the lagoon for fishing rather than to reclaim its land for agricultural land development or industrial land construction.

Before the war, the government had been decreasing rice production, having already neglected domestic agricultural production (United Nations 1955). Immediately, after losing the war (1945), the government had to tackle the food problem, and the project for the lagoon was the strongest candidate for developing arable lands (Akita Sakigake Shimpo 1945-8-26). The MAF proposed an impoldering plan in 1945, to build 18,500 ha of arable lands for the returnees from the former colonies and from battlefields in Asia and the Pacific Ocean who did not have any productive lands or jobs. At that time, the government expected fishery and agriculture to coexist (Akita Sakigake Shimpo 1945-10-19), but it was not clear how. In order to resolve this issue, the APO held a meeting in which staffs of the prefecture were able to propose their ideas. One suggestion was building a partial polder, and another was prohibiting any construction projects in the lagoon that would damage the fishery.

The president then presented to the MAF the APO's internal decision that there would not be any reclamation work in the lagoon (Akita Sakigake Shimpo 1946-3-13).

The prefecture and the national government had refrained from disagreeing during the war, but now, opposing to the APO's decision, the government technocrats insisted on a full reclamation plan. Their opinion had not changed; if anything, it was even stronger.

Coexistence of Fishery and Agriculture

After the war, the lagoon development project was back on the table, with the goal of increasing food production and facilitating the population increase of the whole country. Japan had regained independence politically, but self-sufficiency in food required economic independence. The government and local municipalities laid plans to develop new agricultural settlements throughout the whole country, and they expected 5000 farming families to settle in the Hachirogata polder (Akita Sakigake Shimpo 1955-10-26). The MAF completed a reclamation plan in 1948 (the post-war MAF Plan) that proposed 10,250 ha of reclaimed land, and assumed that 5500 new families would settle on it. This number included 3250 local families (including fishermen) (Morooka 1955). This post-war plan was not intended to actually allocate farmland to local residents, but it does express the government's intention of expanding farmland throughout the country (Akita Sakigake Shimpo 1953-8-18). The MAF published a full reclamation plan as a comprehensive development plan in September 1951 (Akita Sakigake Shimpo 1951-9-10). But the Economic Stability Headquarters of the government requested that it carry out reclamation projects only with the consent of the local fishermen (Akita Sakigake Shimpo 1952-8-29). (Akita Sakigake Shimpo 1952-4-27 & 1952-8-14).

In response, the vice president of Akita Prefecture announced that if there were a sound scientific basis and financial support, the prefecture would accept a project for the construction of a polder of around 3000 ha (Akita Sakigake Shimpo 1952-8-29). With the attitude of APO clear, the MAF were able to establish a research office to investigate the impact of the polder project on the local fishermen (Akita Sakigake Shimpo 1953-1-4). Akita Prefecture established an experimental fishery station in the south part of the lagoon (Akita Sakigake Shimpo 1953-5-13). Thus, the government and Akita Prefecture showed concrete consideration to fishermen as things progressed. Indeed, to preserve a stable social situation for building the polder, it was necessary to set up a democratic development process that gave due consideration to the protection of the fishermen's living rights and coordinated fishing and agriculture.

Local residents organized the Hachirogata Development Council, but the fishery groups on the council still opposed the reclamation plan. They said that the political view of Akita Prefecture regarding the fishery had not been consistent and the prefecture had not done enough to satisfy their request for a crackdown on illegal fishing, or constructed the fishway at the estuary that they had been requesting since the 1930s. The fishermen still distrusted the APO as much as they had before the war (Akita Sakigake Shimpo 1953-5-28) (Akita Sakigake Shimpo 1953-8-18). The development plan for the lagoon could not go forward.

To overcome the impasse, the MAF proposed that the fishermen become farmers of the arable lands in the future polder, and the APO prepared to organize a consultation body to integrate the local opinions (Akita Sakigake Shimpo 1953-7-28). Meanwhile, in August 1953, the government took steps toward creating that arable land. In a cabinet meeting, then Prime Minister Shigeru Yoshida announced a project that introduced Dutch reclamation technology to Japan (Akita Sakigake Shimpo 1953-8-1). Masao Koga, a technocrat of the MAF, went to the Netherlands in 1953 in order to invite Dutch engineers to assist in the polder project. (Morooka 1960a, b) the Dutch Government sent him to Professor Jansen of Delft University of Technology, an expert in the field of land reclamation. Jansen came to Japan in April 1954 and submitted a report to the Japanese Government in August 1954 (Hachiro-Gata Reclamation Bureau, 1969).

Jansen proposed a compromise between the MAC Plan (1920s) and the MAF Wartime Plan: a water area in the southern part of the lagoon, connected to the sea through a sluice gate in the estuary, and a polder in the north, constructed with the technology of the Dutch IJsselmeerpolders in the Netherlands. That is, Jansen actively insisted on a middle way. Akita Prefecture and the local municipalities immediately approved the Jansen proposal (Akita Sakigake Shimpo 1954-8-3). In

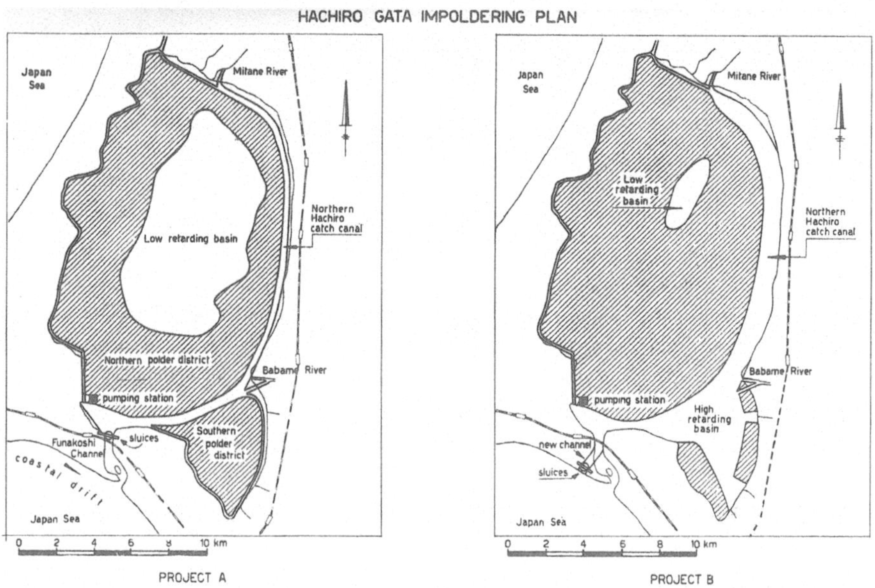

Fig. 8 Two types of polder in the lagoon, proposed by Prof. Jansen in 1954. *Source* Jansen "Some Remarks on Impoldering in Japan," The Ministry of Agriculture & Forestry 1954, p.6; released under a Creative Commons Attribution-NonCommercial-NoDerivatives 4.0 International License

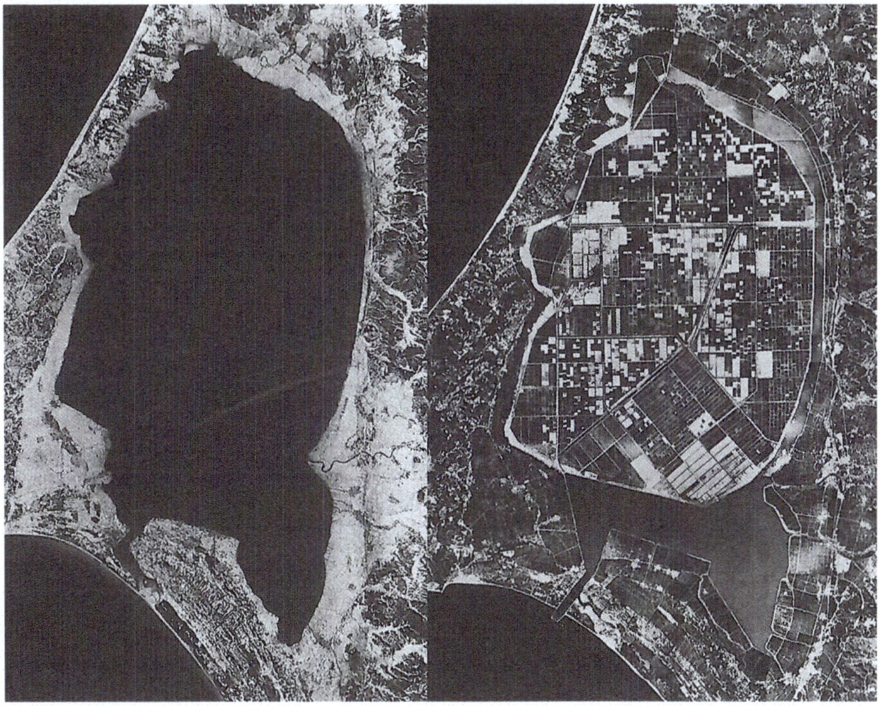

Fig. 9 Comparing the Hachirogata Lagoon before and after (Aerial photos). *Source* General Affairs and Planning Section of the Ogata Mura Village Office; 大潟村役場総務企画課(編) A History of the Rural Development Projects of the Ogatamura Village in the Hachirogata Central Polder;八郎潟中央干拓地「大潟村」における農村集落の建設と村づくりの変遷, 2011, p 7);

addition, local youth associations and women's associations expressed their approval (Akita Sakigake Shimpo 1954-8-23).

The fishermen balked because of the sluice gate. Jansen put it to prevent flood damage and separate the remaining lake from the sea to secure fresh water for the polder (Jansen Pr 1954). But the proposal was introduced in the local newspaper as a possible plan for sustaining the fishery: The map in the newspaper does not show a sluice gate, and readers might well have understood and expected that the Jansen plan would not change the connection between the lagoon and the sea. When the fishermen found out about the water gate, they opposed the Jansen proposal: it was clear to them that it would devastate the fishery. They insisted that landfilling, not draining the lagoon, be the central engine of reclamation work (Akita Sakigake Shimpo 1954-8-3).

The APO had already explained that the fishery would continue with the fishery experimental station in the remaining part of the lagoon (Akita Sakigake Shimpo 1954-5-22). But both the plan and the explanations indirectly indicated that the

Fig. 10 Landscape of the Hachirogata region, seen from the Kanpusan Mountain. Left: the Hachirogata central polder. Middle: the remaining lake. Right: the villages and paddy field around the lagoon (Photo by Yasunori Kitao); released under a Creative Commons Attribution-NonCommercial-NoDerivatives 4.0 International License

Fig. 11 Estuary of the Hachirogata lagoon before the impoldering project, 1949 (Aerial photo). *Source* The Geographical Survey Institute of Japan, Funagawa, USA-M1111-35: https://mapps.gsi.go.jp/maplibSearch.do#1. Last retrieved on the August 15, 2017; released under a Creative Commons Attribution-NonCommercial-NoDerivatives 4.0 International License

future fishery would be a fresh water lake. The fishermen were trying to find a way to reclaim the lands by landfilling at the final stage of the planning of the central polder because they hoped to find a tiny chance to continue their traditional fishing in the brackish lake of Hachirogata lagoon. Opposition to the reclamation project from a standpoint dependent on keeping the brackish water means that it insisted on the natural conservation of the brackish lake.

Nonetheless, the Jansen compromise triggered a debate which led to breaking the fisherman's solidarity. Wealthy fishermen with powered vessels continued to oppose

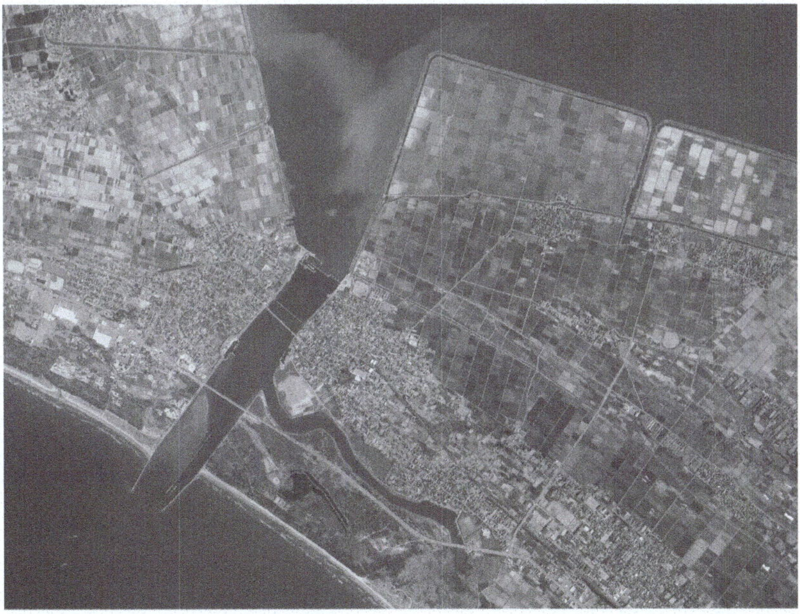

Fig. 12 Estuary of the Hachirogata lagoon after the impoldering project, 2004 (Aerial photo). *Source* The Geographical Survey Institute of Japan, Akita,MTO20046X: https://mapps.gsi.go.jp/maplibSearch.do#1 (Last retrieved on the 15th August 2017); released under a Creative Commons Attribution-NonCommercial-NoDerivatives 4.0 International License

full reclamation. The smaller fisherman who began to understand the limitations of the capacity of the fishing catch in the brackish water had to decide whether to withdraw from the lagoon, and many came to believe that their livelihood would be stabilized by agriculture in the polder (Akita Sakigake Shimpo 1954-8-11). The brackish water disappeared with construction of the polder and the freshwater lake, which ultimately halted the opposition movement by the fishermen (Figs. 8 and 9).

In the end, however, versions of both fishery and agriculture continued at Hirochigata, a way forward made possible by the plan of 1955, drafted by the MAF based on the proposal by Jansen. In this plan, a total of 699 immigrant families were to be settled on 13,289 ha of the polder, and 7245 units of "allocated increased arable lands" (Zoutan in Japanese) are detailed (Morooka 1955). In the same year, the APO made provision for 5000 immigrant families to settle on 13,000 ha of the arable land (The Governor Office of the Akita Prefecture 1956). It can be seen by comparing the number of settlers that the APO plan addressed national policy, while the MAF plan was regionally friendly. The government and prefecture were trying to surmise each other's wishes and to act with consideration for each other's situation, a strategy known as "sontaku." (Unfortunately, sontaku has come to mean pandering to someone in a position of authority.) The plans can be regarded as a mutually beneficial bureaucratic strategy, and they became the basis of a solution for the discrepancies

Fig. 13 Tide sluices of the Hachirogata lagoon. These tide sluices were built in 1961. This changed the remaining lake into a fresh water lake (Photo by Yasunori Kitao); released under a Creative Commons Attribution-Non Commercial-No Derivatives 4.0 International License

between fishing and agriculture in the administrative and regional planning of the Hachirogata development (Figs. 10, 11, 12 and 13).

Conclusion

Today, one aspect of the history of the Hachirogata lagoon has become its entire cultural heritage: Japanese scholars and museums celebrate the land reclamation as a signal achievement, emphasizing physical size, national investment, and technological innovation. Moreover, stories of the promise of the new land, and the future of the people living and working there, all form a kind of positive propaganda. Politically and socially, it has been better to forget the rest of the history, particularly the struggle with the historical past and local fishing traditions.

However, in reassessing the heritage of Hachirogata polder and its region, it is important to see the long debate over how best to use lake and the fishery, and to recognize the finding of a balance of the different viewpoints—between the lake and the polder, fishery and agriculture, traditional and modern—as a historic sociocultural planning achievement. The fishermen's protests ensured that the process included all stakeholders, a message of this heritage that might inform planning going forward. Finally, heritage debates can take this heritage of balance as a model for thinking about narrating history more fully; and planners and politicians can take it as a model for resolving future conflicts over land use during climate change and the shortage of natural resources on the earth.

Acknowledgements This work was supported by JSPS KAKENHI Grant Number 16H02386.

References

General Affairs and Planning Section (ed) (2011) 大潟村役場総務企画課(編), "八郎潟中央干拓地「大潟村」における農村集落の建設と村づくりの変遷; A history of the Rural Development Projects of the Ogatamura Village in the Hachirogata Central Polder, The Village Office of Ogatamura, p,3, pp. 38–41

Hachiro-Gata Reclamation Bureau (ed) (1969) The history of reclamation project of Hachiro-Gata; 八郎潟干拓事業誌, p 11

Heuvel D, van den Heuvel D, Mesman M, Quist W (eds) (2008) The challenge of change: dealing with the legacy of the modern movement. Delft University Press, Delft

Horikiri Z (1942) 堀切善次郎, "治水の急務[Emergency on Flood Control ,河川[The Rivers] Vol. 1, No.4, The Japan River Association, pp 9–10

Jansen Pr (1954) Some remarks on impoldering in Japan, The Ministry of Agriculture & Forestry, p 5, pp 40–41, p 46

Kanamori M (1940) 金森誠之, "National Planning and Development Project of the Hachirogata Lagoon; 国土計画と八郎潟開発;", Civil Division at Akita Prefecture

Kanamori M (1940) 金森誠之, The greater East Asia Co-Prosperity Sphere and Hachirogata; 大東亜建設と八郎潟, The Ministry of Home Affairs of Japanese Government 1940

Kitao Y (2008) The modern agriculture village on the Hachirogata-Polder in Japan: the design process during the economic development period. In: 10th international conference of Docomomo, the Netherlands, pp 449–451

Ministry of Agriculture and Commerce (MAC) (1914) 和蘭における耕地拡張大計; The great planning of expanding Arable lands in the Netherlands, Ministry of Agriculture and Commerce, Japanese Goverenment 1914

Miyata M et al. (2012) 宮田正馗, 他, "ゼロからの自治 大潟村の軌跡と村長・宮田正馗; Self—government: a history of Ogata Mura village and the Mayer Miyata, Kojinsha Publishing; 公人社 2012

Mizutani S (1940) 水谷 鏘, Evocating water resources for productions; 水資源の生産力喚起", Hydro Civil Eng 水利と土木, The Japan River Association, 13(7):2–7

Morooka M (1960a) Memories of the Hachirogata Lagoon; 八郎潟干拓の思い出", The Magazine of National Development; 国土開発, 9(5):40

Morooka M (1955) 師岡政夫, About the planning of the Hachirogata Polder; 八郎潟干拓計画に就いて", Tohokukenkyu; 東北研究, 5(6):14–21, 40–41

Morooka M (1960b) 諸岡政夫, Memory on the impoldering project of the Hachirogata Polder; 八郎潟干拓の思い出, Developing the Hachirogata Lagoon; 拓けゆく八郎潟, The Institute of National Development Projects; 国土開発調査会 ,p 40, p 62

Sawa S (1937) 澤 重民, A study on flood disaster;水害に関する若干の考察", Hydro Civil Eng; 水利と土木, The Japan River Association 11(12):2–7

Tahara J (1942) 田原二郎, A report on the releasing fish into the rivers; 河川工作物に対する魚族放流事業の概況, Hydro Civil Eng; 水利と土木, 1(5):12–15

The Fisheries Experiment Station of Akita Prefecture (1916) "The Report on Using the Water Area of the Hachirogata Lagoon; 八郎潟水面利用調査昱告書", Akita Prefecture, 1916
The Governor Office of the Akita Prefecture (ed) (1956) 秋田県知事室広昱係(編), "八郎潟干拓:米の増産50万石世紀の大事業; Impoldering project of the Hachirogata Lagoon—large scale rice production project in the 20th Century",秋田県八郎潟干拓推進委員会; The Promotion Committee of Impoldering Project of the Hachirogata Lagoon 1956
The Japan River Association (1939) Civil Engineering in the War Footing;戦時下における土木事業(foreword), Hydro Civil Eng; 水利と土木, 12(1):4
Tokusaki K (1938) 徳埼 香, An discussion on expropriation on fishery right; 漁業権の収用について, Hydro Civil Eng; 水利と土木, The Japan River Association, 10(12):35–40
United Nations (1955) Economic commission for Asia and the Far East, "Water resource development in Ceylon, China: Taiwan, Japan and the Philippines", United Nations, 1955, p 70

Local Newspaper

Akita Sakigake Shimpo Newspaper, 1935-5-30 (八郎湖埋立やめて貰いたい;Stop the Reclamation Project in the Hachirogata Polder)
Akita Sakigake Shimpo Newspaper, 1935-5-31 (濱口村が村営で八郎湖の埋立開墾;The Hamaguchi Mura Village starts reclamation project in the Hachirogata Polder)
Akita Sakigake Shimpo Newspaper, 1935-10-23 (開墾か停止; Stop or Start of the Reclamation)
Akita Sakigake Shimpo Newspaper, 1935-6-21 (部課長会議の申合せ事項;The Meeting of the Bureaucrat of the Akita Prefecture Office)
Akita Sakigake Shimpo Newspaper, 1936-9-8 (Promotion Movement for the Babame-kawa River improvement)
Akita Sakigake Shimpo Newspaper, 1937-2-5 (魚族繁殖計画と共に一部を干拓;A reclamation project in the Hachirogata lagoon together with the fishery promotion planning)
Akita Sakigake Shimpo Newspaper, 1938-9-23 (八郎湖魚道改修;An Improvement Plan for the fish-way in the ensure of the Hachirogata lagoon)
Akita Sakigake Shimpo Newspaper, 1939-4-23 (An improvement plan for the fishway in the estuary of the Hachirogata lagoon)
Akita Sakigake Shimpo Newspaper, 1939-4-23 (調査に来る農林省も大乗気;The MAF comes to the Hachirogata lagoon for research)
Akita Sakigake Shimpo Newspaper, 1941-12-10 (宝庫八郎潟の開発;The Development project of the Hachirogata lagoon, the treasure)
Akita Sakigake Shimpo Newspaper, 1942-4-18 (第一に港の完成 八郎潟は潰すな; Building the port is important, but do not develop Hachirogata lagoon)
Akita Sakigake Shimpo Newspaper, 1945-8-26 (土木事業の今後;Land development project in the near future)
Akita Sakigake Shimpo Newspaper, 1945-10-19 (八郎潟の干拓;Impoldering project of the Hachirogata lagoon)
Akita Sakigake Shimpo Newspaper, 1946-3-13 (八郎潟干拓反対決議;The decision on the objection for the reclamation project of the Hachirogata lagoon)
Akita Sakigake Shimpo Newspaper, 1951-9-10 (八郎潟の干拓へ;Toward the development project of the Hachirogata lagoon)
Akita Sakigake Shimpo Newspaper, 1952-4-27 (八郎潟振興調査費;The budget for the economy promotion research on the Hachirogata lagoon)
Akita Sakigake Shimpo Newspaper, 1952-8-29 (農林省4案示す;The MAF proposed four projects)

Akita Sakigake Shimpo Newspaper, 1952-8-14 (Investigation on the possibility of the reclamation project)
Akita Sakigake Shimpo Newspaper, 1953-1-4 (結論を今年中に;Conclusion would be made in this year)
Akita Sakigake Shimpo Newspaper, 1953 − 5−13 (The biggest fish farm in the Tohoku Region)
Akita Sakigake Shimpo Newspaper, 1953 − 5−28 (県の方針ただす;Requesting for the APO)
Akita Sakigake Shimpo Newspaper, 1953-8-18 (八郎潟の利用開発;Using and Developing the Hachirogata Lagoon)
Akita Sakigake Shimpo Newspaper, 1953-7-28 (八郎潟開発同盟発足;Founding the alliance of the Hachirogata development project)
Akita Sakigake Shimpo Newspaper, 1953-8-1 (干拓事業を協力推進-吉田首相から指示;Promoting the impoldering project with the support of the prime minister)
Akita Sakigake Shimpo Newspaper, 1954-5-22 (八郎潟干拓共同案なる;The Collaborative Plan for the impolderig project on the Hachirogata lagoon)
Akita Sakigake Shimpo Newspaper, 1954-8-3(漁民ヤ案に反対表明;Opposition against the Jansen Plan)
Akita Sakigake Shimpo Newspaper, 1954-8-23(全額融資は困難;The loan that covering total cost from the World Bank is not easy)
Akita Sakigake Shimpo Newspaper, 1954-8-11(ヤ案の干拓に賛成;Approval for the Jansen Plan)
Akita Sakigake Shimpo Newspaper, 1955-10-26 (八郎潟干拓を展望する;Viewing the reclamation project of the Hachirogata lagoon)

Yasunori Kitao is Professor at Kyoto Women's University, Japan, and he previously worked at the Kyoto Institute of Technology and Delft University of Technology. His books include *Collaborative Urban Design (2005), Architecture of Ralph Erskin* (2008), *Urban Street Design & Planning* (2013) *and The Architecture in Colombian Environment (2014)*. He currently works on heritage and the built environment that are at once social, cultural, and industrial.

The windmill park in the Noordoostpolder at Westermeerdijk 2015, Wikimedia, released under a Creative Commons Attribution-Share Alike 4.0 International license

Chapter 11
The Noordoostpolder: A Landscape Planning Perspective on the Preservation and Development of Twentieth-Century Polder Landscapes in the Netherlands

Steffen Nijhuis

Abstract The Netherlands has a centuries-long tradition of reclaiming land. In the last century gaining land from water peaked with the IJsselmeerpolders, made possible by technical innovations. The Noordoostpolder (1937–1942), one of the IJsselmeerpolders, is a unique example of a fully designed agricultural landscape of the twentieth century. It is the first Dutch modern polder in which the layout was planned as an integral task, involving all its agricultural, urban, and landscape elements at once, while reflecting the state of the art in design, science, and engineering. Using the Noordoostpolder as an example, this chapter discusses the preservation and development of twentieth-century polders as cultural heritage landscapes. It elaborates a preservation-through-planning approach that takes spatial development with historical landscape structures as a basis. The chapter also briefly elaborates a critical way of understanding the coherence and variation of modern landscapes such as the Noordoostpolder, providing clues for spatial planning by systematically delineating and identifying spatial design principles.

Keywords Land reclamation · Protection through planning · Landscape planning · Industrial agricultural landscape · Noordoostpolder · Heritage landscape

Introduction

Polders are a significant type of water landscape that humans have created in coastal and alluvial lowlands all over the world by reclaiming land from water. Here, water levels are artificially controlled for food production and everyday life. Interaction between humans and water has produced a rich variety of polder landscapes throughout the ages, in the Netherlands ranging from the tenth to the twenty-first century (Nijhuis 2016; Van der Ven 2004). Globally, polder landscapes are under threat due

S. Nijhuis (✉)
Delft University of Technology, Delft, The Netherlands
e-mail: S.Nijhuis@tudelft.nl

C. Hein (ed.), *Adaptive Strategies for Water Heritage*,
https://doi.org/10.1007/978-3-030-00268-8_11

to climate and economic change: increasing flood risk due to sea-level rise, ongoing subsidence due to intense drainage, and rapid urbanization. These valuable cultural heritage landscapes must be safeguarded; knowledge development can provide clues for preserving and transforming them in ways which acknowledge and cultivate their local variation and regional coherence. Polder landscapes are not only important hydraulic phenomena but also spatial constructions and cultural expressions: The polder landscape as one can see and experience it is also a display of tacit knowledge—the kind of knowledge that is difficult to transfer to another person by means of writing it down or verbalizing it—and symbolic values related to land reclamation, water infrastructure, agricultural practice, and landscape planning and design.

This chapter presents a landscape planning approach for protecting and developing twentieth-century polders as cultural heritage landscapes.[1] Landscape planning is concerned with the long-term development and preservation of natural and cultural landscapes, developing and implementing strategic goal-oriented concepts, and allocation of types of land use. It also addresses design aspects such as form and meaning, develops design principles, and seeks to organize a variety of structural landscape elements in a physical, functional, and aesthetic arrangement. Specifically, this chapter introduces an approach that can be characterized as 'preservation through planning,' which takes landscape development with historical landscape structures as a basis. The chapter will also elaborate a critical way of understanding the coherence and variation of these landscapes, providing clues to developers and planners for spatial planning and design by systematically delineating and identifying spatial design principles. The Noordoostpolder, a unique and completely designed polder landscape from the twentieth century in the Netherlands, serves as an example (Figs. 1 and 2).

The next section introduces the Noordoostpolder as the twentieth-century cultural heritage landscape. Then, the chapter elaborates on the protection-through-planning approach to the countryside in the Netherlands, and how this approach is implemented in the region. The section before the conclusion will briefly introduce an analytical framework to identify landscape design principles that can successfully facilitate new development and guide landscape change.

The Noordoostpolder as a Twentieth-Century Cultural Heritage Landscape

Reclaiming land is a centuries-old tradition in the Netherlands. In the last century it peaked in size with the IJsselmeerpolders, made possible by technical innovations. In the centuries before, land reclamation had often been a strategy by which landowners extended their holdings, an investment by rich merchants, or was a public undertaking. In the case of the IJsselmeerpolders, the Dutch government wanted to

[1] This paper is based on an invited lecture presented at the 'International symposium on conservation and utilization of modern industrial heritage; contemporary subjects in the Netherlands, Italy, China, Germany, Indonesia, and Japan.' Tokyo, March 4, 2017, and March 5, 2017.

Fig. 1 The IJsselmeerpolders in the centre of the Netherlands (S. Nijhuis, Delft University of Technology)

safeguard food provision for the nation by providing agricultural land; later, urban development became an important motive. The IJsselmeerpolders—Wieringermeerpolder (constructed 1927–1930), Noordoostpolder (1937–1942), Oostelijk Flevoland (1950–1957), and Zuidelijk Flevoland (1959–1968)—provide almost ten percent of the total arable land in the country today (1,045,000 ha).

In general, there are three important types of land reclamation: (1) impoldered low-lying lands, (2) drained lakes, and (3) lands reclaimed from the sea. During the construction of these polders, open water or waterly lowland is turned into landscapes

Fig. 2 The Noordoostpolder (Paul Paris)

for working and living and therefore includes flood protection, drainage, and often irrigation. Modern polder landscapes as such are not only part of the history of hydraulic engineering and industrial agriculture, but also territorial expressions of the interaction of physical conditions, knowledge, and management in the Dutch cultural and political context of the twentieth century (Fig. 3). The IJsselmeerpolders, lands reclaimed from the sea, are truly cultural landscapes that are defined and shaped by cultural perceptions and practices and, in turn, shape and structure social and cultural experiences (cf. Head 2000). Considering polders as cultural landscapes broadens concepts and understandings of cultural heritage past monuments and sites and exemplifies the interdependence of social, aesthetic, ecological, and economic functional values, and puts polder landscapes forward as an important subject for water and heritage studies.

The Noordoostpolder (NOP) was the first IJsselmeerpolder, drained after the completion of the Afsluitdijk in 1932. The dikes for the polder were built between 1937 and 1941, and the polder was dry in 1942 (RIJP 1986–1990). The two former islands of Schokland and Urk were also included in the polder. The polder has an area of 48,000 ha; it measures 26 km from north to south and 24 km from east to west. The NOP is a unique example of a fully designed landscape of the twentieth century, really a '*Gesamtkunstwerk,*' or a total piece of art with a coherent perspective on architecture, urban planning, landscape architecture, civil engineering, and land-use planning (Fig. 4). It is the first modern Dutch polder in which the layout was planned as an integral task, involving all its agricultural, urban, and landscape elements at once (Minderhout et al. 1955). In general, much had been learned from the drain-

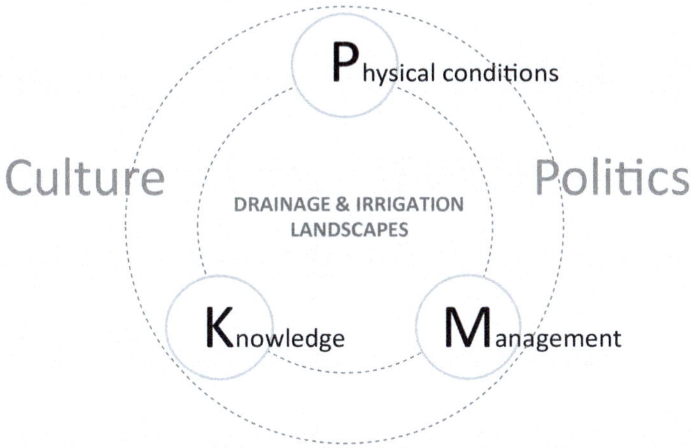

Fig. 3 Polder landscape as territorial expression of the interaction of physical conditions, knowledge and management in the Dutch cultural and political context (S. Nijhuis, Delft University of Technology)

ing of the Wieringermeer. But if one considers the Wieringermeerpolder to be the prototype of modern, twentieth-century land reclamation, then the NOP is the first mature example in the Netherlands. It would in turn become an exemplar for other modern land reclamations, including the Hachirogata Polder (1963–1966) in Japan (HIPO 1969).

The NOP was designed for optimal agricultural production and was also regarded as a scientific, aesthetic, and intellectual project. For the aesthetic aspects of the polder, the publication 'The future landscape of the Zuiderzee polders' (1928) was a great influence (Fig. 5). This manifesto is today regarded as an important milestone in Dutch urban and landscape planning. Here, influential Dutch modernist planning experts such as Dirk Hudig and Theodoor van Lohuizen put forward preferred principles for landscape design, derived from comparing the Beemster, the seventeenth-century polder, to the Haarlemmermeer, a polder from the nineteenth century. They regarded the first as a successful example, and the latter a less successful one. Sociologists and agricultural engineers also helped set up the NOP, paying attention to questions of future population, the role of transport, and the location of population cores. The various scientific movements and professional disciplines came into conflict with one another in the discussion on the ideal design of the NOP (Geurts 1997; Bosma and Andela 1983), so negotiation, planning, and realization played a more important role than ever before.

Landscape planning began to play an increasingly important role in spatial design, as a counterpart to the urban planning of the polder addressing landscape as an important element in planning. Specifically, a landscape framework—based on soil, geography, and landscape history—made space for nature and recreation, enriching

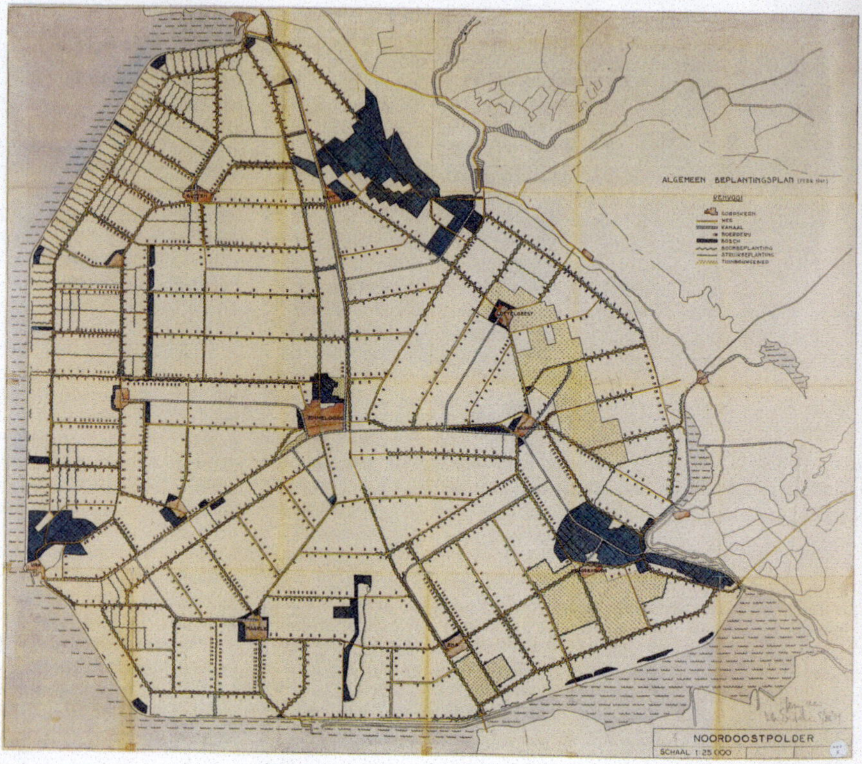

Fig. 4 General planting scheme of the Noordoostpolder, 1947 (Nieuw Land)

the network of roads and villages that urban planners laid over the polder land (as in the Wieringermeer).

Centrality plays a major role at every level in this polder. Emmeloord, with the polder tower (Netherland's highest water tower) as its center, is the center of concentric rings: the ring of villages, a condensed outer ring, and the dike ring. The landscape planners conceived woods and grassland in 'polder rooms' for the edges of the polder; the heart of the polder has a high degree of landscape openness and is used for arable farming. The allocation of the land and the concentric structure of the road system and village pattern all support this articulation of the polder space into 'rings.' At a smaller scale, one finds the concentric structure in the spatial plan of the villages, and in the farmsteads, grouped around the intersection of a polder road and drainage ditch, at the corners of the parcels.

The ten villages in the polder are planned at regular intervals and circle around Emmeloord, the capital of the polder. Architects or consultants employed by the government designed most of the villages in the NOP, laying each out according to the principles of the Delft School, with tradition as the guide. The hierarchical spatial

36

weg 1), waarlangs, achter de onmiddellijk aan den weg staande boomen, de berm opglooit, omdat ze de kruin dekt van een, vrij lagen, scheidingsdijk.

Wegen, omsloten door kreupelhout, zijn in dit gedeelte van het land zeldzaam; ze passen niet meer in het moderne bedrijfsleven. Ze bezitten romantische intimiteit, vooral als ze smal zijn en slingeren 2).

Ook hakhout aan de eene zijde komt voor, [4]. De intimiteit van het besloten laantje gaat dan verloren; het uitzicht aan de eene zijde blijft open, maar doet de afsluiting aan de andere zijde al spoedig als hinderlijk ondervinden. Maar bekoorlijk kan ook zulk een weg zijn 3). Holland heeft er meerdere.

Vredig en landelijk is een beplanting met knotwilgen, vooral aan stillere smalle wegen zonder boerderijen, [5] of alleen met kleinere plaatsen, zelf in knotwilgenhout gedoken.

Beplanting aan eene zijde met zware boomrijen is vooral mooi aan wegen met boerderijen en aan de landzijde van wegen langs water. Op ringdijken komt ze vaak voor, [6]. Ze eischt losheid en schilderachtigheid. In het landschap bekoort ze door haar doorzichtigheid. Jonge eenzijdige beplanting werkt armoedig.

1) Als grensscheiding gelegen tusschen Drechterland en de Vier Noorder Koggen.

2) Een fraai voorbeeld is het smalle weggetje tusschen Hoogmade en Leiderdorp.

3) Het weggetje tusschen Oudendam en Wilnis bewijst het. Hier speelt het water aan de Westzijde van den weg echter een belangrijke rol.

3 Kruisweg Haarlemmermeer

4 Veenkade bij de Dongersluis (Nieuwer Ter Aa)

5 Friezeweg tusschen Nieuwveen en Oude Wetering

37

Verspreide beplanting is schaarsch, maar vol bekoring, [7].

Soms zijn de kruispunten van wegen gemarkeerd door opgaand hout. Een variant daarvan is aan zeer smalle wegen het accentueeren der wegverbreedingen, waar voertuigen elkaar kunnen voorbij komen, door enkele hoog opgeschoten boomen, soms wel, soms niet tusschen lager hout, [8].

De heg is als begrenzing van den weg schaarsch. Ze is te vinden, waar aan den voet van den dijk de sloot ontbreekt, en het aangrenzend land van den weg moet worden afgescheiden. Ook waar de berm van den weg breed is en een sloot eerst op eenigen afstand ligt, [9]. De berm wordt dan als weide of moesgrond gebruikt.

Juist omdat ze zeldzaam is, geeft ze welkome afwisseling. Maar wanneer ze te hoog opschiet, beneemt ze het uitzicht op het land; zelfs een lage heg onttrekt den voor onze wegen zoo karakteristieken berm aan het gezicht, vooral wanneer deze steil is.

Geheel open wegen hebben hun eigen bekoring. Als weg op zichzelf, hetzij met breeden, vlakken berm, die tot rusten uitnoodigt, hetzij hooger gelegen met steiler kanten. Het uitzicht is er ruimer, ongebroken ligt het landschap voor oogen, [10].

De wegen met boerderijen erlangs toonen een rijker karakter met sterker afwisseling (wat niet zegt dat ze reeds daarom schooner zouden zijn), vooral wanneer de boerderijen van ouderen datum zijn, en in oud en vol geboomte zijn

6 Ringdijk van de Beemster bij Purmerend

7 Bij Leidschen Dam

8 Weg tusschen Wilnis en Oudendam

Fig. 5 Pages from ‘The future landscape of the Zuiderzee polders’, 1928

form of the villages, turned in on itself, was intended to serve as a counterpoise to the dislocations of urbanization and industrialization. In elaborating the plans, the designers experimented with architectonic concepts in the spirit of Camillo Sitte that were based on aesthetically pleasing historical examples.

The village of Nagele was an exception. Here, influential Dutch urban planners and architects such as Cornelis van Eesteren, Gerrit Rietveld, and Aldo van Eyck, among others from the Modernist collectives ‘De 8’ en ‘Opbouw,’ came up with their own initiative to design the village and apply their planning ideas (Van Woensel 1999; Van der Wal et al. 1992; De 8 en Opbouw 1952). They structured Nagele systematically and evenly, and they separated the major urban functions—living, working, traffic, and recreation. The central open green space, surrounded by urban facilities and seven neighborhoods, illustrates a new organization of society. The design and construction of Nagele also formed the first turn to the urbanization of the countryside (Fig. 6).

As such the NOP reflects the state of the art in landscape and urban planning, architecture, agricultural and social science, and engineering in the second half of the twentieth century, at multiple spatial scales, from the region to the building. Therefore, the NOP can be regarded as a cultural heritage landscape.

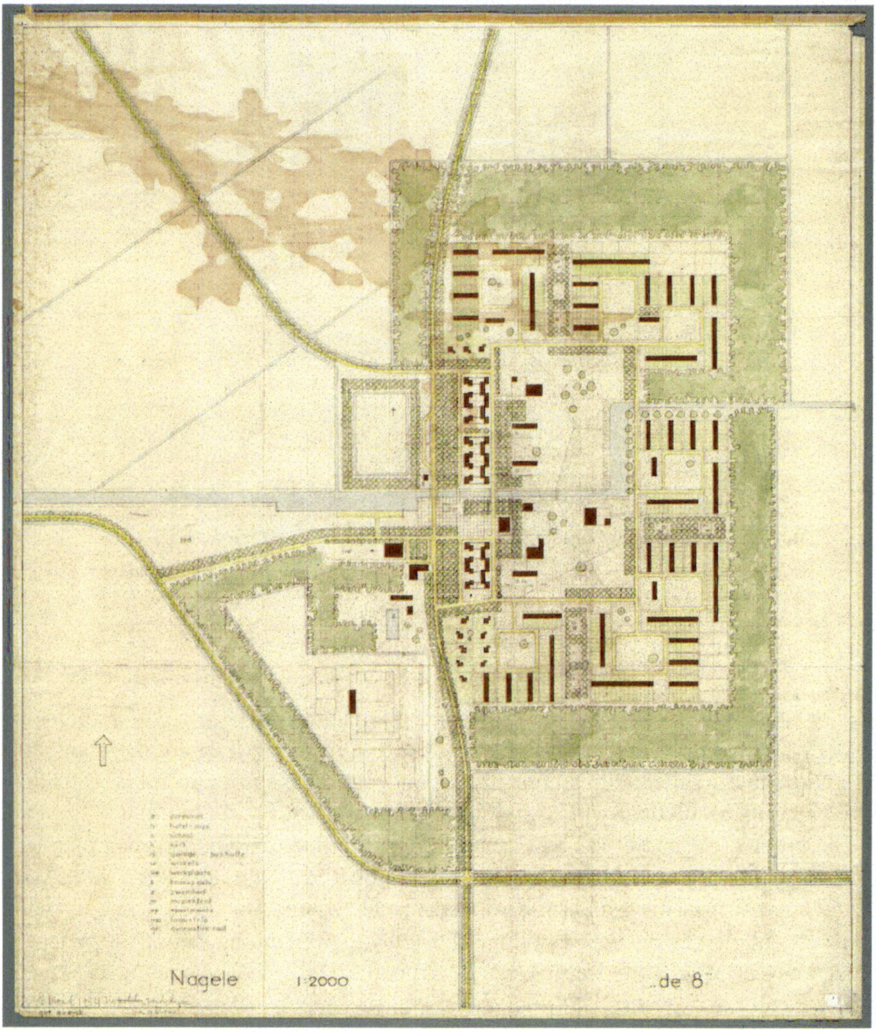

Fig. 6 Nagele, final design by Aldo Van Eyck, 1954 (NAi)

Protecting Polder Landscapes Through Planning

A cultural landscape is defined as an area, as perceived by people, whose character is the result of the action and interaction of natural and/or human factors (Council of Europe 2000). This definition stresses the dynamic character of landscapes. Polders change, as does every landscape with and without human interaction. Sometimes there are profound changes taking place, sometimes less radical changes. A landscape is a dynamic system that is continually transforming under the influence of natural processes and social requirements. Linking the past to the present, the landscape is a layered entity, a palimpsest where traces that time has covered can reinforce or contradict each other; it is a window into a range of histories, chronologies, events, and meanings connecting the traditional and the contemporary, the tangible and the intangible.

Managing and preserving heritage landscapes are very different from protecting a site or monument. The landscape is always open to significant change on a variety of timescales even without human intervention. Its perceptual aspects are just as dynamic as its physical ones: People change their view of the world and thus their landscape for several reasons. In other words, a cultural landscape cannot be preserved unchanged (Fairclough et al. 2008). Preservation of cultural landscapes should focus on the 'management of change,' aiming to create a future in which the past in one form or another plays an appropriate role (Fairchlough et al. 2008).

The landscape of the IJsselmeerpolders is subject to distinctive kinds of change. The NOP is now 75 years old, and most of its layout and buildings are still in place and recognizable. However, it is also a living landscape, where people live, work, and spend their leisure time. Protecting and developing this cultural landscape requires careful attention to this combination. Moreover, rapid urban development, function change, and climate change can level or standardize these rural landscapes, negating their characteristic spatial differences and risking the loss of cultural identity.

The legal instruments (such as the Dutch Monuments and Historic Buildings Act 1988/2012) that aim to preserve the state of monuments and sites do not apply to such living cultural landscapes that continuously change due to agrarian and other economic developments. In the Netherlands, where agricultural systems are the most intensive in Europe, the legal protection of cultural heritage landscapes is extremely problematic. In fact, the Netherlands is one of the few countries with hardly any protected cultural landscapes (UNEP-WCMC 2017). The Dutch government did recently ratify the European Landscape Convention (Council of Europe 2000), which promotes the protection, management, and planning of European landscapes and organizes European cooperation on landscape issues; this illustrates that cultural landscapes are high on the political agenda. And three cultural landscapes were designated as world heritage in recent decades: Schokland and Surroundings (1995), the Defense Line of Amsterdam (1996) and the Beemster Polder (1999) (World Heritage List 2017). However, these are weak forms of protection, since they have hardly any legal basis. More largely, political pressure from landowners, in particular farmers and their representatives, makes it almost impossible to protect landscape with legislation because it leaves little space for future development.

But that does not mean that cultural landscapes have no protection in the Netherlands. During the second half of the twentieth century, the protection of cultural landscapes became part of the political discussion that resulted in important spatial planning instruments embedded in the Dutch Spatial Planning Act. An important milestone in this respect was the 1977 policy document, Vision on Landscape Construction, that pleaded for the integral development of the landscape using its existing structure and its historical situation as a basis in order to ensure that development created a recognizable environment with its own identity. This was followed by the first national mapping of cultural values in the rural area by the national government (CRM 1979).

Another important document was the 1999 Belvedere Memorandum, a national policy document that examined the relationship between landscape history and spatial planning. It considered cultural historic values important to identity and a sense of purpose, a resource against globalization, a source of information and inspiration, and forms of aesthetic, ecological, and economic importance. Conservation through development is its motto. By seeking new uses for historical landscapes and buildings, cities and countries can save them; by using cultural heritage in a frugal and responsible manner, one could invest in developing and strengthening of identity, knowledge, comfort, the business climate, and the potential for tourism (Belvedere 1999). The document proposes that the interaction between cultural history and spatial planning, and the recognition of the difference in perspectives of stakeholders, can create the conditions for discovering a new balance between retaining cultural heritage and developing it (Belvedere 1999). This process of development can help renew citizens' engagement with cultural heritage.

This type of policy document influenced Dutch planning, inviting professional groups, already directed toward the future, to look backward, and encouraging historians, who used to look only backward, to take stock of the future (Bosma 2010). The result was planning instruments that embraced the idea that historic valuable landscapes should be used in such way that preserves them and at the same time gives them a valuable use. But how are these principles put into practice?

Planning Instruments and the Noordoostpolder

Statutory spatial plans play an important role in land-use planning and management in the Netherlands. The Dutch Spatial Planning Act prescribes and regulates statutory spatial plans attending to cultural heritage (and other issues) at the level of the national government, the province, and the municipality (Needham 2007).

The Dutch government can set spatial policy in different ways, but works mainly through national structure plans for a policy sector or broad aspect of spatial policy, and through a Memorandum on Spatial Planning. In 2012, the National Policy Strategy for Infrastructure and Spatial Planning came into force. In this policy document, the NOP as a whole is among the thirty selected areas that jointly exhibit the social dynamics of the reconstruction of the Netherlands after World War II (in the period 1940–1965). The complementary National Policy Strategy for Heritage and Space explains the core qualities of these areas, describing the NOP as large-scale land reclamation with rationally distributed agricultural farmhouses and yards and a functional pattern of roads with associated planting. It also notes the ring of villages around the main core (Emmeloord) as an important characteristic. According to this plan, provinces and municipalities should model new development on these characteristics in their future planning.

Here, the Dutch government focuses on administrative agreements with provinces and municipalities on development-oriented protection zoning. It asks a province to make a Provincial Regional Plan or a Provincial Environmental Plan setting out the desired future development of the province. That plan is self-binding: It binds the province to act in accordance with its own plan and coordinates the actions of the province itself. Because the province is empowered to regulate the action of lower authorities, it can have great consequences for the involved municipalities: It can give or withhold approval to their new Municipal Zoning Plans. For the NOP as a whole, the Environmental Plan Flevoland 2006–2015 is still in effect, offering a policy framework for the municipalities to elaborate on. It puts forward ambitions for enhancing the vitality of the landscape (e.g., new functions), ensuring and improving the quality of the countryside (e.g., strengthening spatial structures), and handling existing situations and functions (e.g., facilitating development). The plan cites the World Heritage site, Schokland, and the villages Urk and Nagele as important basic features of the NOP that needs to be preserved, some historical objects (e.g., pumping stations, bridges), the main water courses, the inner ring road, and the open landscape in the center.

On the municipal level, the Municipal Structure Plan (which is not obligatory) sets out the desired future development of the municipality as a whole, including the framework for the zoning plan(s) and the plans for various sectors of the municipality (e.g., housing, transport, employment). The most important planning instrument is the Municipal Zoning Plan, which sets out the activities which may make take place on the land in a designated area, 'in so far as this necessary for good ordering of land.' This zoning plan is the basis on which a municipality can grant permits for development. It is the only operational plan. The zoning plan can be very general or very detailed and is legally binding. Municipalities are also obliged to make a land-use plan for the land outside the built-up areas, which is to say the rural areas. The 'Rural Zoning Plan Noordoostpolder 2004' is the most important plan for the landscape of the NOP. It is a generic planning framework for the protection and development of this rural area, focused on regulating the main land uses.

The municipality is by law obliged to update zoning plans within ten years. At present, environmental issues in relation to agriculture make it practically impossible for the municipality to update the zoning plan for the NOP. As a step in the process for meeting the requirement, the municipality chose therefore to establish a Management Regulation for the Countryside 2016. The plan specifically mentions several important historical features that developers should take into consideration: the parcellation structure (form, size, and proportion of the plots themselves and their constellations), the farms and yards, and three types of farmhouses. The recently published structural vision Noordoostpolder 2025 provides the basis for the new Municipal Zoning Plan (whenever it can be updated) and is strategic in nature. The structural vision is primarily intended to seduce and inspire stakeholders and to promote initiatives and investments that will contribute to responsible development of the countryside. It also reiterates the enforcement of its cultural–historical and landscape values through descriptions and maps (Fig. 7).

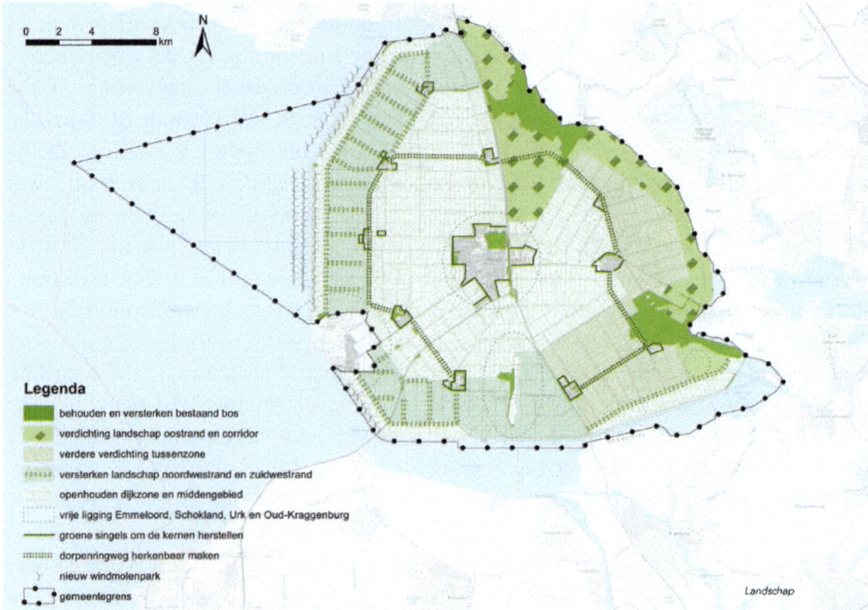

Fig. 7 Landscape value map from the recent Structural vision Noordoostpolder 2025 (Municipality Noordoostpolder, Kuiper Compagnons)

In the Netherlands, such statutory plans protect landscapes such as the NOP to some degree by indicating the landscape character and describing historical landscape structures. However, the plans remain descriptive, only identifying important characteristics in a very general way and providing hardly any clues for how developers or municipalities might make use of them in future plans. Though often perceived as restrictive, the plans do leave room for developers to interpret them—which often results in 'misinterpretations' and destruction of important landscape features, such as the characteristic structures and openness of polders (e.g., encroachment of typical open areas by new farm buildings).

A Critical Design Perspective for Development and Protection

In the NOP, *development* means adapting the polder to agricultural demands (e.g., extra and bigger farm buildings), to housing projects, and to energy transition away from fossil fuels (e.g., wind turbines, water storage). In order to accommodate such development within a cultural heritage landscape, and to facilitate change in an effective way, a design perspective is needed. Design is about changing situations into preferred ones and is future-oriented. Therefore, describing and indicating the

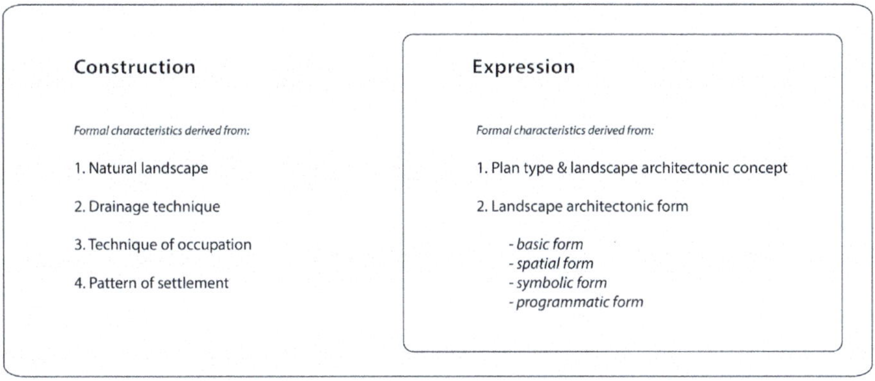

Fig. 8 Analytical framework for the identification of the polder grammar (S. Nijhuis, Delft University of Technology)

landscape character, as the statutory plans do, is not sufficient. Planning instruments should also provide design knowledge to guide and direct landscape change in a responsible manner, while providing clues for spatial development in the form of design principles or—in this case 'polder grammar.' The polder grammar is the set of structural rules and principles that determines the characteristic composition of the landscape: the complexity of the pattern, the morphology, the visual qualities—and with that, the cultural identity of the polder. Knowledge of the polder grammar is the starting point for new changes in the landscape or adding a new design layer. It respectfully guides and directs landscape change.

We have developed an analytical framework to identify and delineate polder grammar systematically (Nijhuis 2016; Steenbergen et al. 2009). It employs historical maps, newly drawn maps, and photographs to identify the constructive characteristics and the expressive characteristics of the polder form. The constructive characteristics are the formal aspects of the historical transformation of the natural landscape into a habitable and exploitable agricultural landscape. The expressive characteristics are the implicit and explicit visual and formal elements, the compositional, aesthetic, and cultural motifs of polder making (Fig. 8). We understand the polder landscape to be a layered entity, defined first by the underlying natural landscape, to a further degree by hydraulic interventions, and finally by the organization imposed by the agricultural development of the land (Fig. 9). Each stage leaves its traces on the form of the polder. Identifying and delineating this complex layering creates a point of contact between the landscape and developers or municipalities dealing with technical and landscape planning questions, such as water management or urbanization, and thus provides clues for their further development of the polder landscape.

Through design experiments based on the polder grammar, on multiple scale levels, one can investigate which elements from the polder landscapes can be changed to adapt them to economic and climate change without damaging the overall integrity of the polder and how to change them. Research by design, exploring possibilities

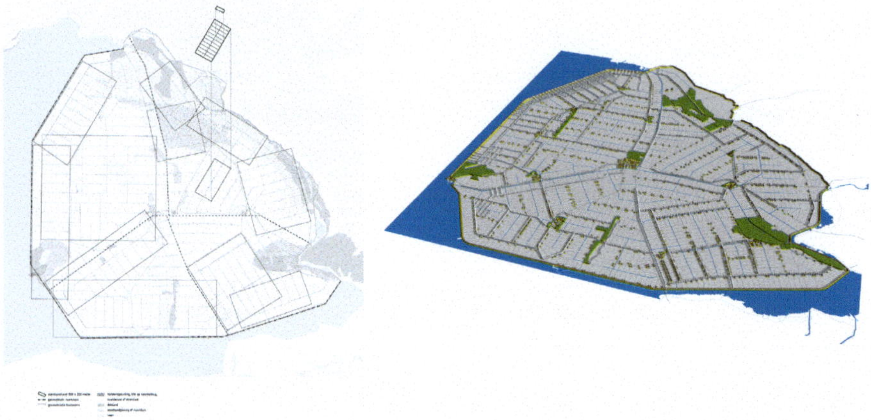

Fig. 9 Some applications of the analytical framework to the Noordoostpolder on the regional scale (S. Nijhuis, Delft University of Technology)

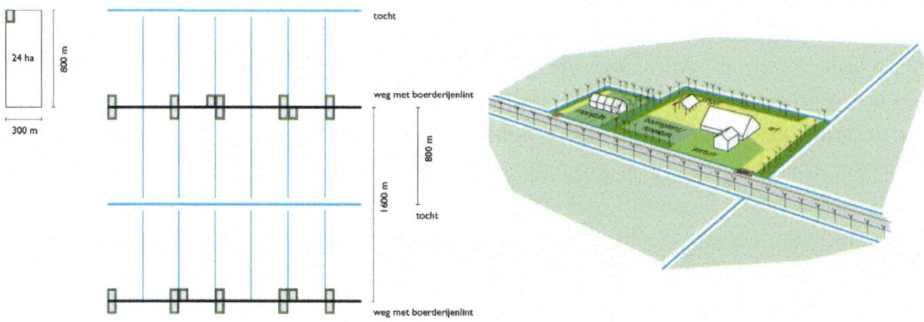

Fig. 10 Typical organization of farms and parcellation in the Noordoostpolder (Feddes 2012)

by spatial design, can help developers, and planners explore the possibilities for spatial development, generating proposals or potential solutions for design problems (Nijhuis and Bobbink 2012). This type of knowledge-based design leads to new, balanced, and coherent polder landscapes with their own identity and spatial qualities. For a full elaboration of the framework see Nijhuis (2016), and application on the NOP and identification of design principles see Steenbergen et al. (2009) and Feddes (2012) (Figs. 10 and 11).

Conclusion

The preservation and development of cultural heritage landscapes like the NOP is a public valuation of tangible (physical) and intangible (social and political) characteristics of the past and asserts a public interest in things traditionally regarded

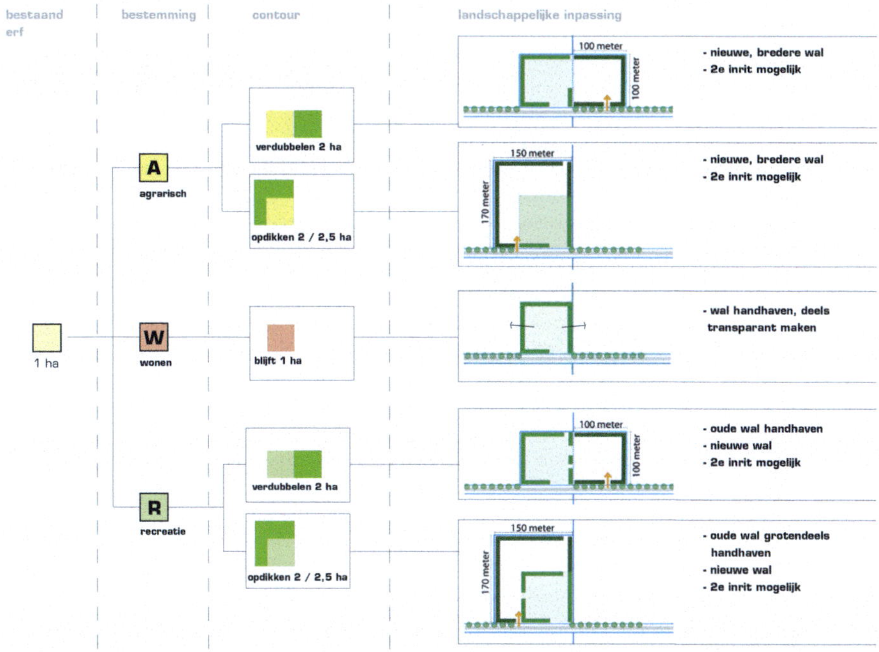

Fig. 11 The development of new farms can be developed based on the design principle (Feddes 2012)

as private. In that respect, heritage landscapes are cultural resources that need to be handled with care. Modern cultural heritage landscapes such the NOP are preserved through governmental planning in the Netherlands, embracing the idea that historic landscape features should be used in such way that preserves them and at the same time gives them a valuable use. This process of protection through planning is more complex and contextual than legal protection. And it is part of a dynamic, political process instead of only belonging to the realm of experts (Renes 2004), which means that more people and organizations are involved. Preservation by planning thus depends on coalitions between partners.

Though the statutory plans in the Netherlands provide an important basis for preservation through planning, with generalized descriptions of important cultural–historical landscape characteristics, the process also calls for augmenting those statutes with design principles in order to make them effective tools for development. Particularly on the provincial and municipal levels, one can find promising developments in this direction, with planners implementing new analytical frameworks for understanding and management as vehicles for translating their ambitions into practical applications. Preservation through planning can build capacity, involve local stakeholders, and stir public debate about the significance of historical landscape features and how to use them in new ways.

References

Belvedere (1999) The Belvedere memorandum: a policy document examining the relationship between cultural history and spatial planning. Sdu Uitgevers, The Hague

Bosma K (2010) Heritage policy in spatial planning. In: Bloemers T (ed) The cultural landscape & heritage paradox: protection and development of the Dutch archaeological-historical landscape and its European dimension. Amsterdam University Press, Amsterdam, pp 641–651

Bosma K, Andela G (1983) The landscape of the IJsselmeer polders. In: Bosma K et al (eds) Het Nieuwe Bouwen, Amsterdam 1920–1960. Delft University Press, Delft, pp 142–174

Council of Europe (2000) European landscape convention. s.n., Florence

CRM (1979) Natuurwaarden en cultuurwaarden in het landelijk gebied: Een inventarisatie. [Natural and cultural values of the countryside: an inventory]. Staatsuitgeverij, Den Haag

De 8 en Opbouw (1952) Een plan voor het dorp Nagele [A plan for the village Nagele]. Forum 7(4/5):172–178

Fairchlough G et al (2008) New heritage, an introductory essay—people, landscape and change. In: Fairclough G (ed) The heritage reader. Routledge, London, pp 297–312

Feddes Y (ed) (2012) Landschapsvisie Noordoostpolder. Landschap van rust en regelmaat. s.n., Utrecht

Geurts A (1997) De Groene IJsselmeerpolders. Inrichting van het landschap in Wieringermeer, Noordoostpolder, Oostelijk en Zuidelijk Flevoland [The green IJsselmeer polders. The design and construction of the landscape in Wieringermeer, Noordoostpolder, Oostelijk en Zuidelijk Flevoland]. Uitgeverij de twaalfde provincie, Lelystad

Head L (2000) Cultural landscapes and environmental change. Arnold, London

HIPO (1969) The record (or history) of the impoldering of the Hachirogata. Hachirogata Impoldering Project Office, Ogata Mura (in Japanese)

Minderhout A, Granpré Molière M, Bijhouwer J et al (1955) De Noord-Oost Polder. Forum 10(1/2):1–52

Needham B (2007) Dutch land use planning: planning and managing land use in the Netherlands, the principles and the practice. Sdu Uitgevers, The Hague

Nijhuis S (2016) Polderscapes: the landscape architecture of the Dutch Lowlands. Fengjing Yuanlin (Landsc Archit) 8:38–57. https://doi.org/10.14085/j.fjyl.2016.08.0038.20

Nijhuis S, Bobbink I (2012) Design-related research in landscape architecture. J Des Res 10(4):239–257. https://doi.org/10.1504/jdr.2012.051172

Renes H (2004) Different methods for the protection of cultural landscapes. In: Palang Hannes et al (eds) European rural landscapes: persistence and change in a globalising environment. Kluwer Academic Publishers, Dordrecht, pp 333–344

RIJP (1986–1990) Wording en opbouw van de Noordoostpolder: Geschiedenis van de ontginning en eerste ontwikkeling (1940-ca. 1960) (4 dln.) [Planning and construction of the Noordoostpolder: history of the reclamation and first development] Rijksdienst voor de IJsselmeerpolders, Lelystad

Steenbergen C, Reh W, Nijhuis S, Pouderoijen M (2009) The Polder Atlas of the Netherlands. Pantheon of the Lowlands. THOTH Publishers, Bussum

UNEP-WCMC (2017) Protected area profile for Netherlands from the world database of protected areas. https://www.protectedplanet.net/country/NLD. Accessed 11 Jan 2017

Van der Ven G (2004) Man-made lowlands. History of water management and land reclamation in the Netherlands. International Commission on Irrigation and Drainage, Utrecht

Van der Wal C, Bruggenkamp JWC, Oterdoom D (1992) De ruimtelijke opbouw van de Noordoostpolder. De jongere bouwkunst en stedebouw in de Noordoostpolder in het kader van het Monumenten Inventarisatie Project (MIP) [The spatial construction of the Nooroostpolder]. Ministerie van Verkeer en Waterstaat, Lelystad

Van Woensel J (1999) Nieuwe dorpen op nieuw land. Inrichting van de dorpen in Wieringermeer, Noordoostpolder, Oostelijk en Zuidelijk Flevoland [New villages on new land. The design and construction of villages in Wieringermeer, Noordoostpolder, Oostelijk en Zuidelijk Flevoland]. Uitgeverij de twaalfde provincie, Lelystad

World Heritage List (2017) Unesco World Heritage List: Dutch properties inscribed on the World Heritage List. http://whc.unesco.org/en/statesparties/nl. Accessed 11 Jan 2017

Dr. Steffen Nijhuis is Head of Landscape Architecture Research, Director European Post-master in Urbanism (EMU), and Associate Professor in the Department of Urbanism, Faculty of Architecture, and the Built Environment, Delft University of Technology. His book publications include: *The Polder Atlas of the Netherlands, Delta Urbanism, and Polder Landscapes of the World.* He currently works on landscape-based strategies for urban development.

The path of the polders in the natural reserve Moëze-Oléron in Saint-Froult, Charente-Maritime, France, Wikimedia, KiwiNeko14, released under a Creative Commons Attribution-Share Alike 4.0 International license

Chapter 12
Europolders a European Program on Polder Landscape, Heritage, and Innovation

Hildebrand P. G. de Boer

Abstract Since the twelfth century, polder landscapes have characterized the Netherlands, in particular, but also have appeared across Europe—vast plains reclaimed from water and repurposed for crops and livestock, farmers and rural communities. Over the centuries, urbanization has brought domestic and international visitors seeking leisure activities in these cultural landscapes. But polders have lain mostly in the shade, as it were, of other landscapes, merely a link between hills, dunes, ocean beaches, and historic cities. The Europolders Program emancipates this characteristic landscape, and strengthens prosperity in it, showing it to be an attractive and interesting territory. The Netherlands has a remarkable hydraulic engineering reputation abroad, not only because of work they have done at home—to endless extraction, reclamation and drainage, irrigation projects, dyke, channel, and harbor works—but also because they brought their expertise to the farthest corners of the world. Polders across Europe were shaped or at least influenced by Dutch (Frisian, Hollanders, Zeeuw) and Flemish people. The Europolders Program focuses on developing a European network of polder landscapes with extensive cultural and natural value. It aims to increase the accessibility, visibility, and awareness of historical polder landscapes, water management, and technological innovation, for the benefit of residents and visitors, and to strengthen regional economies.

Keywords Polders · European landscape · European network · Cultural value · Accessibility · Awareness · Public support · Economic value

H. P. G. de Boer (✉)
Dutch Foundation for Industrial Culture, Amsterdam, The Netherlands
e-mail: hildebranddeboer@hotmail.com

C. Hein (ed.), *Adaptive Strategies for Water Heritage*,
https://doi.org/10.1007/978-3-030-00268-8_12

Introduction

After the Early Middle Ages, around 1000 AD, a new period of economic development in the Netherlands led to a population jump, from about 200.000 inhabitants to more than 800.000 inhabitants in 1300 (Malanima 2010, 5–6). The increasing need for living space and nourishment necessitated major land reclamation in the low, western half of the Netherlands. Fen- and peatlands were drained by draining ditches, which caused the soil to subside, so that extensive construction of dykes became necessary, in combination with sluices, windmills, mechanized pumping works, and drains. Dutch experience in water management techniques, gained during many centuries, diffused to the rest of Europe with travelers: Frisian, Dutch, Zeeuw and Flemish farmers, dyke workers, monks, colonists, engineers, contractors, concessionaires, and capital providers took it to Germany, Poland, England, France, Italy, and Ukraine. Among them, the Netherlands and its people played a historically distinctive role in this land reclamation (Danner et al. 2005, 27–29). Today, many polders still host farmers and rural communities, although city dwellers settling in and around the polders increasingly demand leisure activities, in conjunction with domestic and international tourists celebrating this cultural landscape.

A range of distinctive features makes historical polder systems resilient. They offer active control (water management), accessibility by land and water, multifunctionality for residents and visitors, adaptability to current and future needs (from changes in agriculture and urbanization to climate change), and a distinctive identity (natural and cultural values, combination of cultural landscape and wilderness, cooperation between inhabitants). Yet polders lie mostly in the shade of other landscapes, as it were, merely a link between hills, dunes, ocean beaches, and historic cities. The Europolders Program intends to emancipate them as an important European landscape, and as an attractive and interesting territory for domestic and international tourism. The Program focuses on developing a European network of polder landscapes with extensive cultural and natural value. The aim is to expand its accessibility and visibility while benefitting regional economies. It will ensure that each Europolder will link to a regional network of other historical and innovative sites; each will have an Information Point (for example, a Waterways Museum) explaining each important heritage site or new site related to the polder. In short, the Europolder Program intends to provide a comprehensive view of the story of the polder.

Public support from residents and visitors is a prerequisite for sustainable conservation of this natural and cultural heritage, and for the adoption of policy measures protecting it. Local governments and regional water authorities should apply these measures through inclusive development, with constant awareness of the historical genesis of the cultural landscape, its cultural and economic activity.

Historically, citizens and government in most polders periodically collaborated to guarantee responsible water management, such as keeping watercourses and their banks clean for effective water drainage. This collective work was developed in combination with financial levies by the authorities on citizens, to maintain operations—such as dykes and pumping stations—in an operationally responsible manner.

Such physical and financial efforts are still typical of life in polder landscapes. In addition, polder communities, in cooperation with the authorities, can focus on stimulating awareness by making cultural history recognizable to community members and outsiders alike. Initiatives to restore and reallocate polder monuments—including historic mills, pumping stations, quays, farms, and agricultural production buildings—go hand in hand with an increasing number of hospitality facilities for visitors.

The aim is not only to promote day-trip tourism, but also to stimulate residential tourism as a relevant economic factor. This has become a matter of course in a polder like the UNESCO World Heritage Beemster. The intensive exposure of the Beemster, combined with excellent accommodation options, makes it a valued travel destination and starting point for longer stays in the region. This approach in the Beemster and other polder areas that have already been opened up, such as the French Marais de Poitevin or the Holland Fen in England, will work well in the polder areas in the Europolders Program that are not yet fully accessible.

Earlier Regional Networks

The Europolders Program is based on earlier systems of regional networks of industrial and engineering heritage. An early regional network in the Ruhr area, the *Route der Industriekultur* (Route of Industrial Culture) inspired its basic structure, a tourism route connecting heritage networks that share a theme. It was part of the program for the 1989 *Internationale Bau Austellung* (International Architectural Exhibition), and continuing the IBA's history of displaying new concepts and inventions. In turn, the Route sparked the 1999 *European Route of Industrial Heritage (ERIH)*: it linked eighteen regional networks in Germany, Poland, Austria, Spain, UK, Belgium, Luxembourg, the Netherlands; more than 100 ERIH-Anchor Points; and about 1800 other European industrial heritage sites. Several more regional networks are in preparation, including the German–Dutch *EuregioNetwork Industrial Culture* and the *RotterdamDelta Network* for industrial heritage and innovative industries.

One of the ERIH regional networks, the *HollandRoute* in the Amsterdam Metropolitan Area, was officially opened on 1 July, 2011 by the Commissioner of the King. The HollandRoute takes tourists and recreationists to heritage sites of trade, industry, commerce, engineering, agrarian culture, water management and (military) infrastructure in the Province of North-Holland, all in relation to the local polder landscapes. The HollandRoute currently has six ERIH-Anchor Points, or themed attractions, most of them focusing on water heritage in one way or another: the Heineken Experience, the Dutch Maritime Museum, Zaanse Schans (an open-air museum of a Dutch village), Museum Pumping Station De Cruquius, Museum Steam Tram Hoorn-Medemblik, and the Dutch Steam Machine Museum in the pumping house of the "Vier Noorder Koggen" polder district in Medemblik. Dozens of HollandRoute Points are local historical sites, including windmills and pumping stations. Places to eat, drink, and sleep are Rest Points, all located in historic buildings related to industrial heritage and in the polder landscape. Most Anchor Points, Route Points,

Fig. 1 Anchor Point, HollandRoute, Kromhout Shipyard, Amsterdam (Photograph: De Boer);

and Rest Points are along either the main route or theme routes for cycling, hiking, and boating. Water-related attractions on the HollandRoute include historic waterways, locks, bridges, dykes, shipbuilding and other maritime trade sites, military defense structures, fisheries, and harbors (Figs. 1 and 2). The HollandRoute brings a wider range of visitors than many local attractions, because its associated infrastructure is equally distributed across the Metropolitan Region of Amsterdam. The HollandRoute focuses on both domestic leisure and inward tourism.

In 2013, the Dutch developed a plan to extend the HollandRoute with a new network, the *HollandRoute Polderland Network*. A coherent heritage network across Westfriesland, Waterland, Zaan Area, Amsterdam, and Amstel-Meer, it will showcase the comprehensive, cultural, and environmental polder heritage of the western part of the Netherlands. In a few years, visitors will be able to see and learn about the many features of Holland Polderland in their historical and functional context: dykes, ring canals, drainage ditches, land reclamation, windmills, pumping stations, locks, canals, ditches, bridges, fore polders, peat polders, clay polders, peat extraction, polder roads, railways, tramways, land use, farms, agriculture, urbanism, water authority buildings, polder maintenance buildings, and ecology (Figs. 3, 4, 5 and 6). Making the heritage recognizable to a broad audience of citizens and visitors will strengthen their experience of its history.

History is one side of the coin: the other side of the coin is the future progress that determines the sustainable conservation of the heritage, as well as the implementation

Fig. 2 Rest Point, HollandRoute, Kompaszaal, Amsterdam, former arrival and departure hall of the Royal Dutch Steamboat Company (Photograph: De Boer); released under a Creative Commons Attribution-NonCommercial-NoDerivatives 4.0 International License

Fig. 3 HollandRoute Polderland Anchor Points Cruquius Steam Pumping House (Photograph: De Boer); released under a Creative Commons Attribution-NonCommercial-NoDerivatives 4.0 International License

Fig. 4 HollandRoute Polderland Schermer Museum Mill (Photograph: De Boer); released under a Creative Commons Attribution-NonCommercial-NoDerivatives 4.0 International License

Fig. 5 HollandRoute Polderland, Polder Point, Broekerhaven Boat Lift (Photograph: De Boer); released under a Creative Commons Attribution-NonCommercial-NoDerivatives 4.0 International License

Fig. 6 HollandRoute Polderland, Rest Point Pumping House, Amsterdam Flevopark (Photograph: De Boer); released under a Creative Commons Attribution-NonCommercial-NoDerivatives 4.0 International License

of historic values in new interventions. Think of the possible restoration of former watercourses and quays, paths, polder parcels, fruit gardens, trees, or redevelopment of agricultural, industrial and engineering heritage. New interventions might in many cases be unavoidable, because of climate issues, sea level rise, groundwater silting, economic changes in agriculture, food security, and further urbanization. Heritage counts in this future agenda, because it makes recognizable the ongoing historical dynamics of a polder community and the particular ecological conditions of the polder biotope. Even in the midst of progress, a historical perspective should be valued. Only then it will remain manifest for citizens and visitors, in tangible heritage such as the HollandRoute Polderland Network (see photographs of Cruquius Steam Pomp and Schermer Museum Mill).

From HollandRoute Polderland to Europolders Program

In 2014, the Provincial Government of North-Holland and the HollandRoute Foundation concluded that the HollandRoute Polderland Network could be embedded in a wider European context, raising awareness of polder landscapes and their cultural and natural heritage. They foresaw a European network of polder landscapes, with historical connections to the activities of the Dutch (Frisian, Hollanders, Zeeuw) and

Flemish. Such a network would map a number of historical dynamics with emphasis on the period between 1100 (the earliest land reclamation in Northern Germany) and about 1875 (the latest land reclamation and settlement of Mennonite communities in Ukraine) (Danner et al. 2005, 11). It would study the influx of Dutch capital and the use of financial arrangements to finance various water management projects and explore the role of Dutch and Flemish engineers, investors and settlers in a number of major water management projects, acting as intermediaries between the Low Countries and the host countries. The network would track the Dutch presence, or footprints across the landscape, in sluices, canals, mills, dykes, settlements, and other tangible heritage. The countries involved would be the Netherlands, Belgium, France, Italy, UK, Germany, Poland, and Ukraine. This idea is the heart of the emergent Europolders Program.

As in the HollandRoute Polderland Network, the ambition of Europolders is to interlink the involved polder areas not only on the basis of a common history but also the challenges of the future. Those challenges include sustainably conserving the heritage itself, as well as incorporating historic values in new developments to attract the interest of residents and visitors. The Europolder networks reveal unusual new landscapes to both international tourists and urban holidaymakers; several Europolders are historically related to nearby cities, or even to the urban food supply. In most parts of Europe, tourism and recreation are continuous economic growth factors, even though economic crises; a growing interest in cultural-historical tourism and cultural landscapes confirms the strategic value of the Europolders Program for European cooperation (Fig. 7 see map).

Today, an effective agenda for better regulation and innovation of the polders landscapes should not only focus internally on heritage and water issues, but also look outward to market polders to city populations as a cultural and natural outlet. Polder landscapes in urban areas, such as Hamburg, Gdańsk, Bordeaux, Venice, and Amsterdam can do this easily. The Europolders Program can be a building block for such innovation and for improving tourists' access to the culturally and naturally valuable polder landscape.

The Gdańsk (Danzig) and Malbork (Marienburg) Marshlands, in the delta of Vistula and Nogat, have been plagued by floods over the course of history. The floods of 1540 and 1543 made the area unusable and inaccessible for a long time. The nearby city of Gdańsk was growing, and its city council decided to turn the wetland into agricultural land to secure their food supply. This period coincides with the rise of the Mennonites in Friesland. Foreman Menno Simons joined the Swiss Anabaptists in 1531; in 1536, he broke with the Roman Catholic Church and fled with his followers to Groningen and Eastern Friesland, and later to Schleswig-Holstein to escape prosecution. Many of the Mennonites were farmers looking for land to cultivate.

On 28 November, 1547, the Gdańsk city council granted the village and estate of Reichenberg (Rychemberk) to Philippus Edzema, and granted him permission to settle it with people of his nationality (Dutch) (Global Anabaptist Mennonite Encyclopedia Online). This was the beginning of decades of colonization. Between

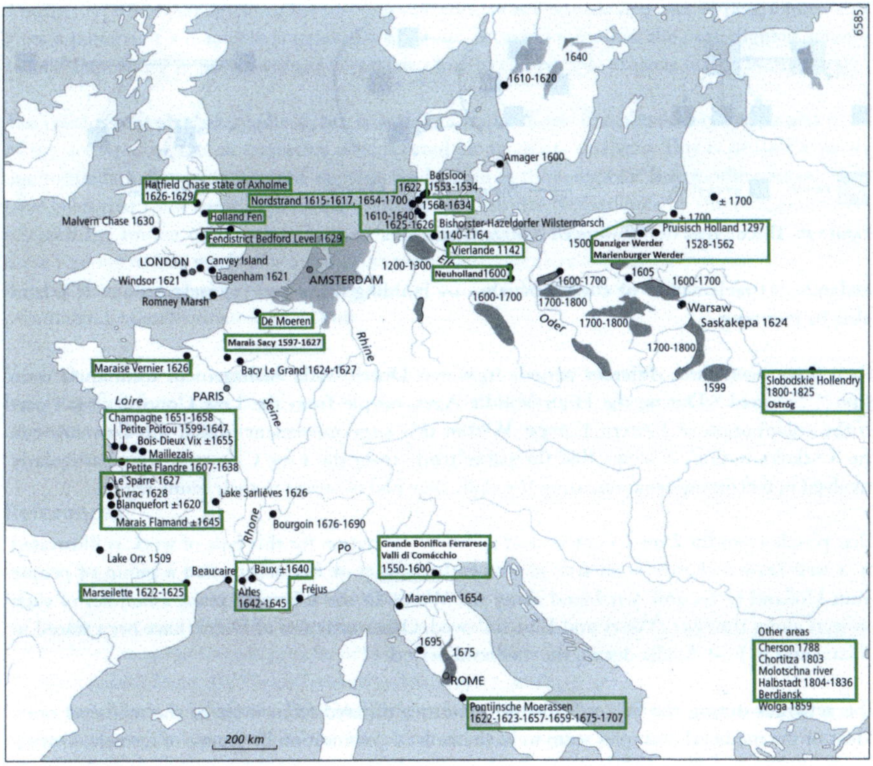

Fig. 7 Locations of the Europolder Areas (within the green lines) on the Indicative List (after: Van Veen 1962, 54). Aspects of Europolders: Example Poland); released under a Creative Commons Attribution-NonCommercial-NoDerivatives 4.0 International License

1547 and 1615 10.000 to 11.000 immigrants arrived in the delta (Dr. Zbigniew Chodyla in Danner et al. 2005, 36). As historian Zbigniew Chodyla later wrote:

> Thanks to the Dutch, the major elements of hydraulic systems were constructed, such as main and local flood control embankments, drainage canals and ditches, dikes and sluices (Fig. 8), ponds, bullock gears, and in particular bucket windmills. In addition, they planted belts of trees and bushes in the fields and introduced braided fences. (Dr. Zbigniew Chodyla in Danner et al. 2005, 42)

In this way, an estimated more than 100.000 ha of agricultural land was developed in the Gdańsk and Malbork area (in all Poland about 255.000 ha) from 1547 to 1800, exploited by the Mennonites as free farmers with the consent of the Polish authorities for grain, cattle, and horses. The colonists built farmhouses and villages with churches and graveyards (Figs. 9, 10 and 11) (Dr. Zbigniew Chodyla in Danner et al. 2005, 37). Some became entrepreneurs in trade and export, or in the processing of agricultural products. Their trade went through the domestic market and export market in Gdańsk.

Fig. 8 Vistula lock at Szkarpawa (Photograph: De Boer); released under a Creative Commons Attribution-NonCommercial-NoDerivatives 4.0 International License

Today, several production buildings are recognizable in the area between Vistula and Nogat, for example, the maltery of the Danziger Aktienbrauerei in Nowy Staw (Neuteich) (Fig. 12), the dairy, and parts of the liquor factory of Heinrich Stobbe in Nowy Dwor Gdański (Tiegenhof). During the nineteenth century, a dense rail infrastructure was established: narrow gauge connections, stations, water towers, and bridges (Fig. 13).

All these tangible elements of heritage can rather easily be used to make the area's fascinating history recognizable to citizens and visitors. The core of this recognition will be the Polder Information Point in the Zulawsky Museum Park, in and around the former dairy at Nowy Dwor, in association with a historical society (*Klub Nowodworski*). From this Point, history can be linked with innovative local developments, with an emphasis on water management and water technology, in association with the Regional Water Management Board Gdánsk (Regionalny Zarząd Gospodarki Wodnej w Gdańsku).

Across the Europolder Program, the historical and innovative aspects of each polder itself, visible in objects and spatial structures, will be part of a regional polder network and connected by different sorts of trails. The Main Connection trail runs between the major attractions of polder heritage. The Main Water Connection offers visitors unexpectedly interesting possibilities in water recreation and provides them easy access to the delta area (Fig. 14). Theme Trails, mainly for hiking and biking,

Fig. 9 Farmhouse in Ostaszewo (Schöneberg, Photograph: De Boer); released under a Creative Commons Attribution-NonCommercial-NoDerivatives 4.0 International License

are attractive routes over structures such as dykes, farmland paths, former railway connections (Fig. 15). Long Distance Trails offer a combination of heritage links (use of the historic infrastructure such as roads and waterways), and packages describing hospitality facilities, such as guides, points of interest, water-sport sites, cafes, restaurants, and accommodation options. Thus, the polder is fully equipped to educate and entertain its residents and growing number of guests, in conjunction with information on websites, apps, print media, and regional and national marketing. In addition, the Europolder network offers information at the Polder Information Points about the historical context of the specific polder within the European network. In this way, the Europolder network is the platform for knowledge exchange between the organizations running the different regional networks, mutually reinforcing the regional polder networks in Europe.

Proposed Europolder Network

The Europolder network will be a partnership among the Netherlands, Belgium, France, Italy, UK, Germany, Poland, and Ukraine. It thus provides a framework for understanding the long interconnected development of European landscapes, and a

Fig. 10 Mennonite church in Cyganek (Tiegerweide, Photograph: De Boer); released under a Creative Commons Attribution-NonCommercial-NoDerivatives 4.0 International License

means for future cultural integration. Dutch professionals were key players in the shaping of European polders. The area in the Province of Noord-Holland (Figs. 16 and 17), for which in 2013 the Master Plan for the *HollandRoute Polderland* was developed (De Boer 2013), will become part of the Europolder network, with its characteristic polders, such as Beemster (1612), a UNESCO world heritage site (Fig. 18), Schermer (1635), and Haarlemmermeer (1852). It will be complemented by the Flemish De Moeren polder (Bert Toussaint in Danner et al. 2005, 141–143).

The Europolder network will include at least two polder complexes from each of these areas (De Boer 2015).

The oldest documented Dutch land reclamation outside the Netherlands took place in the basin of the Weser and Elbe. Polders in France, Italy, Poland, and Ukraine were also built with the support of Dutch engineers.

In 1297, settlers from Holland began draining the area of Preussisch Holland (Paslek) (Fig. 19). In 1547, the first Mennonite settlers came to the Danziger and Marienburger Werder. It was the beginning of 250 years of tremendous Mennonite influence in the region of the rivers Vistula and Nogat (Dr. Zbigniew Chodyla in Danner et al. 2005, 33–55). The French *King Henri IV* (1553–1610) made use of Dutch skills for reclaiming and cultivating wetlands. The Dutch were also involved in reclaiming land from the marshes of the Po Valley (Ferrara, Comácchio, Venetian Territory) in Italy during the sixteenth and seventeenth century. Dutch engineers

Fig. 11 Mennonite cemetery in Stogi (Heubuden, Photograph: De Boer); released under a Creative Commons Attribution-NonCommercial-NoDerivatives 4.0 International License

Gillis van den Houte (Egidio Vandenhoute) and *Everardus van Cortgene* (Everardo Corceine) designed successful drainage concepts (Salvatore Ciriacono in Danner et al. 2005, 151–168). Dutchmen were active in England, especially during the reign of *King James I* (James Charles, 1566–1625) and *King Charles I* (1600–1649) (Dr. Tom Williamson, in: Danner et al. 2005, 103–119). The Holland Fen (Lincolnshire) was drained by several Dutchmen during the seventeenth century. At the end of the eighteenth century, the Industrial Revolution made its appearance in the polder, with the replacement of windwatermills by steam pumping stations and with that the mechanization of water level management.

The area around Chortyzja in Ukaine, populated by Mennonite settlers from 1788 and after, and the area near the Molotschna River, where Mennonite settlers reclaimed land starting in 1803 (Gerlach 2002, 2007) have been identified as future parts of the Europolder network.

The synchronism between Dutch land reclamation activities in and outside the Netherlands is striking, whereby knowledge and experience were exported to areas where more or less similar conditions were expected. From the development of the windwatermill in the fifteenth century to the weather-independent controllable pumping stations of today, Dutch technology has created and maintained perhaps the most artificial type of landscape in Europe. It is also a truly European cultural landscape, due to the continuing international knowledge exchange for management

Fig. 12 Former malt factory at Nowy Staw (Neuteich, Photograph: De Boer);

and security in relation to climate change. The Dutch Union of Water Authorities within and outside Europe also takes an active position in this, both by responding to the interest in the Dutch water board system and in the innovative promotion of sustainable water management.

The Europolders project highlights the polder landscape as a cultural landscape of land reclamation, as a common and living European cultural heritage. It identifies Europe as the first industrial continent, with its industrial and engineering heritage a result of mutual transnational influences. The Europolder network transmits this cultural European dimension to a broad public through transnational cooperation between a growing number of historically valuable and innovatively controlled polder landscapes, as a celebration of our European heritage.

Conclusion

Polders can be considered maritime cultural landscapes, intensively connected with water and influenced by it. A polder is not a natural feature and cannot survive on its own; it requires continuous human maintenance. The attributes that the regional Europolder networks use here make up the basic condition for sustaining the polder

Fig. 13 Narrow gauge railway in Nowy Dwór Gdański (Tiegenhof, Photograph: De Boer); released under a Creative Commons Attribution-NonCommercial-NoDerivatives 4.0 International License

Fig. 14 Main water connection near Drewnica (Schönbaum, Photograph: De Boer); released under a Creative Commons Attribution-NonCommercial-NoDerivatives 4.0 International License

Fig. 15 Quiet trails between Vistula and Nogat (Photograph: De Boer); released under a Creative Commons Attribution-NonCommercial-NoDerivatives 4.0 International License

Fig. 16 UNESCO World Heritage Beemster Polder (1612, Photograph: Stichting Werelderfgoed Nederland); released under a Creative Commons Attribution-NonCommercial-NoDerivatives 4.0 International License

Fig. 17 Schardam, sluice (1592, Photograph: De Boer); released under a Creative Commons Attribution-NonCommercial-NoDerivatives 4.0 International License

Fig. 18 Brandenburg (Germany), Neuholland (Photograph: De Boer); released under a Creative Commons Attribution-NonCommercial-NoDerivatives 4.0 International License

Fig. 19 Dluzyna, near Paslek (Preussisch Holland, Photograph: De Boer);

landscape. The *first (physical) layer* includes historical and current spatial elements that illustrate the story of the formation of the polder. Further, human use of the polder reveals the *second (spatial) layer*, with aspects such as diversity of land use and the built environment. The attributes used here in the course of time make human life in the artificial landscape tangible. The *third (social) layer* concerns the interconnectivity of the polder communities as part of a society, with aspects such as innovation, regulation, economic exchange, mutual connections, and hospitality. The combination of hospitality, heritage, and innovative developments in each polder adds up to substantial value for both the domestic population and external visitors.

The Europolders Program aims to intensify this combination, by creating regional heritage and innovation networks with active stakeholders, and by forming a European platform to further increase the awareness of polder landscapes among a wide audience with many different expectations. The resulting public support will confirm local identities and create conditions for local people to further preserve historical values and making contemporary values recognizable. The Europolders Program is an integrated policy that fosters preservation, transformation, and the adaptive reuse of historic water-related structures to economically strengthen communities and help sustain a unique and historically important landscape type.

References

de Boer HPG (2015) Europolders, a European Programme on Polder Landscape and Tourism. HollandRoute Foundation, Amsterdam
de Boer HPG (2013) HollandRoute Polderland, Landschap, Erfgoed, Toerisme. HollandRoute Foundation, Amsterdam
Danner HS, Renes J, Toussaint B, van de Ven GP, Zeiler FD (2005) Polder Pioneers, The influence of Dutch engineers on water management in Europe 1600–2000. Utrecht
ERIH (2001) The European route of industrial heritage, the master plan. Duisburg
European Route of Industrial Heritage—ERIH e.V.: www.erih.net
Gerlach H (2002) Die Rußlandmennoniten I, Ein Volk unterwegs. Kirchheimbolanden
Gerlach H (2007) Die Rußlandmennoniten II, Westpreußen, Rußland und zurück. Band 2, Kirchheimbolanden
HollandRoute Foundation: www.hollandroute.nl
Malanima P (2010) Energy and population in Europe, the Medieval growth (10th–14th centuries). Institute of Studies on Mediterranean Societies (ISSM)/National Research Council (CNR), Roma
Van Veen J (1962) Dredge, drain, reclaim: the art of a nation. 5th edn. Den Haag

Hildebrand P. G. de Boer studied History of Art and Archeology at the State University of Groningen, the Netherlands. Since 1983, he has worked in heritage conservation, redevelopment, and tourism, focusing on industrial and water-related heritage. Besides his international work for Europa Nostra, ICOMOS, and the International Committee for the Conservation of Industrial Heritage (TICCIH), De Boer is managing director of the Dutch Foundation for Industrial Culture and the founding vice-president of the European Route of Industrial Heritage (ERIH e.V.).

Torenfort Uitermeer, a part of the New Dutch Waterline with a bridge construction, 2008, Bert van As for the Rijksdienst voor het Cultureel Erfgoed, Wikimedia, released under a Creative Commons Attribution-Share Alike 4.0 International license

Chapter 13
Hold the Line: The transformation of the New Dutch Waterline and the Future Possibilities of Heritage

Gerdy Verschuure-Stuip

Abstract The redevelopment of the New Dutch Waterline, also known as the New Hollandic Waterline, was crucial to a change in public appreciation of Dutch military heritage and its connection to landscape design. Starting in 1980, new methods of revitalization combined preservation, renewal, and narrative approaches. At the same time, the work on the New Dutch Waterline changed; a nationally driven project became a series of local interventions. Throughout the effort, it was critical to success to have different actors understand and promote it as a heritage landscape of national importance. The project undertook not only to revitalize individual fortresses, but to enhance regional identity and tourism, a new scale in heritage debates. This chapter shows the importance of understanding and intervening in defense heritage as landscape–as well as individual objects. It also indicates how addressing these different scales can help in future spatial challenges. Finally, it addresses how understanding water heritage can help to tackle the imminent challenge of climate change at the scale of the landscape.

Keywords New Dutch waterline · New Hollandic Waterline · Defense landscape · Heritage management · Landscape planning · Water · Transformation

Introduction

Dutch military heritage is gradually being reused for civilian purposes, as it has lost its active role in military defense and its position and features are no longer classified information. Over the last 20 years, local groups have successfully modernized many military properties, particularly fortresses and bunkers, in order to return them to public use (Fig. 1a, b).

G. Verschuure-Stuip (✉)
Department of Urbanism, Faculty of Architecture and the Built Environment,
Delft University of Technology, Julianalaan 134, NL-2628 BL Delft, The Netherlands
e-mail: g.a.verschuure-stuip@tudelft.nl

C. Hein (ed.), *Adaptive Strategies for Water Heritage*,
https://doi.org/10.1007/978-3-030-00268-8_13

(a)

(b)

Fig. 1 **a** Detail, Honswijk Fortress near Everdingen. Courtesy of the author, (Verschuure-Stuip 2017); released under a Creative Commons Attribution-NonCommercial-NoDerivatives 4.0 International License **b** Honswijk Fortress near Everdingen. Courtesy of the author, (Verschuure-Stuip 2017); released under a Creative Commons Attribution-NonCommercial-NoDerivatives 4.0 International License

Because these transformed structures have been so popular, this modernization is referred to as the renaissance of military heritage in the Netherlands (Hannema 2014). Although the Dutch military past can be regarded as controversial (Verschuure-Stuip 2017), the New Dutch Waterline (NWD) has generally been held in public favor. The name New Dutch Waterline (Nieuwe Hollandse Waterlinie) was introduced and has been used in national and international literature, although in recent years, it has sometimes also been rendered as the New Hollandic Waterline.

The revitalization of properties, and objects such as bunkers and fortresses, as heritage at the scale of the landscape, rather than as individual items was new to the Netherlands. It approached that properties. Almost invisible to an uninformed observer, these landscapes had a military autograph that revitalization made visible in narratives of regional identity.

The integration of strategies inherent in this approach is consistent with current Dutch heritage policy, which holds regional identity to be vitally important to the local and national economy and tourism (Ministerie van OCW 2017). The revitalization of the New Dutch Waterline was first proposed as a pilot project in the Belvedere Memorandum (Feddes 1999), a national interdepartmental policy document which linked heritage to new spatial developments and planning. The main axiom of this influential memo was "renewal through development." Its overarching idea was that heritage should relate to spatial intervention and, for that matter, become a leading factor in it. And policy and professional attention have shifted from object-oriented heritage preservation into development and heritage on a large scale. This chapter discusses how the built military objects were part of large defense lines, as well as how the New Dutch Waterline was transformed at regional and local levels from 1980 on. Further work might profitably address future energy challenges in terms of both objects and landscapes. The account presented here may aid in such an effort.

Large-Scale Military Landscapes

For centuries, it was the duty of the military to protect cities, regions, and even nations against foreign attacks by using characteristics of the landscape such as water, groundworks, soil, and planting to build defense systems. The Netherlands in particular used water defensively, building moats to protect castles and using bodies of water for city fortifications (which featured planted ramparts–massive earthen bodies overgrown with grass and trees). Dikes and roads, were protected by *stellingen*, small-scale military lines. These were often, in turn, part of larger defense structures, which could be as long as 80 kilometers.

For many years, the focus of heritage was at the level of the object. That orientation has recently shifted to include larger structures and their interaction with the landscape (RCE 2009a). This growing knowledge has led to the revitalization of large-scale defense lines such as the Grebbe Line, Maas Line, IJssel Line, Southern Waterline, Peel-Raamstelling, State-Spanish Line, Stelling of Den Helder, Stelling of the Hollandsche Diep and Volkerak, Stelling of the Meuse estuary and Haringvliet, Stelling of the Afsluitdijk, Western Brabant Stelling, and many others (RCE 2009b). A digital map compiled by the State Heritage Department illustrates the variety of defense lines in different periods of time.

Large-scale defense systems can be classified very generally according to the military strategies which employed them as land defense lines or as water defense line*s* (Beek and Kooiman 2004). The largest land defense lines were those that formed the Roman border and the Atlantic wall built along the entire shore of the North Sea during World War II. The most well-known water defense lines are the Old Dutch Waterline (in use 1629–1815), the precursor of the New Dutch Waterline or New Hollandic Waterline (in use 1815–1964), and the Stelling van Amsterdam (built from 1880–1920, decommissioned in 1963). In the last two cases, water was used in the system to prevent the enemy from initial entry; the amount and role of water in each type of defense system varied.

Land defense lines were divided into frontier lines, which controlled an area, and dominance lines, which controlled a specific location in the landscape, such as a road or a dike. The lines also included adjacent fortified cities or fortresses flanked by earthen *schansen*, or casemates (i.e., concrete group shelters) and other military objects. The focus in this system was on the fortified cities or fortresses which were changed and updated many times. Most cities were originally surrounded by brick walls or palisades. The system connected these in order to dominate an area and lakes or swaps were used to direct the enemy to places. In 1672, cities like Bourtange, Coevorden, Zwolle, Deventer, Zutphen, Doesburg, Arnhem, Nijmegen, Grave, Heusden, Geertruidenberg, Bergen op Zoom, Breda, Hulst, and Aardenburg formed the main Dutch defense landline (Will 2002).

During the Dutch Revolt against Philip II of Spain and the Eighty Years' War of 1568 to 1648, cities updated their fortifications with ramparts to withstand the impact of cannon shots. These massive earthen works were created by engineers who used mathematical calculations to produce regularly shaped bastions and courtines, which are still visible in historic inner cities today. They initially surrounded the ramparts with wide, wet ditches to hamper attackers. This system, the *Old-Dutch fortification model,* was improved several times into more ingenious systems of bastions and courtines, with spreads of earthen ramparts to keep the enemy further from the city borders. One updated system was the *Renewed Old Dutch fortification model*; a later one was the *New Dutch fortification model* (Huizinga and Deinema 1994). These different methods of defense can be found under other names in the city patterns of many historic European cities, including UNESCO World Heritage cities.

In short, fortified or frontier cities were the main element of these defense lines, which also used characteristics of the landscape for defense (Huizinga and Deinema 1994). Systems to inundate smaller fields, rivers, and marshland were installed to hinder the enemy, but they also became parts of water transportation, water management, and other non-military infrastructure. These fortification systems were both a blessing and a curse. So when the Dutch Waterline was installed, cities behind that line could start to dismantle their ramparts in order to expand. This dismantling of inner city walls started in 1805 and ended in 1951. In the first period (1805–1813), cities turned ramparts into city parks, as in Haarlem and Leiden. In both the second phase (1854–1874), which started with the Kringenwet, and the third phase (1874–1900), which started with the "Vestingwet" (1874), more cities were allowed to dismantle ramparts. In the fourth phase (1900–1951), the complete system was judged to be outdated and all cities were allowed to dismantle ramparts. Some cities kept their fortification system for cultural-historical reasons or water management issues ('s Hertogenbosch, one such instance, is shown in Fig. 2a–c) (Verschuure-Stuip 2014).

Water defense lines are a typical lowland innovation, based on military experience with the use of water in the sixteenth century. To end the sieges of Brielle (1572) and Alkmaar (1573) during the Revolt against Philip II of Spain, rebels broke the polder dikes of and inundated a polder. Learning from this success, the Prince of Orange, Willem I, ended the long siege of Leiden (1573–1574) by ordering his forces to break the sea dikes near Rotterdam to flood a large part of the province of Holland. These improvised and defensive moves were used by his sons, Maurits of Orange and Frederik Hendrik of Orange, in the introduction of a waterline in 1598, which was installed to block hostile attacks from east of the province of Holland, the economic heart of the Netherlands. However, the plans were not completed at the time. Then, in 1672, the young republic was attacked heavily and the Dutch hastily implemented inundation plans and built the Dutch Waterline (these excluded Utrecht). In valleys and lower parts of the Netherlands, smaller waterlines were created. The seventeenth century Grebbe Linie, located in the lower areas of Gelderland between Soest, Amersfoort, and Woudenberg–and moving up to Wageningen—is one such example.

These water defense lines cleverly combined two systems: a series of adjacent inundation fields to create a vast amount of unbridgeable water and systems of fortresses to defend areas where inundation was not possible: the access points. In peace, the line was almost invisible, as even the fortresses, group shelters, and bunkers were planted with trees and shrubs to hide their existence. During a hostile approach, an ingenious system of waterworks, canals, and sluices would flood these inundation fields to a depth of about fifty centimeters of water to prevent the enemy's entry. This was not deep enough for boats, it was deep enough to hide the ditches of the peat meadow landscape. Soldiers who would try to wade through would fall and drown in their woolen uniforms.

(a)

(b)

Fig. 2 a, b, c Fortification zone of 's Hertogenbosch redevelopment in the last ten years. Courtesy of Verschuure-Stuip 2016; released under a Creative Commons Attribution-Non Commercial-No Derivatives 4.0 International License

(c)

Fig. 2 (continued)

Growing Knowledge of Defense Landscapes

Since 1999, two national policy documents, the Belvedere Memorandum (1999) and the policy letter on the Modernisation of Monuments Care (MoMo) (Ministerie OCW 2009) have stressed the importance of research on large-scale heritage landscapes, including military lines, to understanding regional identity and regional narratives (Ministerie van OCW 2017). In the last few years, Dutch knowledge of these large-scale defense systems has grown rapidly. In 1999, only a few defense lines were identified and mapped in the Belvedere Memorandum (Feddes 1999); that project was completed in 2004 with an inventory of military objects (Beek and Kooiman 2004). In spring 2017, the State Heritage Service (SHS) put a digital map online, classifying defense systems in six periods by time of construction, state organizations, and conflicts (Fig. 3a, b): Spain/Republic (from 1482, different shades of blue), French period (from 1795, yellow), Kingdom of the Netherlands (from 1815, different shades of red), World War I (from 1914, orange), World War II (from 1940, green), and the Cold War (from 1948, purple). Generally, these defense or heritage landscapes have both physical aspects (objects, lines, and areas), and social and mental aspects; they are both tangible and intangible.

(a)

(b)

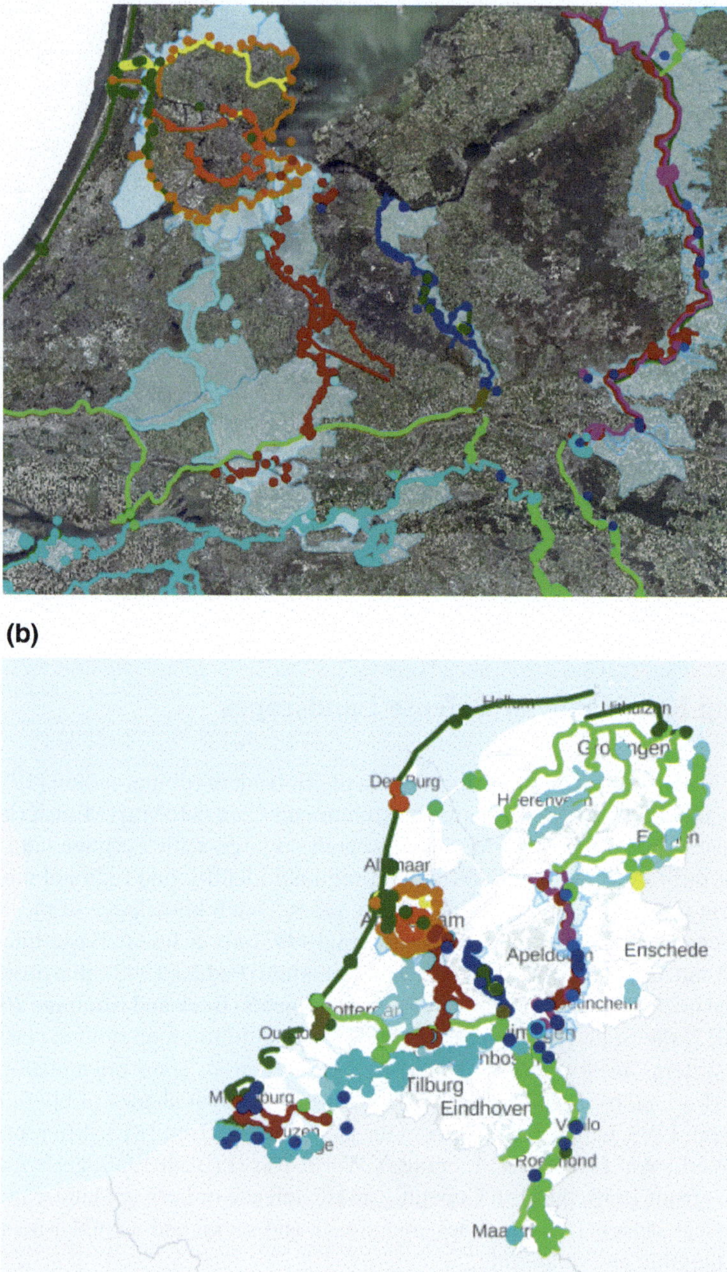

Fig. 3 **a, b** The State Heritage Service created this digital map of all military objects from the sixteenth to the twentieth century. Courtesy SHS 2017; released under a Creative Commons Attribution-Non Commercial-No Derivatives 4.0 International License

Although knowledge of its history and technical use is a good start, it is not enough for reusing this historic line for the future. The challenge of rehabilitating the New Dutch Waterline (NDW) can be summarized: creating public awareness and local participation, growing popular knowledge of the past, securing cultural historic values, embedding the project in various governmental policies, and creating a financial plan and a widely accepted transformation plan (Verschuure-Stuip 2016). The plan needed to map out ways to preserve historic buildings and attract tourism to boost the economy, and to be part of current spatial developments, like water management and tourism. One of the possibilities was to reuse the inundation field for water storage during peak load moments, as part of adapting river landscapes to climate change. The fate of the New Dutch Waterline became a national project because the defense line was spread over four provinces, and several water boards, many municipalities, and other stakeholders were involved. But the most important issue was that the transformation of the New Dutch Waterline was a completely new approach, taking on the heritage of a large landscape.

The New Dutch Waterline was nationally revitalized (1980-now) in a well-documented process of change, from which other heritage development projects could learn. This process of change was described in publications by Raats (2011, 2016), Luiten (2011), and Verschuure-Stuip (2016). The last publication was based on studies in the MSc Landscape elective, Heritage Landscapes (2014–2017) and researched under the name Historic Urban Landscapes in the research group Design & History at Delft University of Technology (Luiten, Verschuure-Stuip 2014).

The New Dutch Waterline

It was the French emperor Napoleon Bonaparte who ordered his Minister of War general Cornelis Krayenhoff and hydraulic engineer Jan Blanken to improve the design of the former Dutch Waterline and to extend it to include Utrecht at the end of the eighteenth century. After Napoleon's defeat, King William I of Orange had his military construct this renewed defense line, the New Dutch Waterline, between 1815 and 1885. The Dutch kept improving the Waterline until the advent of World War II (Brand and Brand 1986; Luiten et al. 2004; Will 2002; Steenbergen et al. 2009; Steenbergen and Van der Zwart 2006; Klinkert 2007).

The New Dutch Waterline ran between the cities of Muiden (the Zuider sea) and Gorinchem (to the tide area Biesbosch) (Fig. 4), and it cleverly used the geomorphology of the landscape: it was situated at the transition of the lower western area and the higher eastern of the Netherlands. It combined two systems. The first system contained a main resistance line, forming the backbone in the defense with a series of adjacent inundation fields plus waterworks (water inlet locks, inundation canals, and dikes) to flood the land as quickly as possible. The second system defended non-

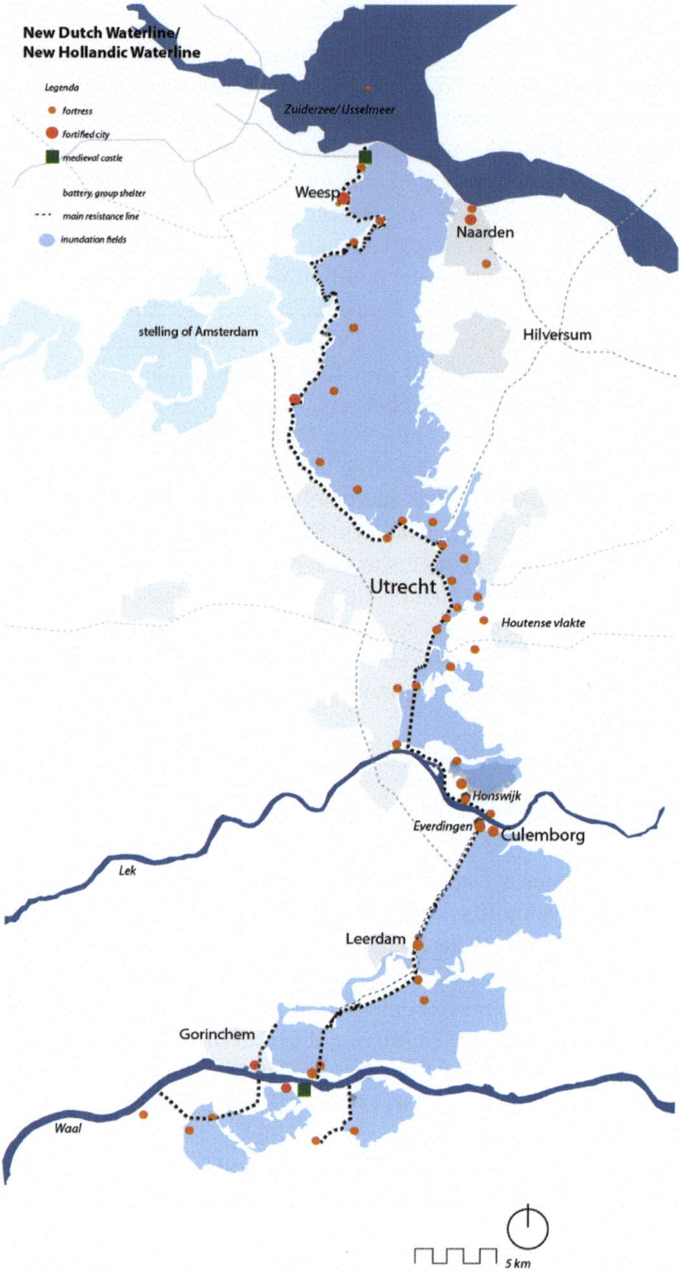

Fig. 4 Outline of the New Dutch Waterline, showing the main resistance line, the inundation fields in times of war and the various fortresses, Werken and Stellingen. Courtesy Verschuure-Stuip 2016; released under a Creative Commons Attribution-NonCommercial-NoDerivatives 4.0 International License

floodable places with military objects: two old castles, several tower fortresses, and seven fortified cities. Due to changes in warfare, this clear outline later incorporated numerous casemates, bunkers for group shelter, and so on. Further military change rendered the defense line useless: during World War II, airplanes simply flew over it. The New Dutch Waterline was officially taken out of national defense in 1961; since then, nature has gradually taken over its fortresses, casemates, and bunkers.

Revitalization of the New Dutch Waterline

The revitalization of the New Dutch Waterline originally focused on landscape planning and restoring objects for tourism and recreation. The shift began in 1980, when renewed attention to the physical remains also triggered the long process of change of these military objects into a large-scale landscape heritage structure (Brand and Brand 1986).

To understand the process of connecting heritage values to future development, the revitalization of the New Dutch Waterline can be divided into six phases (Fig. 5): initiatives (1980–1993), reflection (1993–1997), starting (1997–2003), transition (2003–2008), national implementation (2008–2013), and provincial implementation (2014–future) (Fig. 5) (Verschuure-Stuip 2016).

The first phase started in 1980. At the time, historic buildings were protected under the 1961 Monument laws. Historic buildings were physically reused by preservation and restoration, a field of expertise that was still largely separate from the development of large-scale landscapes and the expansion of cities to fit the needs of a growing population. Some fortresses were named as preserved monuments. This phase included two separate initiatives. First, due to financial cuts, the Department of Defence started to sell military sites in 1980 (Luiten et al. 2004): one fortress went

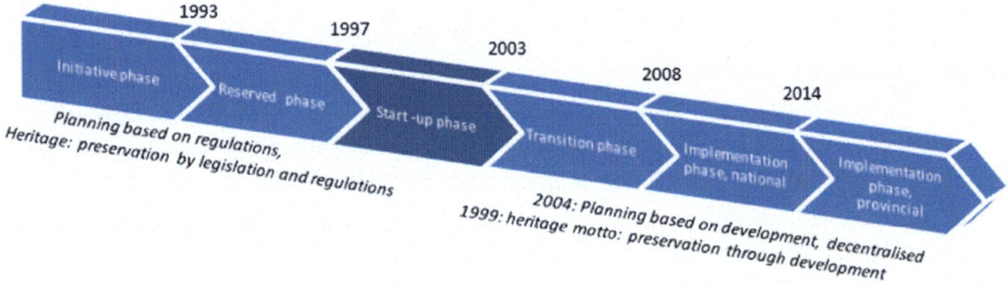

Fig. 5 Six phases in the transformation process of the New Dutch Waterline (1980-now) and changes in spatial planning and heritage management. Courtesy (Nadin et al. 2018); released under a Creative Commons Attribution-NonCommercial-NoDerivatives 4.0 International License

to an individual wine trader, three to the State Forestry Service (*Staatsbosbeheer*), and three to the province of Utrecht. Second, in 1986, an art project and exhibition at Fortress Asperen featured the New Dutch Waterline, then an almost forgotten defense line, and a book was published about its history in 1986 (Brand and Brand 1986). The resulting public awareness was the start of the growing attention to military heritage in the Netherlands. At first, these initiatives were local and small-scale and focused on local use. Then, the province of Utrecht, municipalities, and even ministries became involved through ownership and plans were made which started in 1988 (Raats 2011). In 1990, the planning process came to the national level when the State Heritage Service (SHS), the State Spatial Planning Service, the Ministry of Traffic, Spatial Planning, and Environment, and the Ministry of Agriculture, Nature, and Fishery made joint plans for the future of the New Dutch Waterline. By 1993, three separate, visionary plans had been published, all addressing the natural and cultural aspect of this defense line and highlighting future spatial possibilities; these resulted in governmental attention to cultural aspects of the landscape and national involvement. These plans in turn became part of the Nature Policy Plan on the cultural aspects of the landscape and led to a planning instrument: National (protected) landscapes (Raats 2011). So within only 13 years, an almost forgotten history became a promising multi-disciplinary spatial project (Bosma 2009).

The second phase was one of reflection (1993–1997). Although future plans were presented, which one would think would speed up the transformation into a heritage site, the contrary happened. Due to unclear ministerial responsibilities, the planning project paused at a national level. Meanwhile, however, researchers, experts and amateur researchers within all kinds of organizations, including SHS and historic military foundation Menno Coehoorn (who wrote two books on the subject), continued to collect information on the history and the cultural value of the Waterline: historic overviews, oral histories, drawings, maps, and archival documents. This research on how the line functioned in the past became critically important in the transition phase (Luiten 2011). In 1995, the New Dutch Waterline was placed on the preliminary list of Dutch UNESCO World Heritage Sites for its cultural-historical value (Bosma 2009). A range of actors were involved in this phase: researchers and heritage experts, and attention shifted from the preservation of single objects to understanding the history and cultural value of systems.

In the third phase, revitalization actually started (1997–2003). In 1997, more changes in the ownership of fortresses renewed public interest in revitalizing military objects. The National Government decided to reform the State Forestry Service into an independent organization, and one of their first tasks was renewal plans for its fortresses (Raats 2011). Even more important for our purposes, the 1999 Belvedere Memorandum, highlighted the New Dutch Waterline as a pilot project. Heritage should be preserved by reuse, was its motto, a rather new insight at that time that gave heritage a forward-driven approach (Feddes 1999); and indeed, it accelerated rehabilitation of the New Dutch Waterline. Many professionals were attracted to this new approach.

The transformation was led by the newly funded New Dutch Waterline project bureau, which started to lobby provinces, water boards, municipalities, the public, and landowners for ideas and funding. In 2000, a design competition delivered an overall plan, which became part of the master plan "Line perspective: Panorama Krayenhoff"(Luiten et al. 2004). The basic idea was that the cultural history of this defense landscape would be the "backbone for current and future large-scale spatial challenges" (Luiten, et al. 2004, p 22). To do so, the history of the line was not a set of rules restricting how to preserve. It was presented as an inspiration for urban and landscape quality and for finding methods to connect (local) people to these sites. Designers wrote that it could be developed as a contemporary mega-large-scale green zone, a park for the urbanized western part of the Netherlands, just as nineteenth-century planners turned the earthen ramparts of fortified cities into parks for city folk. Recreation, ecology, and water management could be connected to identity and heritage. And the water system of the historic inundation fields could be used for current water management issues (Luiten et al. 2004). This last value was not actually realized, however.

The master plan was a landscape plan that contained three different maps: a"blue" map (water), "green" map (nature and ecological structures), and "red" map (urbanization and tourism). But it was formulated on a high level of abstraction without pilots, case studies, or in-depth research. This approach encircled the plan with a visionary atmosphere, a positive vibe that inspired many stakeholders, aldermen, and future owners to committed themselves to participating in the revitalization (Raats 2011).

This master plan was finally installed as national policy in 2003. Finances were secured by national funding and through public/private cooperation (PPS). A private party could propose their own plan with matching funding, within a set of clear rules determined by the involved ministries. To ensure the high quality of new plans, a Q-team was installed (Luiten et al 2004). This starting phase was dominated by the state in cooperation with the involved ministries, not by public actors or municipalities.

The fourth phase was one of transition (2003–2008) in which the ideas were worked out in real-life plans, spatially, organizationally, governmentally, and financially. The transformation of the New Dutch Waterline was basically top down, organized by state and provinces and the enthusiastic people of the project bureau in Utrecht and connected to national and provincial spatial policies (Raats 2011). The New Dutch Waterline project bureau divided the entire line into seven areas or envelopes, each with a smaller team of experts. The State Service Rural Landscape (*Dienst Landelijk Gebied*) had the challenging job of getting all of the governmental actors to sign on, including five ministries, five provinces, 25 municipalities, five Water boards, and a fair number of organizations and heritage groups (Colebrander 2009).

For a quick start, it was important to make an iconic design which would attract (inter)national attention (Actieprogramma Ruimte en Cultuur 2005). The cut-through

Fig. 6 Cut-through bunker 599 as part of the water storage and preservation of a bunker landscape. Courtesy Verschuure-Stuip 2014; released under a Creative Commons Attribution-NonCommercial-NoDerivatives 4.0 International License

Bunker 599 served this purpose (Fig. 6). Designed by RAAAF, and initiated by the SSRL and the Culemborg municipality, the design showed a refreshing approach, telling the story of a bunker and also addressing its cultural-historical values and preservation. The closed bunker was cut in half to show its interior and how it was connected by a wooden path to a large pond storing water.

The New Dutch Waterline project bureau focused on economic viability and connection to current large-scale intervention, which was expressed in three main goals: spatial recognizability; the line in head, heart, and hands; and socially and economically sustainable use (PHB NHW 2011).

The fifth phase was the first implementation phase, starting at nationally level but with a growing role for the provinces (2008–2013). The "New implementation agreement Pact van Rijnauwen" was meant to identify the main priorities and organize cooperation between state and provinces (PB NHW 2006). The focus was on the physically reusing military objects and sites in the various envelopes; almost no plans were introduced to address the entire line.

Some funding for the project came from the budget of the New Dutch Waterline bureau but most of it came from the budgets of projects under other national spatial development programs (PB NHW 2011) (Raats 2011). Managers of the New Dutch Waterline discussed how to connect initiatives and budgets, so that both initiatives would benefit (Luiten 2011). For the most part, transformation a top-down process, partly involving municipalities and local actors. In one exception, the Culemborg municipality and a group of local people turned Fortress "Werk aan het Spoel" (Culemborg) into an open-air theater and restaurant.

In the sixth phase, work on the New Dutch Waterline shifted to the provinces (2014–future). Officially, this phase was set for 2012 and 2015, extending to 2020 for new plans. But the national decentralization in spatial planning in the Netherlands, first mentioned in 2004 and implemented gradually, changed this timeline and the revitalization itself. *Government*, or governmental planning, for which the Dutch are famous, changed into *governance*, in which provinces and municipalities as well as local people are planning together (Van der Zande and During 2010). New projects emerged from these new players; plans were more connected to community needs, resulting in growing public commitment in use and maintenance, or "co-creation." In this shift, the provinces Utrecht, Gelderland, North Holland, and North Brabant became responsible for implementing new plans until 2020 (Provinces 2014). In this phase, they revitalized many fortresses, bunkers, community roads, and sluice complexes; preservation methods were combined with modern and robust architecture. An example is the new national New Dutch Waterline museum at fortress Vechten, designed by Anne Holtrop: A ordinary casemate formed the front; the earthen backside was turned into a large patio, with all rooms of the museum were facing it.

In this last phase, the provinces Utrecht, North Holland and Brabant are taking the lead instead of the state. The state will only be involved in the application for monumental status from UNESCO. The New Dutch Waterline will be part of the UNESCO World Heritage site Stelling van Amsterdam.

Challenges for a Sustainable Future

Some of the most important issues of this century—in general and for our built heritage—will be energy transition (and its consequences for our surroundings) and making our landscapes resilient in the face of climate change. Recently, the State Heritage department started to plan to combine heritage with sustainable energy production (Ministerie van OCW 2017). Although plans were initially presented in "Lineperspective Panorama Krayenhoff" to use the inundation fields for occasional high floods, these plans were not realized. Recently, H + N+S landscape architects, and Volharding Breda and RO&AD Architects, based on work in a joint design laboratory, presented ideas for creating renewable energy in the Stelling of Amsterdam and The New Dutch Waterline. Their general idea was to restore the old values of these defense landscapes by creating new values in the fight against climate change. Plans were made on different scales. One of these ideas was to use parts of the inundation fields to store CO2, in order to have room for water storage during peak-level moments and to slow down the oxidation of peat, which causes subsidence. (Peat is a subsoil dehydrated by the digging of ditches in the past and the extraction of water.) Modern "inundation" can help to slow down this process, create a "historic appearance," and make our landscape more resilient. Another idea was to use the cold, wet, and stable inner climate of fortresses for functions which needs cold climates like beer brewing and ICT centers (Dietz et al. 2017).

City temperatures are much higher than those in the rural landscapes, which is called the urban heat island effect (Kleerekoper et al. 2012). In turn, people need their cooling systems more, using more electricity, causing more deaths, and so on. Water around fortified cities and castles, as part of the New Dutch Waterline, might be used to "cool" these urban areas. These flowing historic water bodies with their trees and parks, called green-blue structures, might have a cooling effect during summer and preserve regional identity at the same time.

New green-blue systems have been realized in cities, but the effects of reuse on military heritage need more testing. To optimize their effects, these plans should not be made for every fortress or even city, but used for these large-scale defense lines.

Conclusion

The transformation of the New Dutch Waterline (1980–now) from military large-scale defense line to a line with touristic attention accelerated civilian use of historic military heritage of all types and scales and with all types of methods in the Netherlands. This new attention resulted in research on both large-scale defense lines to objects, and to new interventions on both the regional scale to individual sites. These interventions connected heritage sites to spatial planning, mainly tourism. To tackle future problems, the New Dutch Waterline should also be used to create energy both on the small scale and at the level of the landscape.

The strength of the New Dutch Waterline project that was combining small-scale work on fortresses with plans for large-scale development, although projects at the landscape scale less common than those on individual objects. Individual projects would not have won so much attention when these were only be addressed as objects. Because of its scale, heritage became an important factor. The narrative of this hidden line and the success in the beginning of the transformation was part of the positive awareness.

Moreover, the New Dutch Waterline brought military heritage and industrial heritage, two rather new types of heritage, into fashion for new developments, which combined them with modern architecture. This combination helped the public understand the historical dynamics of larger scale landscapes and acknowledge the value of our military past as part of regional identity.

The next phase, building on the previous shift from a nationally driven process to local participation and co-creation, should tackle one of our greatest challenges: to turn our heritage landscapes into energy landscapes, using the full potential of these large amounts of water (historically used for defensive inundation) and to enhance the green-blue systems around fortified cities and castles. Streaming water, as part of green-blue structures, can help by its cooling capacity. The role of water in our military heritage is changing. In the past, water was used to keep the enemy out. Nowadays, water can help us in our to battle with climate change and energy transition.

References

Actieprogramma Ruimte en Cultuur (2005) Architectuur en Belvedère beleid 2005–2008. SDU uitgeverij, Den Haag

Beek and Kooiman Cultuurhistorie (2004) Het post-militaire landschap 2.0, een overzicht van linies en stellingen in Nederland. Den Haag, pp 9–10

Bosma K (2009) Nieuwe Hollandse Waterlinie, een sculpturaal landschap met een dubbele bodem, in: Steenbergen C, van der Zwart J, Grootens J, Brons R, Colebrander B, Bosma K (eds) Atlas Nieuwe Hollandse Waterlinie, Rotterdam: 010 Publishing, p 55

Brand H (1986) De Hollandse Waterlinie. Veen uitgeverij, Utrecht/Antwerpen

Colebrander B (2009) Inleiding, in: Strategisch Laagland, Atlas Nieuwe Hollandse Waterlinie, Rotterdam: 010 Publishing, pp 84–85

Dietz N, Hugtenburg J, Veul J, Kil A, Koster R, van Ree B, Schouw J (2017) Energielinie, erfgoed in transitie, Amersfoort 2017

Feddes F (1999) Nota Belvedère Beleidsnota over de relatie cultuurhistorie en ruimtelijke inrichting. SDU uitgeverij, Den Haag, pp 6–16

Hannema K (2014) Wedergeboorte van Militair erfgoed, in: ArchitectuurNL 1, pp 18–23

Huizinga S, Deinema J (1994) Vestingsteden. Alphen aan den Rijn, Atrium, pp 3–16

Kleerekoper L, Van Esch M, Baldiri Salcedo T (2012) How to make a city climate-proof, adressing the urban heat island effect. Resour Conserv Recycl 64:30–38

Klinkert W (2007) Water in oorlog, de rol van het water in de militaire geschiedenis van Holland na 1550 In Beukers E (ed) Hollanders en het water, twintig eeuwen strijd en profijt 2. Hilversum: Verloren, pp 486–504

Luiten E (2011) Gereanimeerd erfgoed; Nationaal Project Nieuwe Hollandse Waterlinie als format voor het landschapsbeleid, In: Bulletin KNOB 2011: 6, pp 223–230
Luiten E, Verschuure-Stuip G (2014) Heritage landscapes New Dutch Waterline (Delft University of Technology), Delft
Luiten E, van Hezewijk J, Joosting Bunk E, Witsen PP (2004) Panorama Krayenhoff. Linieperspectief Ruimtelijk Perspectief Nieuwe Hollandse Waterlinie, Utrecht, pp 22–40
Ministerie van OCW (2017) Projectplan Erfgoed telt: naar een toekomstbestendig erfgoedbeleid. Den Haag, pp 1–2
Nadin V, van der Toorn Vrijthoff W, Arjomand Kermani A, Piskorek K, Alewijn N, Mashayekhi A, Verschuure-Stuip G, Brand N, Pendlebury J (2018) The impact of urban planning and governance on the historic built environment (PICH): final report. Delft 2018
PH NHW (2006) Een linie, samen sterk in uitvoering, overkoepelende uitvoeringsprogramma Nationaal landschap Nieuwe Hollandse Waterlinie, Utrecht: projectbureau Nieuwe Hollandse Waterlinie, pp 19–24
Pn NHW (2011) Linie in Bedrijf, ambitie, strategie en uitvoeringsambitie 2020. Projectbureau Nieuwe Hollandse Waterlinie, Utrecht, pp 13–26
Provinces (Gelderland, Noord Brabant, Noord Holland, Utrecht) (2014) Pact van Altena bestuursovereenkomst Nieuwe Hollandse Waterlinie 2014–2020, Utrecht
Raats K (2011) Het gemeenschappelijk offensief voor de Nieuwe Hollandse Waterlinie in de 21ste eeuw, Dillema's bij het samen werken aan de Nieuwe Hollandse Waterlinie. Amsterdam, pp 13–31
Raats K (2016) De herbestemming van de Nieuwe Hollandse Waterlinie, hoe collectieve actie werd bereikt in een geschakeerd landschap. Groningen 2016
RCE (2009a) Aanwijzingsprogramma Nieuwe Hollandse Waterlinie, Amersfoort
RCE (2009b) Erfgoedbalans 2009, Archeologie, monumenten en cultuurlandschap in Nederland, Amersfoort, p 89
RCE (SHS) Landschap in Nederland, kaart van verdedigingswerken. https://landschapinnederland.nl/bronnen-en-kaarten/militaire-landschapskaart. 21 Aug 2017
Steenbergen C, van der Zwart J (2006) Strategisch laagland, Digitale atlas Nieuwe Hollandse waterlinie, Rotterdam: 010 Publishing
Steenbergen C, van der Zwart J, Grootens J, Brons R, Colebrander B, Bosma K (2009) Atlas Nieuwe Hollandse Waterlinie, Rotterdam: 010 Publishing
van de Zande A, During R (2010) Erfgoed en Ruimtelijke planning; beleid, wetenschap, instrumenten en uitvoering. SDU uitgeverij, Den Haag, p 19
Verschuure-Stuip GA (2016) Military brownfields. The New Dutch Waterline. In: Bagaeen S, Clark C (eds) Sustainable transformation of military terrains, Routlegde: Taylor and Francis, pp 143–158
Verschuure-Stuip GA (2017) Musserts' Wall in Lunteren, design approaches for perpetrator-victim heritage in: Atlantis pp 36–41, 28 oct
Verschuure-Stuip GA, Labuhn B (2014) Urbanisation of former city fortifications in the Netherlands 1805–2013. In: Brebbia CA, Clark C, Defence sites, II, Heritage and future, Vol 143, Southampton: Witt press, pp 247–248, pp 245–256
Will C (2002) De Hollandse Waterlinie. Matrijs, Utrecht

Gerdy Verschuure-Stuip is an architect and assistant professor of landscape architecture at Delft University of Technology. She publishes and teaches on both landscape architecture and heritage issues in the Netherlands. Research topics include country estates and their connection to the landscape, planted avenues, and military heritage.

Part IV
River and Coastal Planning

Pilgrims plunge into the water holy Ganges river in the early morning, Wikimedia, Piyush, released under a Creative Commons Share-Alike 4.0 license

Chapter 14
'Absent–Present' Heritage: The Cultural Heritage of Dwelling on the Changjian (Yangtze) River

Andrew M. Law and Xi Chen

Abstract Drawing upon post-structuralist theories of heritage and Derridian theory in particular, this chapter examines the idea of 'absent–present' heritage: heritage that is non-existent, but whose trace remains in the present in a social, memorial and sometimes physical way. Exploring this theoretical approach to heritage, this chapter examines cultures of dwelling on and alongside the Yangtze River. Empirically, it examines the histories and contemporary status of floating fields, sampans and tracking on the Yangtze; treating these social material entities as absent–present heritage, it ends with a discussion of augmented reality as a new digital technology that could allow for the conservation of this heritage. It is also suggested here that such conservation practices might offer space for critical reflection.

Keywords Absent–present heritage · Dwelling · Floating fields · Sampans · Tracking · The Yangtze

Introduction

In the following writing, we view heritage as a series of multiple cultural thematics and/or an assemblage; heritage is a becoming, a constellation of social relations where powerful and powerless agents, quotidian actors and conservation institutions come together to demark particular social and material sites. Heritage is not a given but is the product of a long series of relations and relationships that enable it to exist. This chapter rejects the idea that there is anything like an asocial heritage which exists beyond the meaning that humans give it (beyond time and space). Moreover, by taking a mainly processual ontological stance, we also suggest that what is regarded as heritage is also subject to change over time and space; heritage

A. M. Law (✉)
Newcastle University, Newcastle upon Tyne, England, UK
e-mail: andrew.law@newcastle.ac.uk

X. Chen
Brunel University, London, England, UK
e-mail: Xi.Chen@brunel.ac.uk

C. Hein (ed.), *Adaptive Strategies for Water Heritage*,
https://doi.org/10.1007/978-3-030-00268-8_14

is only heritage for the time being and what is considered heritage at one moment might be disregarded as heritage in another. Drawing upon the material aspects of a theory of heritage developed by Harrison (2013), we also contend that heritage cannot be solely understood as a socially constructed concept; indeed, as Harrison suggests (particularly through the work of Barad 2003, 2007), heritage must be understood as a mutual becoming of meaning (discourse) and materiality (Harrison 2013). From the inscription of ancient buildings to the inscription of Tango on the Intangible World Heritage list, heritage involves materialities (of architecture, of bodies). Concepts of heritage are subject to power relations which select, filter and/or negate multiple discourses of history. The act of demarcating a time and space as 'heritage' is therefore to acknowledge and to give value to particular histories and spaces over others.

This chapter unpacks these theoretical ideas of heritage through the notion of absent–present heritage: heritage which is non-existent, but whose presence remains in the present. Specifically, we explore those cultures of dwelling on and alongside the Changjiang or Yangtze River.

In recent years, a number of articles, papers and books have been written with concerns for the 'loss' of the natural and cultural heritage of the Changjiang or Yangtze River in China (see for instance, Turvey 2008; Zhang et al. 2008; Courtney 2017a); the loss of natural heritage has been associated particularly with the depletion of wildlife within and around the river, including the catastrophic extinction of the Baiji dolphin, which has gained international attention (Turvey 2008). Alongside these debates, scholars have reported on the extensive social and physical damage done to particular parts of the Yangtze as a result of the Three Gorges Dam and other interventions into the environment (see for instance Qing 1998; Murray and Cook 2002; Hvistendahl 2008). Some of this academic commentary has focused on the problematic mass resettlement of people in the excessive reconstruction of the landscapes of the river (Murray and Cook 2002: 103; International Rivers Network report 2003).

In evaluating these various environmental shifts, this chapter supports current calls for the conservation of nature, natural settings and communities within and around the Yangtze. But it also argues that more needs to be done to document and protect the 'absent–present' cultural heritage of the river; drawing upon post-structuralist and Derridian theory in particular, this chapter defines absent–present heritage as that heritage which relates to legacies, ghosts, memories, remembrances and shadows of practices; in other words, this chapter deals with heritage which has ceased to exist in the present, but whose traces haunt the present. Specifically, this writing examines a series of intangible dwelling practices associated with the river. While many of these practices now cease to exist, an absent–present culture and heritage of dwelling remain on and alongside the Yangtze, as many original traditions and practices of dwelling have been reconfigured and re-established within new cultural and economic frameworks. Drawing upon alternative conservation approaches, it is claimed then that more needs to be done to 'conserve' these legacies; however, rather than simply reviving the past, as an act of synthetic nostalgia, it is contended that new forms of digital and augmented reality based conservation (associated with historical texts and particular sites along the river) might allow for new forms of memorial conservation.

A History of Environmental–Human Decline on the Changjiang (Yangtze) River

The environmental problems of the current Yangtze River—including flooding, pollution, the loss of wildlife and the displacement of communities—do not have their roots in contemporary history (although the contemporary moment has certainly exacerbated environmental issues on and alongside the river); rather, the river and its surrounding environment have been subject to climate change and intense human interventions that can be traced back to prehistory and the medieval period in particular (Turvey 2008; Zhang et al. 2008; Courtney 2017a). As Turvey has suggested: 'a medieval economic revolution more than a thousand years ago led to further increase in population growth, deforestation and intensive rice farming across the Yangtze Basin; some cities had by now already reached populations of over a million inhabitants and timber shortages began to be reported from across the region' (Turvey 2008: 19). Turvey notes that the Qing dynasty (1644–1912) saw a 'population explosion during the eighteenth and nineteenth centuries' with the result that many flood plains were converted into agricultural land (Turvey 2008: 19). Thus, by the late imperial period, the Yangtze area suffered increased soil erosion with sediment running off the land and into the river; environmental processes which, in turn, expanded the delta; the area also underwent significant changes to rainfall and water quality (Turvey 2008: 19). Courtney has claimed that the 'ultimate cause[s]' of the great flood of 1931—which claimed an estimated 2 million lives—were 'excessive deforestation, wetland reclamation and the over-extension of river dyke networks' (Courtney 2017a). Thus, longitudinal patterns of human intervention into the Yangtze have had long-term impacts on the environment well into the present. [See the work of Turvey, Courtney and others. Courtney recently asserted that flooding along the Yangtze might also be regarded as a form of heritage in itself (2017b)].

Another cause of great change in the environmental conditions of the Yangtze has been intense deforestation, particularly of the upper Yangtze Basin (Waugh 2003: 293; see also Murray and Cook 2002: 100–104). As Wisner et al (2004) have pointed out, since '1985 it is estimated that forest cover in the Yangtze Basin has fallen by 30 per cent' (Wisner et al 2004: 181). One result is that 'soil erosion has increased, so that silt levels have increased markedly' (Murray and Cook 2002: 100; 104). Cumulative deforestation has been connected to flooding, including the major flood of 1998; as Kram et al. (2012) have contended 'The flood, China's worst in 44 years, drowned more than 4000 people and rendered 14 million homeless… [B]ecause 85% of the Yangtze River Basin had been logged, monsoon rainfalls coursed relatively freely towards the river' (Kram et al. 2012: 15).

As well as flooding, there has been a general shift in the environmental and cultural landscapes of the Yangtze; in her extensive photographic work along the Yangtze (2000–2003), artist and photographer Linda Butler has commented on the extensive 'transformation of the landscape': 'buildings, roads, bridges and whole cities had begun to sprout high on the hillsides. Simultaneously, the old cities and villages were being dismantled and the materials—the floor slabs, rebar and bricks—were reused' (Butler 2004: 2). The growth of industrial pollution, fishing, netting, damming and boating–vessel traffic has had a negative effect on the water quality and biodiversity of the area (Turvey 2008: 35–40; Hays 2009; Dandan 2014; The Guardian 2014; World Wildlife Fund 2017; Water Policy International nd.). Turvey has pointed in particular to the onslaught of 'huge quantities of untreated water, loaded with industrial and agricultural pollutants, sewage and other waste products' which have been continuously pouring into the river (Turvey 2008: 35). Though it is difficult to know when this started, he suggests that by '1985, wastewater emissions along the river totalled almost 130 billion tons' (Turvey 2008: 35). The results have been catastrophic, with the death of marine animals (including the Baiji dolphin) and inestimable risks to human health. As Turvey attests, 'levels of stomach cancer and cancer of the oesophagus, thought to be caused by drinking polluted water and eating polluted Yangtze fish, are on the increase in communities along the river' (Turvey 2008: 205; see also Guang 2010).

But of all the negative human interventions on the Yangtze, the Three Gorges Dam (constructed in the 1990s) has been perhaps the most severe. The Dam has elicited strong reactions from environmentalists, activists, academics and locals living alongside it (Qing 1998; Chetham 2002; Butler 2004; Hvistendahl 2008). Environmentalists have noted that it has increased the potential for landslides, earthquakes and the risk of water-borne diseases (Butler 2004; Hvistendahl 2008). Observers have also pointed to the immense human costs of the project; as Hvistendahl (2008, but writing before the Dam was constructed in 1994) has explained: 'To date, the government has ordered some 1.2 million people in two cities and 116 towns clustered on the banks of the Yangtze to be evacuated to other areas before construction, promising them plots of land and small stipends—in some cases as little as 50 yuan, or $7 a month—as compensation' (Hvistendahl 2008). The development of the dam has also submerged human settlements—at least 1000 cities, villages and towns—as well as farmland and natural resources (Hays 2011).

From Representational to Non-representational Heritage Theory: The Challenge of 'Absent–Present' Heritage

In recent years, post-structuralist inspired theorists have examined the idea of heritage (and conservation) in more processual terms. Rather than debating over how to 'represent' the heritage of people, societies and nature, writers such as Harrison (2013), Hillier (2013) and Pendlebury (2013) have started to view heritage as a becoming and an assemblage. Drawing upon actor–network theory (ANT) (Latour 1996, 2005), assemblage theory (Deleuze and Guattari 1987; De Landa 2006) and the theory of agential realism (associated with Barad 2003, 2007), Harrison (2013) describes heritage 'as a strategic socio-technical and/or bio-political assemblage composed of various people, institutions, apparatuses (*dispositifs*) and the relations between them' (Harrison 2013: 35). Inspired particularly by the theory of agential realism, Harrison also contends that heritage can be profitably understood as a mental (discursive) and material entity; heritage does not only occur 'in the minds of humans… but involves a range of material beings who co-produce heritage as a result of their own affordances or material capabilities' (Harrison 2013: 113).

While post-structuralist assemblage theories of heritage provide a new conceptual lens, more needs to be done to understand the temporal 'presentism' of heritage. Heritage is often defined through notions of what is tangibly or intangibly 'present' in the now (especially in the listing strategies of UNESCO), with less attention paid to the 'heritage' that is absent, or that heritage which has a shadowy status in the present. There is a paucity of research on the way that some forms of heritage haunt the present. The problem with the presentism of heritage is that it often negates the traces and ghostly legacies of some cultural (or material) sites in the quest to identify and delineate other sites that are easily recognisable in the now/present. Here, the concept of hauntology in the work of philosopher Jacques Derrida is particularly interesting; in *Spectres of Marx* (1994), Derrida discusses the haunting of the neo-capitalist and/or neo-liberal moment by the ghosts of Marx. Writing in the early to mid-1990s, Derrida's point here is that despite attempts by interlocutors—such as Francis Fukuyama—and others to discuss the end of history and the triumph of capitalism (see Magnus and Cullenburg 2006), the present is always already haunted by its other (in this respect Marx). Derrida's concept of hauntology has implications for ideas of time, memory and the legacies of history more generally. As Loevlie observes: 'To live is to be haunted. Our "here and now", our material presence, is never stripped, bare or alone. Neither is our subjectivity. We are always caught up in invisible and intangible webs of the past, of the Other, of the future, of death' (Loevlie 2013: 337). In thinking through these issues, theorists of heritage have begun to critically think through the concept of 'absent heritage' and the need to presence the absence (Harrison 2013; Micieli-Voutsinas 2016). Harrison once again has provided insights into some of these theoretical challenges in his suggestion that 'Absent heritage… has developed as a significant global cultural phenomena in which the visual and aesthetic language of heritage conservation is applied to the conservation of voids or absent spaces to maintain an "absent presence"' (Harrison 2013: 169).

Hiller indirectly deliberates over the importance of absent heritage in her discussions of the Newmarket Saleyards and Abattoir in Melbourne, and of the current memorialisation of cows in these sites. She argues for a new form of 'hot heritage' conservation practice that brings life to multiple times and histories and different narratives of the abattoirs. Moving beyond a sanitised and safe memorialisation of the abattoirs, Hillier calls for an alternative politics of remembering which encourages conservationists to represent the 'anguished face of the cows' in their journey towards their slaughter (Hillier 2013: 865). Thus, Hillier contends that we must view heritage 'not as an assemblage of passive objects, but as having a "life" of its own and "characteristics of its own, which we must incorporate into our activities in order to be effective, rather than simply understanding, regulating and neutralizing it from outside"' (Hillier 2013: 868).

Drawing upon these various positions, we use a strategy of remembering to give meaning and indeed value to the past and the remaining and sometimes unacknowledged contemporary practices of human dwelling on and alongside the Yangtze River. While the history and cultures we describe have predominantly disappeared their ghosts and traces exist in the present; these traces deserve to be part of contemporary conservation efforts to protect the 'heritage' of the river.

Although it is literally impossible to cover the long history of the many cultural settlements that have dwelled on the Yangtze, in the historical discussions that follow we draw upon a selection of writers from the nineteenth century, including colonial authors, travellers, academics, geographers and missionaries, all of whom actively sought to chart and document the many activities that took place on the Yangtze. In some instances, we also draw upon Chinese sources, including writing by Chinese historians. We counter hegemonic discourses running through the colonial texts, but explore their interesting and detailed insights into the life of the people and the cities that reside next to the Yangtze. The following historical analysis is therefore highly selective, in order to illustrate a historical culture of dwelling on and alongside the Yangtze.

An early historical commentary from the southern Song dynasty (1127–1279) scholar, administrator and poet, Lu You (1125–1209) documents a culture of living on the river. After a trip to Lushan Mountain, on a tributary of the Yangtze, Lu You made the following notes:

> Moving out into the big river, we met a raft made of wood and measuring over ten or more chang across [a chang is 3.3 m] and over fifty chang long. There were thirty or forty houses on it, complete with wives and children, chickens and dogs, mortars and pestles. Little paths ran back and forth and there was even a shrine—I've never seen anything like it. The boatmen tell me that this is actually a rather small raft. The big ones sometimes have soil spread over the surface and vegetable gardens planted, or wine shops built on them. They are unable to enter the coves but travel on the big river (Lu You quoted in Lynn 1979: 65–66).

In the 1840s, Évariste Huc also refers to these floating rafts or 'floating Islands' in the second volume of his *Journey through the Chinese Empire*:

> We passed several floating islands, those curious productions of Chinese ingenuity, which no other people seemed to have of. These floating islands are enormous rafts, generally constructed of bamboos, which resist the decomposing influence of the water for a long time. Upon the raft is laid a tolerably thick bed of vegetable soil; and thanks to the patient labors of a few families of aquatic agriculturalists, the astonished traveller beholds a whole colony lying on the surface of the water – pretty houses with their gardens, as well as fields and plantations of every sort (Huc 1855: 96)

Zhang, in *Coping with Calamity* (2015), (an environmental history of central China), points to a series of '*Fengtian*' or 'floating fields' (in the Qing dynasty) around the areas of Mianyang, Hanchuan and the 'residents of the lower reaches of the Han

River' (Zhang 2015: 119); in this work, Zhang states that these floating fields or Jiao rafts 'were common in this region and can be understood as a way of dealing with frequent floods' (Zhang 2015: 119). As scholars have pointed out, then, these rafts were common right up until the middle of the twentieth century (Ball 2016: 29). In addition, an enormous number of junk boats or sampans (a mutated expression from the Mandarin san ban, 三板 meaning three planks) and wupans (from the Mandarin Wu ban, 五板 meaning five planks) were not simply a means of transport, but were literally homes for families and sites of economic activity (Chetham 2002: 66–67); as Chetham reports for 'over a thousand years, junks offered a lively variety of services to other boats and towns in remote areas all along the Yangtze' (Chetham 2002: 66). Many of the larger junks were also theatres where 'Actors and musicians performed operas, acrobatics and magic acts' (Chetham 2002: 66); furthermore, junks served as shops, restaurants, tea houses and grocery stores, and some 'sold vegetables and pigs to passing boats' (Chetham 2002: 67). From sing-song sampans, young women serenaded 'hardy boatmen' who spent their money 'in the free and easy way peculiar to the sailing-ship sailors of bygone days' (Chetham 2002: 67 quoting Cornell Plant 1921; see also Ball 2016: 136). Junks came in enormous variety and were connected to local geographic areas along the Yangtze; as the very large tome *The Junks and Sampans of the Yangtze* (1971) (written by a retired river inspector of the Chinese maritime customs) attests, the vessels were based on a 'main structural principle of junk design, which depends for its strength upon a system of bulkheads interspersed with frames or timbers' (Worcester 1971: 29), and often the sails were made from either 'canvas or cloth' (Worcester 1971: 58). Junks were often decorated to suit their owners 'religion, mythology and symbolism' (Worcester 1971: 31). Worcester also points to the sheer craftsmanship embodied in the junk and its navigability (see also Ball 2016: 137) (Fig. 1).

When junk boat crews met rapid waters, they also employed trackers, teams of men (sometimes as many as 200) who took them over the rapids with bamboo lines and harnesses (Worcester 1971: 51). This was tough and dangerous work. Trackers were often directed by the beat of a drum from the junk (Worcester 1971: 52) and as Ball suggests some of the trackers were even whipped by gangers to '"encourage" the exertions' (Ball 2016: 37). [Tracking has a long history and might well have its origins in the Tang era (618–907) or even earlier (see Van Slyke 1988: 121; Ball 2016: 36)] (Fig. 2).

Eastern Sichuan boatmen—boat labourers, rowers and trackers—also developed a rich culture of work songs, or haozi, which Chabrowski identifies 'as tools for their work and a means of group integration' (Chabrowski 2013: 15). The songs allowed the boatmen to coordinate their movements, mentally map the geographies of the river and express their sorrows over their menial work and even their fears of death (Chabrowski 2013).

To be sure, a great deal of the cultural heritage of dwelling on the Yangtze has eroded. While sampans do in fact remain on the Yangtze—Butler (2004:2) points to the continued existence of 'Boatmen in their sampans'—they appear to be very different in terms of their form and function (we will discuss this later). Writing in 1967, Worcester reported that the steamer had a 'great effect on the junk' as did

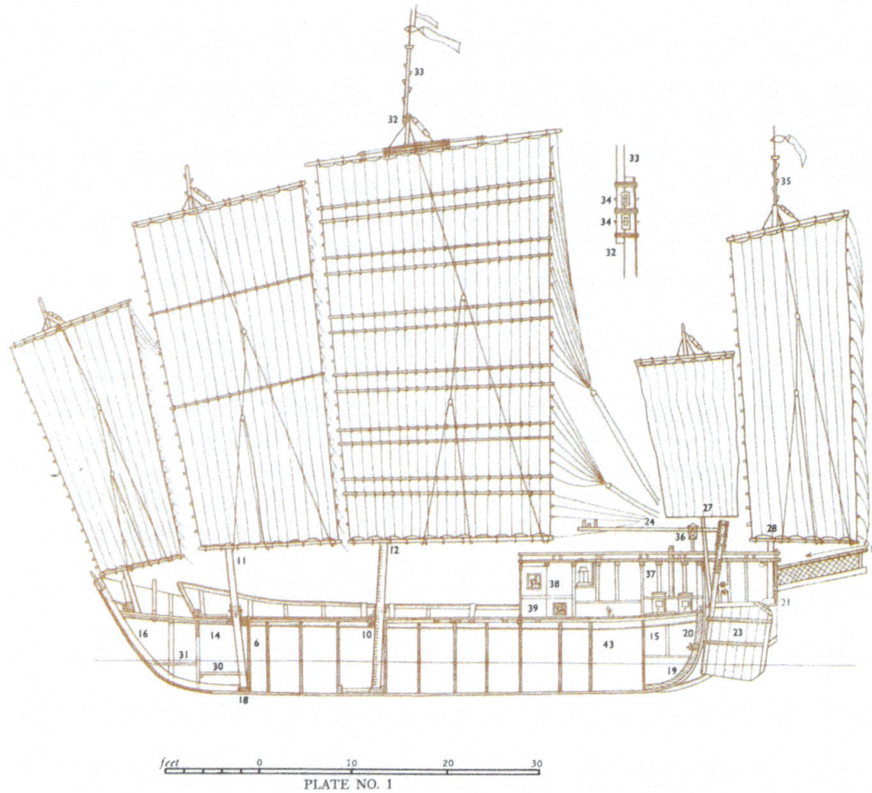

Fig. 1 Picture of one of the designs for a junk from the Yangtze Estuary (and Shanghai area) known as the 'The Sha-Ch'uan or Kiangsu Trader'; Worcester (1971) plate no: 1: page 163; released under a Creative Commons Attribution-NonCommercial-NoDerivatives 4.0 International Licence

a 'large-scale programme of road construction throughout the country, [which]... conveyed the trucks of Mr. Henry Ford and Mr. Dodge' (Worcester 1967, published in 1971). More recently, the Three Gorges Dam has restricted much small shipping in area (Chetham 2002: 249). Many of the practices associated with tracking ceased in the 1950s when the rapids were dynamited, and there is only small evidence of it being practised by locals today. (Ball 2016: 37). In 1998, Ding Qigang reported some evidence for the practice of tracking in the modern era (Qigang 1998: 88). However, commentators such as Butler suggest that tracking probably continued until the 'last of the sailing vessels left the river in 1972' (Butler 2004: 146).

However, in her work *Yangtze Remembered* (2004), Butler depicts a tracker named Mr. Wang (see Fig. 3), who worked on the river between 1936 and 1972. Butler's text hints at the endless loss that has taken place along the Yangtze, but shows that the ghost of tracking still remains. Mr. Wang is said to be able to remember 'the chants the trackers sang to help them pull in unison' (Butler 2004: 19). Later on, Butler suggests that Mr. Wang is the last of his kind, the last remnant of a rich culture

Fig. 2 Image of trackers from Worcester (1971); page 5 entitled 'Hauling a junk over the Yeh T'an. The extra tracking hawser can be seen, as can the sampan which supports the bamboo rope; released under a Creative Commons Attribution-NonCommercial-NoDerivatives 4.0 International Licence'

of practice and dwelling now almost out of memory: 'after he retired and his fellow tracker friends had died, occasionally he rowed himself into the current of the river, singing the old tracking songs and hearing them echo from the cliffs' (Butler 2004: 146).

Here, we now turn from the past to the present; and as well as western travel writing and photography, we also draw upon Chinese sources including talks by Chinese officials, tourist agents and a local cultural photography club. Our own research shows that although the practices of sampan sailing and tracking are no longer the traditional daily work of everyday people, to a large extent these practices have continued for the benefit of tourists. Traditionally, sampans along the Shennong Stream would transport goods daily between Dongba Town on the south bank and Yandu River Town (沿渡河镇) on the north bank branch of the Yangtze River. As the undersecretary of the Dongba Tourism Bureau of Hubei Province has stated, people 'living in the north rel[ied] on the "peapod boat" [a local name for the sampan; see Fig. 4]. They need[ed] industrial goods like salt, cloth, matches and soap from Dongba Town and their agricultural products can also be sold to the south area, such

Fig. 3 Picture of Mr. Wang; used with permission from Butler (2004): plate 6; page 19; released under a Creative Commons Attribution-NonCommercial-NoDerivatives 4.0 International Licence

as corn, pigs and sweet potatoes'. Traditionally, more than 100 sampans worked on this water traffic route (Jiang 2016).

Each peapod boat was about 13 m long and around 2 m wide and took six people to operate: four in the front and two in the rear. When it reached shallow water, the boatmen would get out and tow the boat up the river with bamboo ropes (tracking). After the construction of the Three Gorges Dam, the water rose and the practice of sailing and tracking, to transport goods, ceased. But in recent years, the traditional practices of sailing and transporting goods were replace by a new sampan culture indeed, approximately 600 sampans and 700 trackers (Wuhan Evening Paper 2017) have been serving tourists along the river. Alongside regular tracking displays, at one site members of the Tujia ethnic minority community (one of the 56 ethnic groups said to make up the multi-ethnic Chinese state) perform the tracking naked. Traditionally, the Tujia trackers often worked naked in this region because their clothes would get wet and gain weight as the trackers worked in and alongside the river; specifically, their clothes and the bamboo ropes used to haul the boats would chaff against their skin. Today, nakedness is a tourist attraction: 'naked trackers' have attracted and fascinated visitors and tourists alike, especially foreigners and shutterbugs (Changjiangshangbao 2010). Whilst the authors of this chapter, have not conducted any ethnographic work into the status of the naked trackers, we are cognizant of the possibility that the Tujia ethnic group might be subject to particular kinds of hegemonic ethnic majority discourse. As Xie and others have pointed out, Chinese ethnic minorities have often been subject to a 'Sinocentric' or 'Han gaze'

Fig. 4 Image of a peapod boat. Photograph taken by Jiafa Xiao and used with his permission. Photograph taken 12 September 2010; released under a Creative Commons Attribution-NonCommercial-NoDerivatives 4.0 International Licence

which positions ethnic minorities as exotic and/or primitive (Xie 2011: 101–103; see also Liu 2013) (Fig. 5).

In a travel video created by Chen (2016), the local tour guide Fuyan Zhang, who belongs to the Tujia ethnic minority, explained that the boat workers in the Shennong Stream area now earn their livelihood in three ways: they row the peapod boats and conduct tracking performances for the tourists; they fish and plant; and as the income from tracking performances is relatively low, young people go to the city to work (Chen 2016). Rather than skills and practices that get handed down to the next generation, sampan boating and tracking remain with the older generation. Currently, the oldest sampan boatmen and trackers are more than 80 years old, and the youngest are approximately 50 years old (Jiang 2016). Argubaly the new performative simulacra tourist version of the sampan-tracking heritage of the Shennong Stream is the trace or ghost of something that *was*: a whole way of life that is now lost (Figs. 6 and 7).

Similarly, Chabrowski suggests that tourism has affected the song (song as in music, as opposed to Song dynasty heritage) heritage of eastern Sichuan boatmen: 'the rapidly expanding local tourist industry [has also] participated in returning songs to prominence by making them attractive, accessible and "suitable for" the broader middle-class tastes of the *gaige kaifang* ("reform and opening") generations of consumers' (Chabrowski 2013: 15). At the same time, Chabrowski has noted that their songs have been appropriated by the 'new political centre of Chongqing Municipality (Chongqingshi) in order to construct a distinctive local culture integrated into the

Fig. 5 Naked trackers. Photograph taken by Jiafa Xiao and used with his permission. Photograph taken 12 September 2010; released under a Creative Commons Attribution-NonCommercial-NoDerivatives 4.0 International Licence

framework of a whole encompassing Chinese culture as an "intangible heritage"' (Chabrowski 2013: 15). Ultimately, these initiatives have 'broken the link between the workmen and their culture' (Chabrowski 2013: 15).

Conclusion

Life on the river has changed so much that the simple reconstruction of these practices through touristic endeavours might fail to capture the very varied and intimate habits and ways of living associated with this rich heritage. How then to conserve it? As we shall suggest now, one an augmented reality conservation approach might be one option.

Augmented reality heritage is taking off in a range of settings, from museums to actual heritage sites (Reading 2003; tom Dieck and Jung 2017; Unger and Kvetina 2017), and promises to easily connect people to large historical and memorial resources. Tom Dieck and Jung have defined augmented reality as 'the digital overlay of information on users' immediate surroundings, using devices such as mobile phones or head-mounted displays (HMD) and smart glasses in particular'

Fig. 6 Image of tourists being carried along by the trackers. Photograph taken by Jiafa Xiao and used with his permission. Photograph taken 23 April, 2017; released under a Creative Commons Attribution-NonCommercial-NoDerivatives 4.0 International Licence

Fig. 7 Picture of the naked trackers passing some washerwomen, staged by the Baren Photography Club (巴人摄影俱乐部), a social club that is separate from the local government. Photograph taken by Jiafa Xiao and used with his permission. Photograph taken 12 September 2010; released under a Creative Commons Attribution-NonCommercial-NoDerivatives 4.0 International Licence

(tom Dieck and Jung, 2017: 111; see also Buhalis and Yovcheva 2015). AR browsers allow visitors visiting heritage spaces to use digital devices such as smart phones and/or tablets to access visual or written information; AR markers, actual objects or images in the real world, then trigger 3D virtual information on digital devices.

One of the earliest cultural heritage sites helped by this technology was the ancient temple of Olympia in Greece, where researchers formulated the ArcheoGuide AR system (Buhalis and Yovcheva 2015). Today, the digital technology company Moptil (working in partnership with academics) offers tablets to visitors to the site (and a range of other sites including the Acropolis, Delphi, Knossos, the Asclepieion in Kos and Lindos), allowing them to experience a 3D panorama of an assortment of reconstructed remains and imagined ancient figures (virtual avatars) associated with the space. Their website boasts that 'users have the opportunity to see in Real Time ancient Greeks walking and talking. Also interiors of temples are enriched with human activity in order to reveal the use of every temple' (Moptil 2017). In the Yangtze area, these technologies could be used to offer historical information at a number of sites, or AR 3D display markers could be placed in a selection of real-world sites along the river (and maybe actually on the river) to convey augmented images of floating fields, sampans and trackers. Likewise this technology could be used to generate images of towns, river banks or landscape lost to industrial development.

Such technologies might also be useful in strategies aimed at getting locals, visitors and perhaps even policy makers to reflect on the deep cultural histories, crafts, morphologies and practices that have always already haunted the river and the people who have lived on and adjacent to it. Such technologies might therefore allow for both hot experiences and critical and political reflection. But ultimately these technologies, which could allow us to view spectral images of the past, also hint at a way of representing a heritage that is absent, but whose ghostly trace haunts the present.

Acknowledgements The authors would like to thank the Naval Institute Press, Linda Butler and Jiafa Xiao for allowing us to use the images in this writing; our thanks also to Chris Courtney for his critique of this paper at the conference: 'Risks, to Life, heritage and Community on the Yangtze River' in Newcastle, 4–6 December 2017.

References

Ball Philip (2016) The water kingdom, a secret history of China. The Bodley Head, London

Barad K (2003) Post humanist performativity: toward an understanding of how matter comes to matter. Signs: J Women Cult Soc 28(3):801–831. https://doi.org/10.1086/345321

Barad Karen (2007) Meeting the universe half-way: quantum physics and the entanglement of matter and meaning. Duke University Press, Durham and London

Buhalis D, Yovcheva Z (2015) Augmented reality in tourism: 10 unique applications explained. Digital Tourism Think Tank, Bournemouth University, Dorset

Butler L (With a foreword by Winchester, S) (2004) Yangtze remembered: the river beneath the lake. Photographs and text by Butler L. Stanford University Press, Stanford California

Chabrowski II (2013) "Tied to a boat by the sound of a gong": world, work and society seen through the work songs of Sichuan boatmen (1880s, 1930s). Ph.D. thesis, European University Institute. https://doi.org/10.2870/13065
Changjiangshangbao (2010) People [ONLINE]. Available at: http://env.people.com.cn/GB/11099545.html. Accessed 18 Dec 2017
Chen S (2016) Chanjiang Sanxia Ziyouxing Wuxia Shennong Xi (長江三峽自由行 巫峽 神農溪). [Online Video] 21 April 2014. https://www.youtube.com/watch?v=8dAkUeR-ef8. Accessed 23 May 2017
Chetham Deidre (2002) Before the deluge, the vanishing world of the Yangtze's three gorges. Palgrave, Basingstoke
Courtney C (2017a) Central China Flood, 1931. http://www.disasterhistory.org/central-china-flood-1931#_ftn4. Accessed 23 May 2017
Courtney C (2017b) The heritage of flooding in the Yangtze Valley. In: Unpublished paper delivered to 'Risks to life, heritage and Community on the Yangtze'; international 3 day conference, held at the Confucius Institute, Newcastle University (4–6 Dec 2017)
Dandan N (2014) The taste of Extinction. Global Times Published, 25 May 2014. http://www.globaltimes.cn/content/862072.shtml. Accessed 17 Dec 2017
De Landa M (2006) A new philosophy of society: assemblage theory and social complexity. Continuum, London
Deleuze G, Guattari F (1987) A Thousand plateaus, capitalism and schizophrenia. Continuum, London
Derrida J (1994, 2006) Spectres of Marx. Routledge, London (new edition)
Guang L (2010) 'The story of fishing villages in Jiangsu province, China; Bearing witness to the human cost of water pollution', Greenpeace International. http://www.greenpeace.org/international/en/news/features/Chinese-river-fishing-community/. Accessed 17 Dec 2017
Hays J (2009) Yangtze River. http://factsanddetails.com/china/cat15/sub99/item460.html. Accessed 17 Dec 2017
Hays J (2011) Three gorges dam: benefits, problems and costs. http://factsanddetails.com/china/cat13/sub85/item1046.html. Accessed 23 May 2017
Harrison R (2013) Heritage critical approaches. Routledge, London
Hillier Jean (2013) More than meat: rediscovering the Cow Beneath the face in urban heritage practice. Environ Plan D: Soc Space 31(5):863–878
Huc ÉR (1855) Journey through the Chinese empire, vol 2. Harper, New York
Hvistendahl M (2008) China's three Gorges Dam: an environmental catastrophe? Even the Chinese government suspects the massive dam may cause significant environmental damage, available from Scientific American. http://www.scientificamerican.com/article/chinas-three-gorges-dam-disaster/. Accessed 23 May 2017
International Rivers Network report (2003, January) Human rights dammed off at the Three Gorges, an investigation of resettlement and human rights in the Three Gorges Dam project. A report published by the International Rivers Network, Berkeley
Jiang Q (2016) Chinanews. http://www.chinanews.com/cul/2016/12-20/8099790.shtml. Accessed 23 May 2017
Kram M, Bedford C, Durnin M, Luo Y, Rokpelnis K, Roth B, Smith N, Wang Y, Yu G, Yu Q, Zhao X (2012) Protecting China's biodiversity: a guide to land use, land tenure, and land protection tools (Smith N (ed)). The Nature Conservancy, Beijing
Latour B (1996) On actor network theory: a few clarifications. Soziale Welt 47. Jahrg., H. 4(1996):369–381
Latour B (2005) Reassembling the social: an introduction to actor-network theory. Oxford University Press, Oxford
Liu T-K (2013) Re-constructing cultural heritage and imagining wa primitiveness in the China/Myanmar Borderlands. In: Blumenfield T, Silverman H (eds) Cultural heritage politics in China. Springer, New York, NY

Loevlie EM (2013) Faith in the ghosts of literature. Poetic hauntology in Derrida, Blanchot and Morrison's Beloved. Religions 4:336–350

Magnus B, Cullenburg S (2006) Editor's introduction, in Spectres of Marx, Derrida, J. vii–xii. Routledge, London (new edition)

Micieli-Voutsinas J (2016) An absent presence: affective heritage at the National September 11th Memorial & Museum, Emotion, Space and Society. https://doi.org/10.1016/j.emospa.2016.09.005

Moptil (2017) Olympia 3D–with augmented and virtual reality.http://moptil.com/sites_olympia/. Accessed 8 Feb 2018

Murray G, Cook IG (2002) Green China, seeking ecological alternatives. Routledge, London

Pendlebury J (2013) Conservation values, the authorised heritage discourse and the conservation-planning assemblage. Int J Heritage Stud 19(7):709–727

Plant Cornell (1921) Glimpses of the Yangtze Gorges. Kelly and Walsh, Shanghai

Qigang D (1998) What are the three gorges resettlers thinking? In: Qing D (ed) The river dragon has come! The three gorges Dam and the fate of China's Yangtze River and its people. M. E. Sharpe, London

Qing Dai (ed) (1998) The River Dragon has Come! The three gorges Dam and the fate of China's Yangtze River and its people. M. E. Sharpe, London

Reading Anna (2003) Digital interactivity in public memory institutions: the uses of new technologies in Holocaust museums. Media Cult Soc 25(1):67–85

The Guardian (2014) Wild Chinese sturgeon on brink of extinction in polluted Yangtze. https://www.theguardian.com/world/2014/sep/15/wild-chinese-sturgeon-on-brink-of-extinction-in-polluted-yangtze. Accessed 17 Dec 2017

tom Dieck MC, Jung TH (2017) Value of augmented reality at cultural heritage sites. J Destination Mark Manag 6:110–117

Turvey S (2008) Witness to extinction how we failed to save the Yangtze River Dolphin. Oxford University Press, Oxford

Unger J, Kvetina P (2017) An on-site presentation of invisible prehistoric landscapes. Internet Archaeol 43. https://doi.org/10.11141/ia.43.13

Van Slyke LP (1988) Yangtze, nature, history and the river. Addison Wesley Publishing Company, Reading

Waugh D (1998, 2003) The new wider world. Nelson Thornes, Cheltenham

Water Policy International (nd) Yangtze River. http://www.thewaterpage.com/yangtze.htm. Accessed 17 Dec 2017

Wisner B, Blaikie PM, Cannon T, Davis I (2004) At risk: natural hazards, people's vulnerability and disasters, 2nd edn. Routledge, London

Worcester GRG (1971) The junks and sampans of the Yangtze. Naval Institute press, Annapolis, Maryland

World Wildlife Fund (2017) The Yangtze. https://www.wwf.org.uk/where-we-work/places/yangtze. Accessed 17 Dec 2017

Wuhan Evening Paper (2010) Chinanews [ONLINE]. Available at: http://www.chinanews.com/cul/news/2010/04-01/2202594.shtml. Accessed 23 May 2017

Xie PF (2011) Authenticating ethnic tourism. Channel View, Bristol

You L (Date unknown) '14th May' translated by Watson B, 1979 in 1997. In: Madeleine L (ed) Yangtze River: the wildest, wickedest river on earth. Oxford University Press, Oxford

Zhang J (2015) Coping with calamity: environmental change and peasant response in Central China, 1736–1949. University of British Columbia Press, Vancouver

Zhang Q, Gemmer M, Chen J (2008) Climate changes and flood/drought risk in the Yangtze Delta, China, during the past millennium. Q Int 176–177:62–69

Andrew M. Law is Senior Lecturer in Town Planning in the School of Architecture, Planning and Landscape, Newcastle University, United Kingdom. He has published several books, including *Tudoresque* (Reaktion Books) with Professor Andrew Ballantyne, and articles in a range of journals. His recent research includes the use and abuse of history, historical imaginaries, nostalgia and heritage in China.

Dr. Xi Chen is a Postdoctoral Research Assistant at Brunel University. He has been teaching and practicing alternative architectural research across the UK and China. His research interests lie in rethinking cities and citizenship through participatory design and critical spatial practice. He is keen on exploring the role of the architect/designer in the building of collaborative networks across the professions, government, academia and the public.

The Zhujiajiao water town, China, 2009, Wikimedia, Anna Frodesiak, Public Domain

Chapter 15
Neglected and Undervalued Cultural Heritage: Waterfronts and Riverbanks of Alblasserwaard, The Netherlands

Arie den Boer

> *One of the best things about water is the look and feel of it.... It's not right to put water before people and then keep them away from it.*
>
> —William H. Whyte (1975).

Abstract Alblasserdam is a Dutch dyke village dating to the thirteenth century, with its earliest houses built along the embankment of a major dyke. Most of its history is closely related to shipping and shipbuilding. The village center had a harbor for inland ships and a navigation lock; in the hinterland, industrialization created several yards where workers built many types of vessels: simple wooden rowing boats, wooden ships, and steel ships. Other yards related to shipbuilding—a steel mill, a construction yard for railway infrastructure—rose there too. The material used in these yards came in by ship and ferry. Today, the local ferry has been replaced by bridges and a tunnel. Sites once hosting major shipyard now hold housing. Halls that were used to build minesweepers for the Dutch Navy are now used for building beautiful yachts. Cranes and old buildings alike have disappeared, and new areas have become available for redevelopment. Five objects on the waterfront of the Noord exemplify the connections of history to possible transformations in the future, and the question of safeguarding the area's history as cultural heritage: the site of the Nedstal steel factor, the historic bridge (and an art installation proposed for it), the shipyard of van de Giessen de Noord, the Oude Werf yard, and the Mercon Kloos site. Two citizen initiatives seek to restore and manage cultural heritage in the Alblasserwaard and the river Noord. The analysis shows that cultural heritage has gotten more attention from public and private stakeholders and civil society over time.

Keywords Citizen participation · Water heritage · Public space · Riverfront · River art · Alblasserwaard

---

A. den Boer (✉)
Erasmus University Rotterdam, Rotterdam, The Netherlands
e-mail: arie.den.boer@planet.nl

C. Hein (ed.), *Adaptive Strategies for Water Heritage*,
https://doi.org/10.1007/978-3-030-00268-8_15

Introduction

Until 1977, urban riverfront areas around the world hosted factories, industrial plants, warehouses, harbor facilities, and railways. Typically, they were significantly separated from the surrounding urban fabric (Estevens 2005). But with the evolution of maritime technology and of post-industrial cities, port cities gradually lessened their dependence on port activities. Modern shipping required more space, deeper water, and more extensive port facilities. Riverfronts lost their port functions. Today, seafronts, riverfronts, and lakefronts that were once the industrial birthplace of many cities have become prime locations for new development. Many reimagine a working waterfront as a multi-use public gathering place, and a golden opportunity for a city to redefine itself (Whyte 1975). During recent decades, port cities have restructured themselves and their waterfronts, and many scholars and practitioners have written about riverfront improvement, regeneration, and redevelopment in many cities (Harms 2008). The HafenCity project in Hamburg is an interesting example of possibilities for waterfront development (HafenCity Hamburg 2017).

Along with the city's Kontorhausviertel (or "Trading House District") and Chilehaus ("House of Chile"), the Speicherstadt, a historic warehouse district, has been included on the list of UNESCO World Heritage Sites and is a must-see on every journey to Hamburg (HafenCity Hamburg 2000). The planning for HafenCity was guided by several goals: to reunite the city with the waterfront; to revitalize the waterfront to create jobs and revenue; to benefit port, city, and state; to provide parks, plazas, walkways, and public open space at the water's edge; and to respect the historic character of the waterfront (Harms 2007).

In contrast, Alblasserdam is a neglected and underdevalued site of cultural heritage. It is one of the landscapes surrounded by water in the Rhine-Meuse-Scheldt delta, the Alblasserwaard, and it lies along the tidal river Noord on the west side of the water between Lek and Beneden-Merwede from Rotterdam to Dordrecht. This paper presents an overview of historical and heritage-related narratives of its waterfront, of current debates over its fate, and some future options. It aims to show the complexity of heritage, with its mixture of historical values, economic considerations, safety issues, river art development, and much more. As such, this paper does not claim to offer a clear example of how to do things in heritage policy, nor an analysis deeply anchored in theories of heritage and change. It is a simple but important message is that heritage policies and debates are inherently messy; many agents with many claims and interests discuss what they consider to be valid steps to take when developing an area. This article also shows the complexity of the decision-making process and concludes by considering how various models of negotiation might take the analysis further.

The Alblasserwaard as Focus of Waterfront Heritage

Most tourists to the Netherlands have heard of the canals in Amsterdam, the capital of the Netherlands, or the riverbanks of Rotterdam, one of the largest harbors in the world. Not so many, however, have heard about the Alblasserwaard and its dyke ring, which nature, water management, building, and industrialization have shaped since 1277. Its embankments protect the Alblasserwaard from inundation both from water from the sea at high tides and the flooding of the rivers Rhine and Meuse from Switzerland, Germany, and France; the local saying is that the 'water-wolf could strike from two directions (Louwe Kooijmans 1974). Complaints about the Alblasserwaard and Alblasserdam being neglected go a long way back; Boersma said that no writing had left the printing press that described in detail the history of the dyke city that stretched along the Noord (Boersma 1939). But today, a company called DEAL! Drecht Cities, the central marketing and promotion agency of the six Drecht Cities, is finally promoting Alblasserdam (Deal Drecht Cities 2018).

Alblasserdam is the main village on the west side of the river Noord, and it celebrated its 700th anniversary in 1999. The publication "Alblasserdam 1299–1999," written under supervision of the municipality, says that 1299 may actually be a conservative estimate, as it is the date of a document that describes founding father Nicolaas van Souburgh as already in the region (De Reus 1998).

Alblasserdam is named after the peat river Alblas and the dam that was built where the Alblas connected to the Noord, a larger river (De Wit 2009). Originally, the dam had three culverts to control the water level in the lower-lying ward. At the end of sixteenth century, these culverts were converted to a navigation lock with wooden point doors. The city grew on the dyke and in the center of the dyke ring. On May 10, 1940, the center of Alblasserdam was bombed and destroyed; the townspeople rebuilt it in the Delft School style, a movement that promoted architectural traditionalism in between 1925 and 1955.

Most of the city's history is closely related to shipping and shipbuilding. The village center had a harbor for inland ships and a navigation lock; in the hinterland, industrialization created several yards where workers built many types of vessels: simple wooden rowing boats, wooden ships, and steel ships. Other yards related to shipbuilding—a steel mill, a construction yard for railway infrastructure—rose there too. The material used in these yards came in by ship and ferry.

Recently, these yards have seen huge changes. The steel factory Nedstaal went bankrupt and the site of Mercon Kloos, a railway infrastructure company, has been acquired by a project developer, Whoonapart (Poldervaart 2017a, b). The van der Giessen de Noord shipyard is now a shipyard called oceAnco. The land of the Oude Werf yard has been used for houses and apartments. A location on the riverbank is marked for a jetty to be used by the waterbus, which connects Dordrecht with Rotterdam with Papendrecht, Hendrik-Ido-Amacht- Alblasserdam, Ridderkerk, and Krimpen aan de IJssel.

Alblasserwaard is radically changing, raising many questions about what to do with the areas that have changed, that will change, and that might well need to change.

Five objects on the waterfront of the Noord exemplify the connections of history to possible transformations in the future, and the question of safeguarding the area's history as cultural heritage: the site of the Nedstaal steel factor, the historic bridge (and an art installation proposed for it), the shipyard of van de Giessen de Noord, the Oude Werf yard, and the Mercon Kloos site. Stakeholders are discussing—and thus shaping—the town's present and the future, with current views becoming history over time and cultural heritage standing as witness and proof of the past. In our participatory society, stakeholders use their citizenship networks to put up a vote, a voice, or an initiative to give direction to—or at least influence—the decision making process on the transformation of the urbanized riverbanks in the region.

The River Noord and Alblasserdam

In 1277, the dyke ring of the Alblasserwaard (Fig. 1) was fitted with locks at Kinderdijk/Elshout, and Alblasserdam to be able to control the water level in the Waard's polders: Polder Blokweer, Polder Het Nieuwland, Polder Kortland, Polder Souburg, and Vinkenpolder. Farming and fishing villages like Alblasserwaard rose around the locks, later adding shipbuilding and iron works. Transport overland was restricted to a small network of local tracks until the nineteenth century. Rivers had to be crossed with a ferry, as fixed crossings were not created until the twentieth century (Van Groningen 1992). The bridge over the river De Noord replaced the ferry between Alblasserdam and Oostendam in 1939, part of a 1927 national road plan (rijkswegenplan) (Autosnelwegen 2018). Originally, that bridge was part of highway A15, but with increased traffic pressure it became a bottleneck. A tunnel under the river at the same crossing was the solution, and in 1992, the Noordtunnel was inaugurated. The original bridge now carries local traffic and hazardous materials not allowed through the tunnel. It is still an icon of Alblasserdam and the river Noord and is clearly visible from land, water, and air.

Shipyards were built on both sides of the Alblasserdam harbor. Behind the earliest yard (De Oude Werf) and to the north of the village, a second yard (Mercon Kloos) for railway infrastructure material was built in 1843. In 1936, a steel mill (Nedstaal) was built on the existing yard of Van de Giessen de Noord on the south side. Every shipyard has its own history, written by amateur historians.

In 1950, the Oude Werf was sold to the shipbuilding magnate Cornelis Verolme. Pictures from that year show a shipyard that still has a saw mill. He continued the company, renaming it Verolme Shipyard Alblasserdam (V. S. A.) in 1957. Verolme's redevelopment approach was economically driven, without much attention to the past or heritage. Both intangible and tangible assets of the past disappeared during his tenure. For example, in March 1952, the log pond was filled to become a plot for redevelopment, and later that same year, the sawmill was removed for the same spatial development plan (Alblaserdam net 2018). De Hollandse Molen foundation had been afraid that the sawmill would fall victim to industrialization and luckily found a municipality that was interested in it; the dismantled sawmill, named "Ons

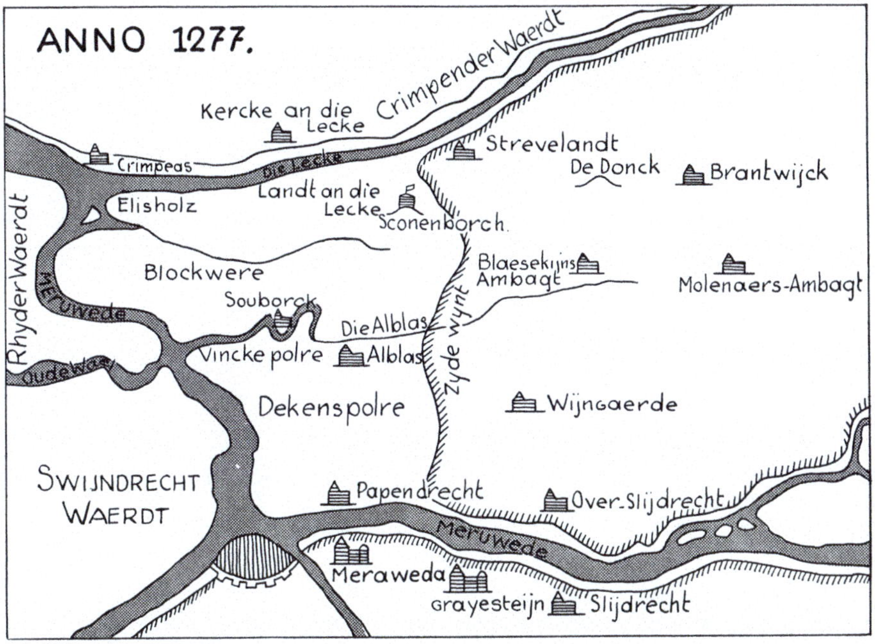

Fig. 1 Alblasserwaard in 1277, with the dyke ring at Zyde wynt (Zijdeweg) (Boersma 1939); released under a Creative Commons Attribution-NonCommercial-NoDerivatives 4.0 International License

Genoegen" (Molendatabase 2018) went to Haarlem. In 1973, parts of it were used to rebuild the "De Kat" mill in Uitgeest after it burned.

In 1977, the Verolme yard was connected with a bridge over the harbor to the Van de Giessen de Noord shipyard. That yard was now used to build subsections for the main yard. In turn, van de Giessen de Noord sold the Verolme yard in 2000 to a project developer and the municipality. The area was no longer needed for shipbuilding and was available for redevelopment. The former Verolme site was the first major building area to expand the built-up area of Alblasserdam.

The municipality collaborated with a project developer, Alblasserwerf CV (Bouwcombinatie Alblasserwerf 2004) and an architectural firm, HVE Architecten (HVE Architecten 2001) to plan 403 homes (60% apartments and 40% houses) in a new residential area, Alblasserwerf, and to plan accompanying public space (Municipality Alblasserdam 2012). The redevelopment was hit by the financial crisis, which raised concerns about financial viability. It was adapted accordingly: in order to finance the project, more houses had to be built on the Verolme location. Some attention to heritage could be kept. A 2001 study recommended keeping part of the building slip foundation, with an area of plank bridges symbolizing the now-filled log pond. But when the private partner ran into problems and sold out, the municipality reconsidered the plans, guided by BBN advisors, a Dutch real estate and construc-

Fig. 2 Shipyard Van de Giessen de Noord at the end of the twentieth century. The last two tower cranes were demolished April 23, 2009. *Source* Homoet 2017; released under a Creative Commons Attribution-NonCommercial-NoDerivatives 4.0 International License

tion consultant. (BBN Adviseurs 2009). This was a second reconsideration of the plan, adding more space to make it economically and logistically feasible. Today, the redevelopment is in its final stage, with the new waterbus jetty located on the De Noord river, close to the former yard at the Jan Smit kade (quay), named after the founder of the shipyard (Homoet 2018a). In 2000, there were plans to expand the Van der Giessen de Noord shipyard, which built luxury ferries (Fig. 2). But after the attack on the World Trade Centre in New York on September 11, 2001, orders suddenly dropped. In September 2002, the company reorganized, shedding a third of its staff, but despite many good contacts with shipping companies, no orders came in. As a result, on August 25, 2003, the management of van der Giessen de Noord sadly announced that the company was closing its gates (Homoet 2018b).

In 1936, the Nederlandse Kabelfabriek (NKF) founded a steel factory, Nedstaal, in Alblasserdam; it also founded another similar steel company, Corus Ijmuiden (its current name). Nedstaal was important to the early development of the municipality of Alblasserdam, especially in terms of the employment it offered (Telegraaf 2014). But it had financial problems in the early twenty-first century and finally closed on January 31, 2017 (Rotterdam District Court 2014). The municipality board and council met confidentially with the Drechtsteden Regional Development Company (ROM-D) on March 13, 2018 to discuss the redevelopment of the site.

The Mercon Kloos shipyard was built in 1843 near Oost Kinderdijk. It has adapted to the economic development of the Netherlands and general market needs, building large items from ships to rail infrastructure material. In August 2004, it moved its activities to the former property of the shipyard Van de Giessen de Noord in Krimpen aan den IJssel.

Thus, over time, more stakeholders were involved in the planning and redevelopment of historic sites, more attention was paid to sites' surroundings, and planners took more criteria like heritage into account.

Cultural Heritage on the Riverbanks

The forelands of the river Noord were developed for industrial activity, especially shipbuilding and steelmaking. Companies chose this location for economic and logistical reasons, including the price of land, access to the river, and the availability of cooling water. The shipbuilding industry, like other industries, continuously modernized production procedures and products to meet global competition. Facilities had to change or become obsolete. Whether they become an active part of the area's history in the shape of industrial, cultural, or other heritage depends on all kinds of discussions, actions, and interests. Five objects on the waterfront of the Noord (Fig. 3) exemplify the connections of history to possible transformations in the future, and the question of safeguarding the area's history as cultural heritage: the Nedstaal site, a proposed art installation on the bridge, the shipyard of van de Giessen de Noord, the Oude Werf yard, and the Mercon Kloos site.

Nedstaal went bankrupt on January 31, 2017 (Fig. 4). The company owned the site with FN Steel; now, the 36 hectare area and its buildings are owned by Ruigenhil Vastgoed B. V. Nedstaal's assets were removed in April, including the river crane,

Fig. 3 Alblasserdam riverbanks: 1. Mercon Kloos; 2. Alblasserwerf (former 'De Oude Werf transformed to housing); 3. OceAnco (former 'van de Giessen de Noord'; 4. Bridge over river Noord; 5. Nedstaal. Photograph Cees van de Wal Fotografie and Film; released under a Creative Commons Attribution-NonCommercial-NoDerivatives 4.0 International License

Fig. 4 Aerial view of FNSteel and Nedstaal. Land is owned by Ruigenhil Vastgoed. Photograph Cees van de Wal Fotografie & Film; released under a Creative Commons Attribution-NonCommercial-NoDerivatives 4.0 International License

an icon of the waterfront; the plants themselves were removed by different parties over the course of the year (RHVG 2009).

Not yet removed is the pollution inherited from the integrated steel mill and its steel melting plant; no plan for remediating the site has yet been made. Meanwhile the presidium (the local legislature) has consulted with the other council members of Alblasserdam to prepare for issues that might arise. These consultations were seen as useful after the experience with the spatial redevelopment of the Verolme location in Alblasserdam and similar locations across the Netherlands. The local government wanted to have a clear playing field both economically and socially, and more control in the planning phase free of the market and civil society.

For a redevelopment of this size, the municipality hired ROM-D, whose goal is to strengthen and expand the regional economy, including developing and selling new commercial property, revitalizing existing (outdated) commercial property, developing housing projects, raising the region's profile through promotion, and regional business development. It sees to it that companies will settle at locations right for them and for the region (ROM-D 2010).

The bridge is a symbol of the river and the municipality of Alblasserdam. Recently, the River Art Rotterdam and Drechtsteden Foundation (Stichting River Art Rotterdam & Drechtsteden) asked artist John Körmeling to develop an attraction on the bridge, and he created Motel Kinderdijk, an actual hotel with rooms overlooking river Noord (Fig. 5). This project also has the support of the Economic Development Board of Drechtsteden. On May 8, 2017, the foundation formally handed the proposal for the hotel to the mayors of Alblasserdam, Hendrik-Ido-Ambacht, and Papendrecht.

Fig. 5 Maquette of bridge art over the Noord: Motel Kinderdijk. Photograph Gemeente Alblasserdam. *Source* Het Kontakt 2017; released under a Creative Commons Attribution-NonCommercial-NoDerivatives 4.0 International License

To proceed, the foundation will also need approval from Rijkswaterstaat, part of the Ministery of Infrastructure and Environment.

The shipyard once owned by van de Giessen de Noord is now owned by oceAnco. From the river, one sees a modern shipyard with a new dry dock, 156 m long by 52 m wide. Much of the old shipyard is gone, including the tall cranes that once distinguished the town's skyline. Only the marine hall remains; here, where workers once constructed polyester minesweepers and minehunters, a new company builds polyester yachts. The Heritage Commission of the municipality of Alblasserdam and the local Historic Society of West-Alblasserdam first objected to demolishing the two remaining tower cranes, but on March 13, 2009, withdrew the objection due to the high cost of keeping them (Homoet 2017, Municipality Alblasserdam 2009). Other huge shipbuilding cranes disappeared from the region when shipyards modernized, closed, or merged with other shipbuilding companies.

The history of this part of the riverfront is told not in structures but in pictures, posters, and other artworks. Working for the Green Heart Foundation (Stichting Groene Hart), two Dutch artists, Marry Teeuwen de Jong and Roel Teeuwen, have designed a new way to maintain the tower cranes as work of art (Teeuwen 2016) a project called Haven Zuid to be located next to oceAnco (Municipality Alblasserdam 2007) (Figs. 6a, b).

The recent financial crisis delayed the project, and it waits for a project developer. Recently, the community of Alblasserdam has selected a work of art by artist Rosalinde van Ingen-Schenau to symbolize the region and serve as a meeting point for tourists (Municipality Alblasserdam 2017). Financed by the municipality and also

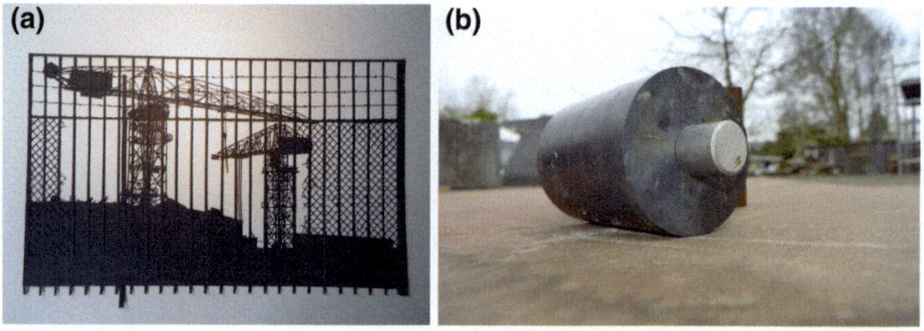

Fig. 6 **a** (Marry Teeuwen)—view through the fence, **b** (Roel Teeuwen)—counterweight of a crane; released under a Creative Commons Attribution-NonCommercial-NoDerivatives 4.0 International License

Fig. 7 **a** and **b** Anamorphosis with aldermen Arjan Kraijo and Dorien Zandvliet Photograph Gemeente Alblasserdam; released under a Creative Commons Attribution-NonCommercial-NoDerivatives 4.0 International License

the province of Zuid-Holland, *Anamorphosis* was placed in the harbor area at the end of 2017. These works of art are a sort of cultural heritage (Fig. 7).

Similarly, the Oude Werf shipyard is only remembered in pictures and books, and on the Internet (Homoet 2018a). The redevelopment of the site started in March 1952 with filling up the mill pond and dismantling the saw mill (Molen Database 2018). The yard had been owned by Verolme (Verolmetrust 1960) and was sold to van de Giessen de Noord in 1977; in 2000, the yard was acquired by Alblasserdam and a project developer. The area has been redeveloped for houses and apartments overlooking the river. From the river, the visitor can see not only the housing but the jetty for the waterbus that connects Rotterdam and Dordrecht. Near the waterbus stop are other artworks (Municipality Alblasserdam 2014), as part of the town's art route that showcase the identity and the diversity of this port city (Alblasserdam Kunstroute).

In March 2006, the Alblasserwaard city council told the city board to rewrite a plan to develop the Mercon Kloos location (Fig. 8) (ChristenUnie Alblasserdam 2006). But the highest judicial institution in the Netherlands, the Raad van State, decided on

Fig. 8 Mercon Kloos location from the air. Photograph Cees van de Wal Fotografie & Film (Van der Zouw 2015); released under a Creative Commons Attribution-NonCommercial-NoDerivatives 4.0 International License

Wednesday, November 29, 2010 that the original project of the construction of 300 houses could not proceed. One issue was that the municipality of Alblasserdam had not taken into account the heavy traffic on the Kinderdijk route when planning the new houses, which would bring the number of traffic movements above the maximum threshold. The municipality proposed alternative routes, but the judges decided that better proposals were required. The project was further delayed by the financial crisis. Recently, Mercon Kloos has entered a new phase with the land purchase by project developer Whoonapart (Poldervaart 2017a, b). First meetings with the press and stakeholders have taken place, and early plans now take into account the village landscape, cultural heritage, and heavy traffic. Clearly, the planners have learned from earlier efforts.

Some of the challenging questions and issues remain. What can be kept from the existing iconic buildings depends on the current state of the buildings, especially their steel frames, which have not been maintained. This project is still in development.

Whoonapart and KuiperCompagnons are keen to involve all stakeholders, and have organized several public meetings to this end. The stakeholders have a range of requirements and demands:

Rijkswaterstaat requires a distance of 25 m between the riverside and housing, mainly for safety, as the Noord is one of the busiest rivers in the Netherlands.

The Waterboard Rivierenland requires that the built-up area be a certain height, because the existing embankment is too low. The Waterboard has decided to reinforce the river side of the embankment because it cannot be made higher due to existing housing.

The Cultural Heritage Agency of the Netherlands (Rijksdienst voor Cultureel Erfgoed) and Foundation World Heritage Kinderdijk (Stichting Werelderfgoed Kinderdijk) have requirements about the skyline of any new buildings visible from the Kinderdijk mill complex. The current inhabitants along the embankment would like to keep their view to the river. Finally, potential tenants for the hotel, the restaurant, and the museum are making bids based on their business models. Whoonapart is including all of these stakeholder demands in their request for a permit for the development of the Mercon Kloos location. The effort spend in the preliminary phase should shorten the next part of the process of obtaining the required permits from the authorities at local, regional, and national levels.

These five examples show the various ways in which the Dutch preserve cultural heritage for our present and our future. Contemporary art is put in place for improving liveability of the riverfronts and for remembering the past. More attention is paid to the historical tangible and intangible values of buildings, landscapes, and views of sites to be redeveloped.

Networks and Processes

Today, riverfronts are celebrated by the state, citizens, civil society, and the market as desirable spaces for sustainability and liveability of neighborhoods, villages, towns, and cities. This celebration generates self-organized initiatives by citizens and civil society within the scope of redevelopment plans of the riverfront landscape. So, as is shown with the riverfront and the forelands of the Noord, more and more actors are getting involved in the spatial development of waterfronts.

Once a center of industrial activity, the Alblasserwaard river area is now a mix of industrial activity, housing, and leisure, with heritage values as a continuous thread through all actions. All actors operate in networks bringing together social, human, and information capital to achieve their different objectives; they pursue processes that link them with each other, whether in agreement or in ongoing discussions. In line with recent concerns about citizens' involvement in the Netherlands (Van Dam et al. 2014), this section highlights three citizen initiatives restoring and managing cultural heritage in the Alblasserwaard and the river Noord.

First, the Water Triangle is one of seven defined *heritage lines* in the province of Zuid-Holland (Province Zuid-Holland 2015a, b). A heritage line is a geographic line connecting various points on the map to create a coherent whole in harmony with a common historical narrative. These points could also frame a linear feature, but they need a common story. The aim of naming objects as a heritage line is to make the cultural landscape more visible. Other lines include the fortifications marking the Roman frontier, the former transport routes along the many canals, and former defense structures in water. The Water Triangle heritage line connects the iconic towns Biesbosch, Dordrecht, and Kinderdijk and the area in the triangle they form. Much has been invested in the Unesco Mill Network Elshout/Kinderdijk, a group of windmills in an exceptional human-made landscape that illustrates the centuries-long

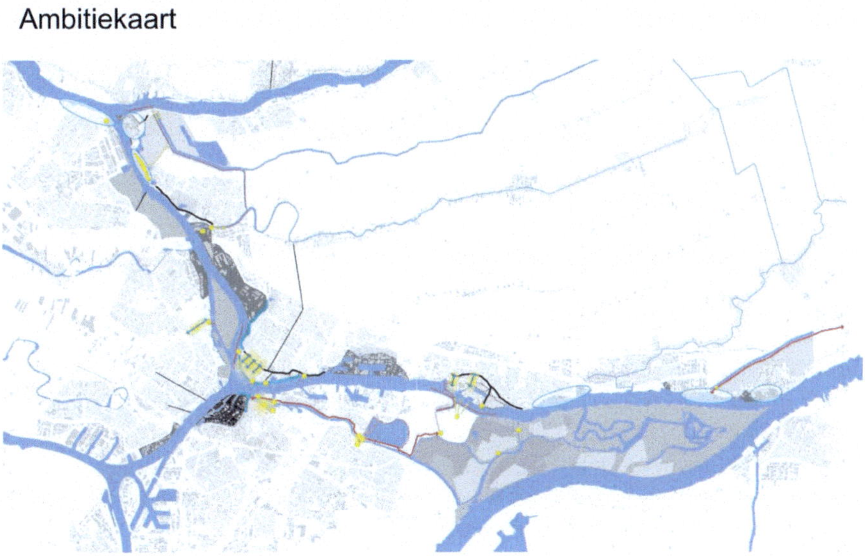

Fig. 9 Ambition map of heritage line water triangle (Waterdriehoek) Province South Holland; released under a Creative Commons Attribution-NonCommercial-NoDerivatives 4.0 International License

battle of the Dutch people to drain parts of their territory and protect them against further inundation. Now a tourist attraction, all the major elements of the complex system have survived (Unesco 1997).

The question of which items to include in a heritage line is one of provincial policy making, part of which is supposed to emerge from bottom-up pressure. Stakeholders who want to discuss heritage and might propose activities to preserve or create heritage meet three to four times a year in heritage committees and they advise the province which projects to subsidize. Typically, September meetings are reserved for judging project proposals from members of the Water Triangle heritage line (Fig. 9).

A second citizen initiative is the RiverArt project, which uses art to connect river towns. It aims to realize 25 art projects in 2025. It was started by the Economic Development Board and is also a project of Dordrecht (which is also part of the Water Triangle heritage line) and Rotterdam. It is an integrated area development project to reinforce riverfront quality with attention to heritage and public space. The first art project is the Floating Forest in the Rijnhaven of Rotterdam (in Dutch *dobberend bos*), a collection of twenty trees floating water on recycled sea buoys in the Rijnhaven, a downtown harbor basin. The installation is a green spot in the port, calling attention to art, innovation, and sustainability (Port of Rotterdam 2016).

The third citizen initiative is a vantage point on the Groote Zaag in Krimpenerwaard: the fourteen meter high Beaver bridge, designed by Ruud Reutelingsperger of the artists' collective Observatorium. The bridge, a structure of steel tubes, is inspired by Beaver castle. From it, you have a panoramic view over The Saw (De Zaag), Rotterdam, the port, and the river (Zuid-Hollands Landschap 2018).

Conclusion

This chapter is a first step in discussing the neglected and undervalued culture heritage of riverfronts in the Alblasserwaard and what people are doing to develop and preserve them. A next step, which could help a deeper analysis of the case study of the Alblasserwaard and its many stakeholders, would relate to the process of valuing this heritage to ideas of negotiation; the IAD model by Ostrom (2005), the CLEAR model by Lowndes (Lowndes et al. 2006), the process of bonding and bridging by Putnam (2001) and of linking by Szreter (Szreter and Woolcock 2004), and the civic voluntarism model of Verba (Verba et al.1995). Possible linkages can be established between these different ways of working. The IAD model from Ostrom could define boundaries of the "action arena" with actors and situations, whereas the CLEAR model could define required and preferred capacities of participants. Putman and Szreter allow us to define the processes that take place with networks and between networks. Bonding takes place between like-minded actors, bridging not-like-minded actors with equal power positions, and linking not-like-minded actors in unequal power positions.

The conceptual model (Fig. 10) shows that the local government has rules to obey and depends on the social, physical, and economic conditions of its locality. The spatial bonding process takes place between directly involved actors.

As discussed, much has been already invested in works of art along the river banks to make the river areas and the region more visible and attractive to investors and tourists. Former construction plants are turned into housing areas, including the jetty for the waterbus and options for hotels, a restaurant, and a museum of local water management. The new owner Whoonapart says that he looks for best practices on the Internet to deal with his property and seeks to listen to all stakeholders.

This chapter considered two shipyards, one steel mill, one construction plant, and a traffic bridge in a municipality on the river bank of the river Noord: witnesses of the past, the present, and the future of water and heritage in the region. In the past, the economic value of the property was the main factor valued by owners and municipalities. Citizens and civil society had a vote and maybe a voice, but economic value (within the applicable laws and rules) settled most decisions. Now, the game is different. With the changing economic value of the area, and also changing values concerning the area's history and future, and with the public having a stronger voice, citizens' initiatives, and involvement are more important to the process of valuing and transforming property and land along the river. This new complexity is highly valued by municipal authorities, but also brings along a need for more scientific research on transforming forelands from industrial landscapes in ways that preserve that same industry and related identity within a diversity of voices on water and heritage.

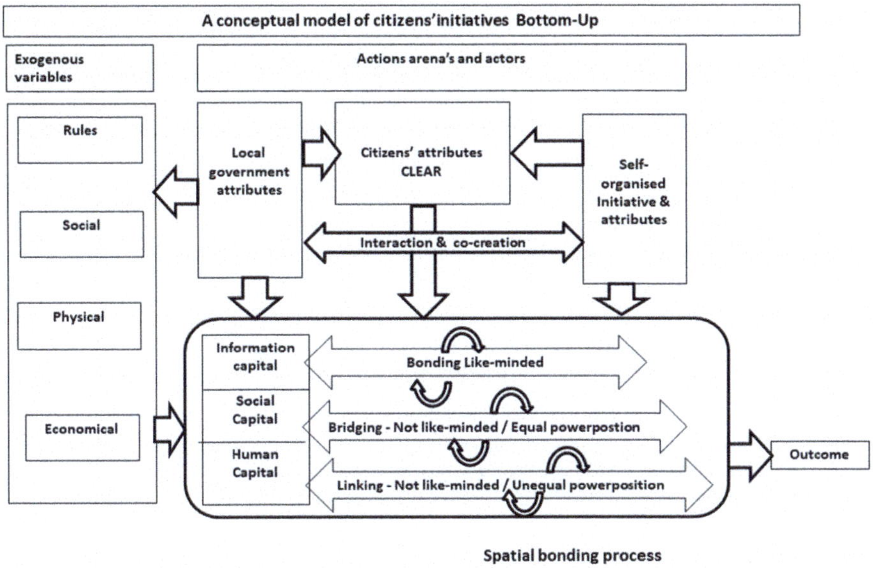

Fig. 10 Conceptual model of citizens'initiatives (bottom-up); released under a creative commons attribution-Non Commercial-No Derivatives 4.0 International License

References

Alblasserdam.net (2018) Retrieved 1 January 2018, from https://www.alblasserdam.net/column/2015-08-25/cornelis-verolme-blijft-voor-eeuwig-aan-alblasserdam-verbonden-.html

Autosnelwegen (2018) Openingsdata nieuwe autosnelwegen. Retrieved 29 June 2018, from https://www.autosnelwegen.nl/index.php/geschiedenis/5-1940-1945-tweede-wereldoorlog

BBN adviseurs (2009, March 30) Herontwikkeling Scheepswerf Verolme Alblasserdam. Retrieved 29 June 2018, from http://www.bbn.nl/gebiedsontwikkeling/projecten/507/herontwikkeling-scheepswerf-verolme-alblasserdam.html

Boersma P (1939) Alblasserdam's heden en verleden, Europese Bibliotheek–Zaltbommel

Bouwcombinatie Alblasserwerf (2004) Alblasserwerf. Retrieved 29 June 2018 http://www.boele.nl/projecten/alblasserwerf-de-hellingen-alblasserdam

ChristenUnie Alblasserdam (2006, November 3). Mercon Kloos. Retrieved 25 Sept 2017, from https://alblasserdam.christenunie.nl/k/news/view/136109/543294/tegenstem-begroting-christenunie.html

De Reus, M.J.C. (1998) Alblasserdam 1299–1999. Gemeente Alblasserdam, ISBN: 90-76540-13-6

De Wit S (2009) Dutch Lowlands morphogenesis of a cultural landscape. Uitgeverij SUN and author, Amsterdam

Deal Drecht Cities (2018, Februari 22). Alblasserdam. Retrieved 29 June 2018, from https://www.dealdrechtcities.nl/drecht-cities/alblasserdam/

Estevens A (2005) Waterfronts rehabilitation. Retrieved 4 Oct 2017, from http://www.isocarp.net/Data/case_studies/636.pdf

HafenCity Hamburg (2000) Architecture. Retrieved 4 Oct 2017, from https://www.hamburg.com/sights/architecture/hafencity-hamburg/

HafenCity Hamburg (2017) Masterplan. Retrieved 4 Oct 2017, from https://www.hafencity.com/en/glossary/glossary-i-m.html#masterplan

Harms H (2007) Changes on the waterfront—transforming harbor areas comparison and evaluation of waterfront developments in two contexts. In: San Francisco Bay Area and Hamburg, Germany

Preliminary Presentation at Shrinking Cities Conference, University of California, Berkeley Feb 2007, p 14

Harms H (2008) Changes on the waterfront—transforming Harbor Areas: comparison and evaluation of waterfront developments in two contexts: San Francisco Bay Area and Hamburg, Germany, IURD Working Paper Series, Institute of Urban and Regional Development, pp 37–48, UC Berkeley

Het Kontakt (2017) Retrieved 20 October 2017, from https://www.hetkontakt.nl/regio/alblasserwaard/algemeen/119603/motel-kinderdijk-kunstidee-voor-brug-over-de-noord

Homoet C (2018a) Oude Werf. Retrieved 29 June 2018, from http://www.scheepsbouw-alblasserdam.nl/oude-werf.html#

Homoet C (2017) Torenkraan. Retrieved 20 Oct 2017, from http://www.scheepsbouw-alblasserdam.nl/torenkraan-23-April-2009.html

Homoet C (2018b) Werf de Noord. Retrieved June 29, 2018, from http://www.scheepsbouw-alblasserdam.nl/torenkraan-23-april-2009.html

HVE Architecten (2001) Verolme terrein Alblasserdam. Retrieved 1 Jan 2018, from http://hve-architecten.nl/projecten/verolme-terrein-alblasserdam/

Louwe Kooijmans LP (1974) The rhine/meuse delta: Four studies on its prehistoric occupation and holocene geology, Leiden, Analecta Praehistorica Leidensia 7

Lowndes V, Pratchett L, Stoker G (2006) Local political participation: the impact of rules in use. Public Adm 84(3):539–556

Molen Database (2018) Molen Ons Genoegen. Retrieved 1 Jan 2018, from http://www.molendatabase.org/molendb.php?step=details&nummer=866

Municipality Alblasserdam (2009, March 13), Torenkraan. Retrieved 28 June 2018 from https://www.alblasserdam.nl/albdam/up1/ZetddqbIwC_Behoud_torenkranen_helaas_te_duur.pdf. Retrieved June 28, 2018

Municipality Alblasserdam (2012) Alblasserwerf. Retrieved 29 June 2018, from https://www.alblasserdam.nl/werk-in-uitvoering/bouwprojecten/alblasserwerf

Municipality Alblasserdam (2007, March 24) Haven Zuid. Retrieved 4 Sept 2017, from https://www.alblasserdam.nl/albdam/up1/ZydikabIqC_120.pdf

Municipality Alblasserdam (2014) Kunstroute. Retrieved 24 Sept 2017, from https://www.alblasserdam.nl/albdam/up1/ZuidaqoJC_kunstroute.pdf

Municipality Alblasserdam (2017, May 22) Kunstwerk. Retrieved 24 Sept 2017, from https://www.alblasserdam.nl/actueel/nieuwsarchief/may-2017/kies-een-kunstwerk-voor-het-nieuwe-havengebied

Ostrom E (2005) Understanding institutional diversity. Princeton University Press

Poldervaart J (2017a, September 11) Nieuwe bestemming voor het gebied Mercon-Kloos. Retrieved 20 Oct 2017, from https://www.alblasserdam.net/nieuws/2017-09-11-10003-nieuwe-bestemming-voor-het-gebied-mercon-kloos.html

Poldervaart J (2017b, September 26) Ontwikkeling Mercon Kloos terrein. Retrieved 22 Oct 2017, from http://www.alblasapart.nl/

Port of Rotterdam (2016, March 17) The port has gained a floating forest. Retrieved 31 Mar 2018, from https://www.portofrotterdam.com/en/news-and-press-releases/the-port-has-gained-a-floating-forest

Province Zuid-Holland (2015a) Erfgoedlijn Waterdriehoek. Retrieved on 25 Sept 2017, from https://www.zuid-holland.nl/@9212/nieuwe/

Province Zuid-Holland (2015b). Zuid-Holland: slimmer, schoner en sterker. Retrieved 28 June 2018, from https://www.zuid-holland.nl/publish/pages/10694/zuid-hollandslimmerschonerensterkerhoofdlijnenakkoord2015-2019.pdf

Putnam RD (2001) Social capital: measurement and consequences. Isuma Can J Policy Res [Internet]. 2:41–51

RHVG (2009) Ontmanteling installaties.Retrieved 28 June 2018, from http://www.rhvg.nl/themas-en-projecten/actueel/23-ontmanteling-installaties

ROM-D (2010) Drechtsteden. Retrieved 29 June 2018, from https://www.rom-d.nl/drechtsteden

Rotterdam District Court (2014, October 20), Netherlands steelmaker Nedstaal files for bankruptcy. Retrieved 29 June 2018, from https://www.metalbulletin.com/Article/3391545/Netherlands-steelmaker-Nedstaal-files-for-bankruptcy.html

Szreter S, Woolcock M (2004) Health by association? Social capital, social theory, and the political economy of public health. Int J Epidemiol 2004(33):650–667

Teeuwen R (2016, Februari 29) Kunstenaars laten mogelijkheden met oude torenkranen zien. Retrieved 29 June 2018, from http://www.alblasserdamsnieuws.nl/wordpress/2016/02/29/kunstenaars-laten-mogelijkheden-met-oude-torenkranen-zien/

Telegraaf (2014, October 17). Staalproducent Nedstaal failliet. Retrieved 28 June 2018, from https://www.telegraaf.nl/financieel/894917/staalproducent-nedstaal-failliet

UNESCO (1997). Mill Network at Kinderdijk-Elshout. Retrieved March 13, 2018, from http://whc.unesco.org/en/list/818

Van Dam R, Salverda IE, During R (2014) Strategies of citizens' initiatives in the Netherlands: connecting people and institutions. Crit Policy Stud 8(3):323–339

Van der Zouw H (2015, August 25) Cornelis Verolme blijft voor eeuwig aan Alblasserdam verbonden. Retrieved 29 June 2018, from https://www.alblasserdam.net/column/2015-08-25/cornelis-verolme-blijft-voor-eeuwig-aan-alblasserdam-verbonden.html

Van Groningen CL (1992) De Alblasserwaard. Waanders Uitgevers, Zwolle/Rijksdienst voor de Monumentenzorg, Zeist 1992

Verba S, Schlozman KL, Brady HE (1995) Voice and equality: civic voluntarism in American politics. Harvard University Press

Verolmetrust (1960, October 11) Geschiedenis. Retrieved 9 Feb 2018, from http://www.verolmetrust.nl/geschiedenis/

Whyte WH (1975) The social life of small urban spaces. Retrieved 5 Jan 2018, from https://www.pps.org/about

Zuid-Hollands Landschap (2018) De Zaag en Stormpoldervloedbos in de Nieuwe Maas. Retrieved 29 June 2018, from https://www.zuidhollandslandschap.nl/gebieden/de-zaag-en-stormpoldervloedbos

Arie den Boer is External Ph.D. candidate at Erasmus University Rotterdam, the Netherlands. His research interests include citizen participation, urban redevelopment in harbor and waterfront areas, and water-bound heritage. His articles include "Obstacles and changes during the decision-making process from citizens' initiative to a plan of reconstruction" (RLICC 2018) and "Is reconstruction of a lock worthwhile by enhanced insights?" (RLICC 2015).

Villages on the Water in Brunei, Wikimedia, Bernard Spragg, released under Creative Commons CC0 1.0 Universal Public Domain Dedication

Chapter 16
Room for the River: Innovation, or Tradition? The Case of the Noordwaard

Sander van Alphen

Abstract Dutch river management has changed dramatically over the past two hundred years. Approaches to flood management, however, date back many more centuries. This chapter compares the state river management program, Room for the River (undertaken from 2006 to 2015), to historical flood management strategies and techniques. We consider which measures used in this program might be useful in addressing the possibly accelerated pace of climate change and rise in sea level. We focus on the cultural water landscape of the Noordwaard. Formed in response to large river discharges occurring in 1993 and 1995–and the precautionary evacuation of two hundred fifty thousand inhabitants which accompanied them–Room for the River consisted of more than thirty projects and eight measures. In a departure from past policies, most measures took place in the winter bed, or floodplain, of the river or behind its dike. Another innovative characteristic was its dual objective of ensuring safety and contributing to spatial quality–which the program defined as the balance between hydraulic effectiveness, ecological robustness, and cultural meaning and aesthetics. The program dedicated a special team to monitor progress on cultural issues. Room for the River represented a paradigm shift from the flood management policies of the recent past. Yet, some of the measures it recommended, such as digging a bypass and raising mounds in the Noordwaard, also recalled into practice much older approaches. This cutting-edge program made visible and returned to use traditional ways of flood management.

Keywords Heritage · Cultural heritage · Floods · Noordwaard · Water management · Rivers · Spatial quality · Ruimte voor de rivier · Polder · Depoldering

---

S. van Alphen (✉)
Vrije Universiteit Amsterdam, Amsterdam, The Netherlands
e-mail: scvanalphen@gmail.com

C. Hein (ed.), *Adaptive Strategies for Water Heritage*,
https://doi.org/10.1007/978-3-030-00268-8_16

Introduction

The Dutch landscape is characterized by river deltas, through which large rivers like the Rhine, the Meuse, and the Scheldt meander to the sea. Because twenty-five percent of the country lies below sea level, approximately two-thirds of the country is at risk of flooding when storms come inland from the sea or rain causes rivers to overflow. Between 1200 and 1953, the country experienced about one hundred and thirty floods (Van Veen 1962). Floods are an important part of the national collective memory: history, literature, and folklore often referred to the management of floods as the "battle against the waterwolf," casting the devouring action of wind-driven water on soft shores as wolfish destructiveness (Te Brake 2002). Today's landscape reflects this long tradition of flood management.

In 1953, one thousand eight hundred thirty-six people drowned and seventy-two thousand people lost their homes. This calamity led to the Delta Act and the construction of the Delta Works (from 1958 to 1997). These steps reduced the probability of flooding in the southwestern river delta to 1:4000 per year (Bijker 2007). During this time, the southeastern part of the Netherlands was twice endangered, in 1993 and 1995, by large discharges of river water. The first case caused thousands of inhabitants of the province of Limburg, located along the Meuse River, to be evacuated, resulting in restoration costs of over two hundred fifty million guilders. The second case caused two hundred fifty thousand people to be evacuated amid operations conducted day and night to strengthen the dikes.

The high water levels of 1993 and 1995 renewed awareness on the vulnerability of the country to sea and river flooding. Moreover, few Dutch trusted their traditional flood protection. In initiating the ten-year program Room for the River (hereafter, RftR), the national government aimed to protect the Netherlands against extreme river discharges over the long term and, at the same time, to contribute to spatial quality–which it defined as the balance between hydraulic effectiveness, ecological robustness, and cultural meaning and aesthetics. In the light of such a history, this article explores the following questions: How do historical flood management strategies and techniques compare to RftR? What measures from RftR might be effective in addressing possibly accelerated climate change and rise in sea level?

Dutch approaches to water management have been internationally recognized for centuries. This has included international activity as well as work conducted in the Netherlands itself–as early as 1113, the Dutch were involved in reclaiming the Weser valley near Bremen in Germany (Van de Ven 2003). This legacy is visible today. By 2018, seven water-related sites in the Netherlands had received UNESCO World Heritage status, a development which recognizes the Dutch testimony to the ongoing battle against the sea, architectural ingenuity, orderly design, large-scale implementation, and their representation of different strategies and techniques in the development of Dutch water management (Van Rotterdam 2015). In the face of climate change and rising sea levels, adaptive water management has proved again and again to be an important intervention. The long history and innovative character of Dutch water management heritage may well offer inspiration for the future.

A Short History of Dutch Flood Management

Although the oldest surviving written records of Dutch water management date from the eighth century BCE, the earliest physical remains date to the fifth century BCE (Van de Ven 2003). These artificial mounds, known in Dutch as "terp" or "wierde," which can be found in the northern provinces of Friesland and Groningen, served as a refuge for cattle during high waters (Van Dam 2010). Often located on top of existing hills, a t*erp* is constructed of sea clay. Over time, a terp would be raised and expanded so that an entire village could fit on top. By the middle of the nineteenth century, however, farmers had dug into and excavated many of these mounds for their fertile soil (De Langen et al. 2016). Terps, it is worth noting, offer three solutions that work in tandem: higher ground for evacuation from high water; dikes, or raised ramparts, along bodies of water, to block water from coming on land; and dams to provide the higher ground that prevents flooding from occurring in the first place.

About two thousand years ago, inhabitants of the Netherlands constructed the first dikes to protect their agricultural land (Projectbureau Belvedere 2010). By the eleventh century, dikes were built along great rivers and tidal inlets in the southwestern and northeastern provinces. Currently, the Netherlands has over seventeen thousand, five hundred kilometers of dikes and relics (Lascaris 2012).

In contrast to dikes, dams are sited in the water. A dam closes off the water behind it; water level can then be regulated by a sluice or pumping device. The dam works even when the water level outside it rises. Thus, dams not only block the water but also regulate the amount of water and strength of its flow. The early seventeenth century Haarlemmer Sluice–which continues to connect the Nieuwendijk and the Haarlemmerdijk in the center of Amsterdam–is one such sluice dam. Dams also provide paths across bodies of water and were often where ships distributed their goods. As a result, the names of some Dutch cities: Amsterdam, Edam, and Rotterdam still end with the word *dam*.

From about 1200 on, the character of dikes would change, now they would become part of efforts to reclaim land from the sea–largely for agriculture. However, an area could only be reclaimed if it had enough sediment and if it only flooded at high water levels. Even after the arrival of this new kind of dike, living on mounds remained prevalent to minimize the impact of floods (Van Dam 2010). According to Petra van Dam, Professor of Water History at the Vrije Universiteit Amsterdam, raising dikes compartmentalized the landscape, a strategy which also reduced the speed, extent, and damage of a flood. A compartmentalized landscape could offer people more time to evacuate themselves and their cattle, depending on how far away dry ground was.

The diking and draining of marshes were so extensive that soil gradually dried out and descended toward the groundwater beneath. The rate of resulting subsidence is estimated at approximately fifteen to twenty centimeters per century (Van de Ven 2003). As a result, large parts of western and northern Netherlands have entered an ongoing spiral of "dredging or drowning" (Te Brake 2002).

Although the Romans had already built hydraulic engineering works in the Rhine–Meuse river area–including sluices, culverts, and dams–the Dutch viewed rivers as unmanageable natural phenomenon until the eighteenth century (Lascaris and de Kraker 2013). Since authorities did not believe conditions could be remedied, they focused on fighting their symptoms (Van der Ham 2004). From the end of the eighteenth to the beginning of the nineteenth century, spillways and bypasses were commonly used to direct surplus river water. Diverting rivers became a popular measure because it could be performed in a relatively short time. At times, as was the case after the flood of 1809, it was important to act quickly.

While the first state plans to manage rivers were presented at the beginning of the seventeenth century (Van der Woud 2006), only individual interventions were carried out. These often merely protected private territory. Contemporaneously, state water engineers and river experts, Christiaan Brunings and Rudolph Tutein Nolthenius, blamed the poor state of river beds and dikes, as well as numerous irregularities in the river system in the first half of the nineteenth century on these unwise remedies which, in turn, led to reduced flow capacity and regular floods (Van Heezik 2007).

Under the influence of the Enlightenment and the Industrial Revolution, a new worldview gradually but decisively developed, characterized by infinite confidence in the power of human reason and the desire for a socially engineered society (Van Heezik 2007). An elite body of engineers at the time possessed the required technological expertise to carry out state-level river management. However, it would be almost one hundred years before the Netherlands would benefit from their understandings (Van der Woud 2006).

New social conditions, including political stability, technological development, and economic growth, would eventually enable the Dutch to modernize river management during the second half of the nineteenth century. The transition was precipitated by an immediate influence: continuing pressure from the German shipping trade industry required both improving the navigability of Dutch rivers and reducing the number of near-floods (Van der Woud 2006). This resulted in another national paradigm shift, in which people and even the law newly considered the general interest to be more important than the private.

The state now adopted a new technique of river management, normalizing the water level, or creating a less irregular course for the summer beds of rivers. Techniques used here included dredging shoals and constructing groynes perpendicular to the river bank in order to create regular width and a steady flow pattern. Diverting rivers was seen as a last and far less favorable remedy (Van der Ham 2004).

With these strategies at hand, better control of rivers would be realized over the next hundred years. Figure 1 is taken from the *Spankrachtstudie* (2004) and was drawn up prior to RftR. Red boxes in the table indicate the number of floods that have occurred along the Rhine River since 1500. As is shown, floods took place at about every ten years. From 1850, when normalization was introduced, the number of floods decreased significantly until the end of the twentieth century, when the occurrence of river floods almost completely stopped. Rivers did, nevertheless, flood in 1926, 1993, and 1995. This last flood triggered the formation of RftR.

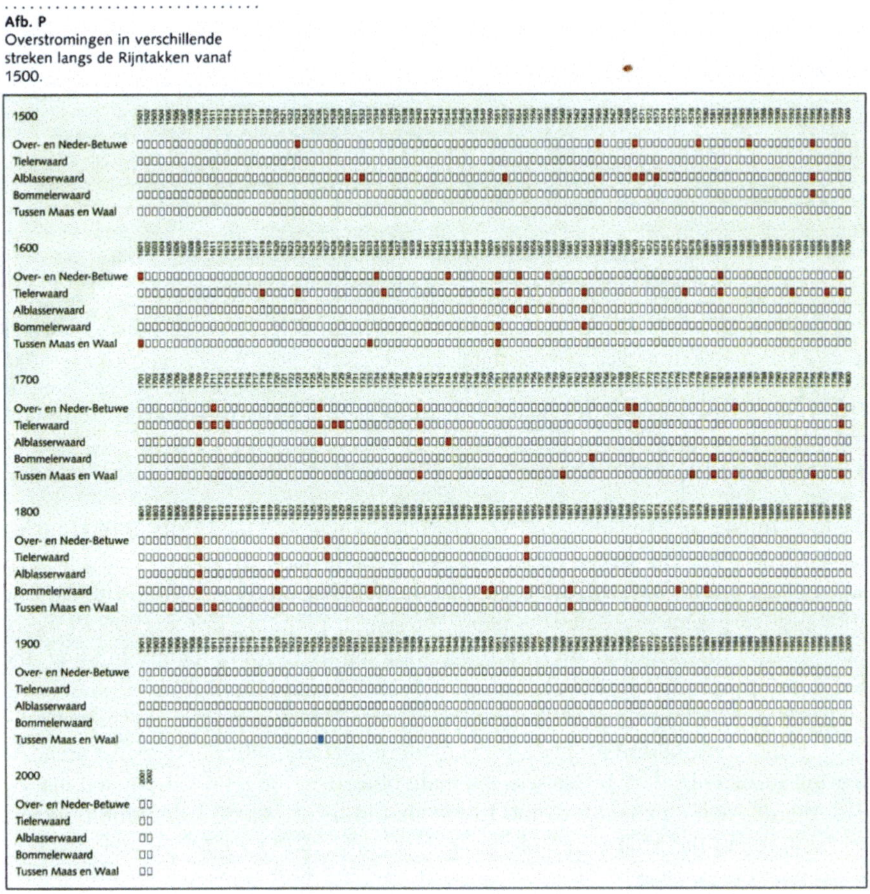

Fig. 1 Floods along different streams of the Rhine River branches from 1500 on/to the present. *Source* Van der Ham (2004). Afleiden of opruimen: de strijd om de beste aanpak tegen het rivierbederf. p. 56; released under a Creative Commons Attribution-NonCommercial-NoDerivatives 4.0 International License

The Dutch also improved the river area to benefit shipping and freshwater supply during the twentieth century. Interesting architectural examples are the three concrete weir structures along the Lower Rhine, which possess graceful curves that can absorb great pressure.

Room for the River

The high water levels which occurred along the Meuse and the Rhine at the end of the twentieth century illustrated the Netherlands' flood vulnerability. In response to

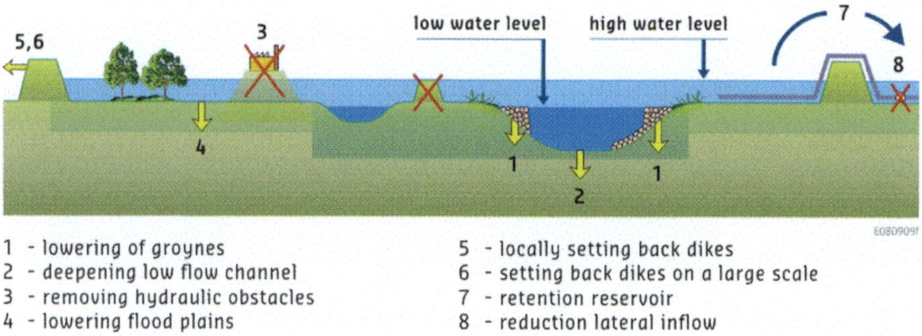

Fig. 2 Room for the river: plenty of possibilities. *Source* Silva, W., Klijn, F. and Dijkman, J.P.M. Room for the Rhine branches in the Netherlands, what the research has taught us. Arnhem: Deltares (WL)/Rijkswaterstaat RIZA. p. 56; released under a Creative Commons Attribution-NonCommercial-NoDerivatives 4.0 International License

this, the Minister of Transport and Water Management and the Minister of Housing decided, in 1996, that additional measures were needed to adapt to adverse climate change and to provide the river area sustainable protection from floods. The national government presented its plans for spatial planning in the 1996 *Planologische Kernbeslissing* (hereafter PKB-4). Although parts were legally binding, the PKB-4 was primarily symbolic.

The PKB-4 had two goals: ensuring the required level of protection against river flooding and contributing to the improvement of spatial quality in the river area. As ensuring sufficient safety levels was the main objective, the report stipulated that by 2015, the discharge capacity of the Rhine should be at 16.000 m^3/s (cubic meters per second) and the Meuse at Borgharen 3800 m^3/s (RPO 2006). However, it was argued that, in response to climate change, normative discharge capacity might have to be raised in the next century.

In total, the PKB-4 identified over 30 locations and proposed a basic set of eight measures to achieve the dual objectives of RftR. The eight measures are shown in Fig. 2.

Most of the measures were proposed for the winter bed of the river (lowering floodplains and dike displacement) or the area behind the dike (locally setting back dikes and retention reservoirs). Although these are not shown in Fig. 2, the PKB-4 also included more traditional dike improvement techniques (encompassing the raising, widening, or strengthening of dikes). Further measures, such as digging bypasses and depoldering, are also absent from Fig. 2.

During the 1970s, the ecological movement often regarded traditional flood protection measures as a threat to the dynamic character of the estuary (Disco 2002). Furthermore, the drastic way in which the state raised the height of river dikes without public consultation–often removing houses–led to resistance from local populations.

However, 1975 would be a turning point. That year, local citizens successfully fought against the dike reinforcement at Brakel in court. Eventually, Plan Ooievaar (1986) would demonstrate that ensuring a sufficient safety level could go hand-in-hand with spatial quality.

The RtfR program also reflected the importance of spatial quality to local citizens and their support of river management. It adopted this as its second objective. In order to ensure that the planning process gave sufficient attention to spatial quality, a Q-team was appointed in 2006. It advised the RftR program office (hereafter, RPO) on conserving and reinforcing spatial quality in water safety, noting that spatial quality is a poorly defined concept that cannot be quantified (Klijn et al. 2013). Spatial quality was, therefore, described as the balance of three values: hydraulic effectiveness–protecting the land from flooding; ecological robustness–building natural processes that require little maintenance; cultural meaning and aesthetics–enhancing existing landscape qualities (Klijn et al. 2013).

To foster these values, the Q-team opted for a formal peer-review approach. However, project leaders and designers of each project were first given the opportunity to raise issues and receive coaching informally. Next, the team visited each project at least three times; after this, key recommendations and a general judgment were presented to the local, regional, or private parties in charge, accompanied by a request for a response. Based on the opinions of the Q-team and initiators' responses, the RPO formulated a provisional statement on the preferred option of the final design. The Vice Minister would then make the final project decision.

In the evaluation process, "cultural meaning" included cultural history, and referred both to human interaction with the water and physical characteristics of the river over time. To research it, the Q-team had to learn about geomorphological origins and current morphological dynamics. The Q-team advocated that planners take a selective approach, in which the past did not lead but co-determined the planning process. The evaluation procedure followed here is in contrast to that at a UNESCO World Heritage Site, where the role of the past plays a more decisive role. However, in the case of the RftR program, other interests, like the agricultural and recreational industry, were also at stake. The role of the past became more subordinate in the planning process. The advice elicited is now being implemented. The various stakeholders are involved in the process at the early stage of plan preparation for dike improvement and river widening projects within the framework of the national Delta Programme (Ministry of Infrastructure and the Environment 2014).

Case Study: The Noordwaard

> The sea had been calm for already a couple of years. Perhaps, that's why few had paid attention. But the sea proved unpredictable and unreliable. What was predicted, and feared by many, happened on the eighteenth of November 1421: shortly before the day of the holy Elisabeth, the sea struck relentlessly. It penetrated the land, leaving nothing but destruction behind (Van der Ham 2003).

Fig. 3 Saint Elizabeth's Day Flood. Master of the St Elizabeth Panels, c. 1490–c. 1495. *Source* Rijksmuseum, Amsterdam; released under a Creative Commons Attribution-NonCommercial-NoDerivatives 4.0 International License

The St. Elizabeth's Flood in 1421 (shown in Fig. 3) marks the Noordwaard's origin and the beginning of this polder's natural process of accretion. It now consists of nearly two thousand five hundred hectares in the Netherlands' southwest. But the Noordwaard would develop very gradually, not becoming fully formed until 1980. For many centuries, land would only be reclaimed from the water by natural events

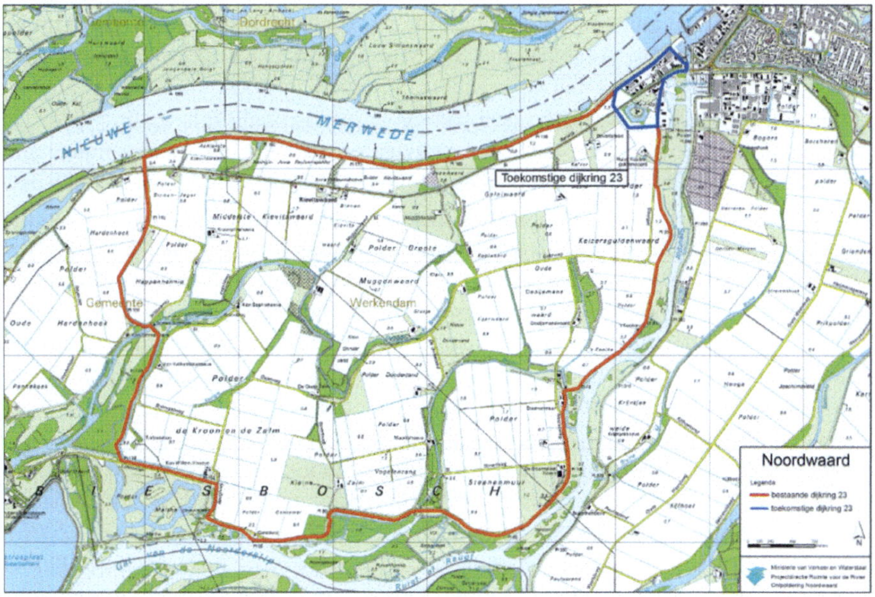

Fig. 4 Future dike ring 23. Existing ring in red; future ring in blue. *Courtesy of/Source* Ministry of Transport, Public Works and Water Management, and Room for the River programme office, 2010. Toelichting Rijksinpassingsplan: ontpoldering Noordwaard. p. 27; released under a Creative Commons Attribution-NonCommercial-NoDerivatives 4.0 International License

such as the accretion of tidal flats, or the growth of shoals and willow trees. Only in the eighteenth and nineteenth centuries did locals conduct the first small-scale reclamations of tidal marshes, surrounding increasingly larger parts of the marshes with dikes (Ministry of Transport, Public Works and Water Management, and RPO 2010). At the onset of the twentieth century, communities merged more and more polders, adjusted the water system, and consolidated agricultural land on a large scale. As it moved from osier fields to grain production in the first half of the twentieth century, the area lost the ability to temporarily store water, as grain required larger farm plots and better drainage. Through such means, a single large polder was created in 1980: the Noordwaard (Marijnissen 2012).

Two particular interventions influenced the creation of the Noordwaard: the construction of the New Merwede Canal (constructed from 1861 to 1874) and the closure of the Haringvliet Dam (in 1970). The New Merwede was dug out to maintain enough drainage capacity on the Upper Merwede River. At the same time, the canal's dike would form the Noordwaard's northern border (shown in Fig. 4). Because of these structures, the Noordwaard could, to a great extent, be drained during the twentieth century. For its part, the closure of the Haringvliet Dam was part of the Delta Works and ensured that the Noordwaard was no longer affected by tidal action.

Land reclamation and consolidation in the Noordwaard reached its zenith in the beginning of the 1980s. At that time, the Noordwaard consisted of large patches of land, with a little water between them. However, in 2006, the central area of the Noordwaard polder would be designed as an occasional high water bypass (in a manner similar to the river diverting activity of the first half of the nineteenth century). This would increase the discharge capacity of the New Merwede and, as a direct result, lower its water level at Gorinchem by thirty centimeters and protect the urbanized area around Rotterdam and Dordrecht against floods. Consequently, most of the Noordwaard would no longer be behind a dike; the likelihood of flooding there (as opposed to flooding of urbanized areas) would therefore increase significantly. By reconnecting the embanked polders to a river and reintroducing water and flood dynamics, the Noordwaard partly lost its function as permanent agricultural polder and was effectively depoldered.

In comparison to the depoldering of the Hedwigepolder in Zeeland, creating a new design for the Noordwaard was a relatively smooth process. There, far more opposition from the local population prevailed (Burgers 2014). A relatively small number of houses, about seventy-five, were sited in the Noordwaard (RPO 2006). Along with regional and local governments, the Ministry of Infrastructure, Public Works, and Water Management (hereafter, RWS), investigated several options for increasing the discharge capacity of the New Merwede. Residents were also given the opportunity to participate in decision making. Eventually, thirty people took up the offer (RPO 2006). Of these, twelve testified at the Council of State (Marijnissen 2012).

Most residents of the Noordwaard were unhappy that they had to suffer for the safety of the inhabitants of Dordrecht and its surroundings. Another point of discussion was the insecurity faced by those left behind (Marijnissen 2012). Of the twenty-five farmers who participated in the decision making, fifteen had to leave the area because there would not be enough agricultural land for them in the newly designed Noordwaard. Of the remaining fifty families, about half had to move, because their homes would be in danger of flooding (Metz and Van den Heuvel 2012). Taking this critique into consideration, authorities included evacuation routes in the new design; and the municipality of Werkendam developed a roadmap for evacuation.

In many cases, RWS would sit down at the "kitchen table" with residents to look for individual solutions. In addition, families were offered the opportunity to sell their homes at market value (before RftR); and farmers were helped in their search for new farmland. Those that still wanted to live in the Noordwaard could get compensation for moving to higher ground or taking measures for better protecting against high water. For example, the design of the new (depoldered) Noordwaard included the construction of ten accommodations on top of a newly raised terp (Department of Waterways and Public Works 2011). Furthermore, a dike ring could close some houses off from the water.

Within this new dike ring (marked in blue in Fig. 4) are a residential area, a business park, and the Steurgat fortress, a feature of the New Dutch Waterline. In order to preserve residents' view of the fortress, a new, innovative type of dike was chosen, a wave-resistant dike which could be lower than older dikes. In addition, the impact of waves was further diminished by planting an eighty-meter-wide willow forest on a gentle slope in front of the dike (Department of Waterways and Public Works 2011). The dike realized was 0.7 m lower and eleven meters less wide than a traditional dike would have been (Ministry of Transport, Public Works and Water Management and RPO 2010).

After deciding that the Noordwaard should be a high water bypass, the state still had to choose between constructing one large bypass or a comprehensive set of several branches. Although one large bypass offered benefits, in that it would be cheaper to construct and maintain, it would be less attractive to recreation–a feature which residents sought. The option of a comprehensive set of several smaller branches was preferred, as it served the most interests: safety (better management of flood risks), housing (attractive living environment), nature (more tidal creeks and intertidal zones), and cultural history (recreating old patterns of the river landscape).

In order to allow the river flow to enter the Noordwaard, the dike was lowered to two meters above sea level over a distance of two kilometers (from the Steurgat fortress to Kievitswaard). When the water level exceeds that two meters, which occurs every winter, river water is able to flow over the lowered dike section into the Noordwaard. In addition, four new northeastern gateways and two new southwestern gateways, along with pumping stations between them, provide sufficient capacity to achieve the hydraulic task of lowering the water level at Gorinchem by thirty centimeters (Bureau Noordwaard 2006).

The new design of the Noordwaard (shown in the map in Fig. 5) is based on the landscape structure of 1905, the period after the completion of the New Merwede and before the large-scale land consolidations. For landscape architect Robbert De Koning, a strong cultural–historical concept offered many possibilities for connecting with the project assignment and for creating a cohesive and recognizable landscape. Due to land consolidations in the twentieth century, many historic elements (such as creeks, polders, terps, osiers, and farms) had disappeared and, consequently, the landscape had lost its cohesion (Bijker 2007). The new design has stored many old waterways at their former location as a method of rectifying the loss and has maintained the original allotment pattern as much as possible. In addition, the return of the creek system has also provided opportunities for nature development.

Conclusion

The case of the Noordwaard represents a paradigm shift in flood management that was developed, ironically, by reviving much older traditions. Living on terps, a form of flood protection dating from the fifth century BCE, has been re-initiated, as has the method of diverting, which involves digging bypasses to drain surplus river water.

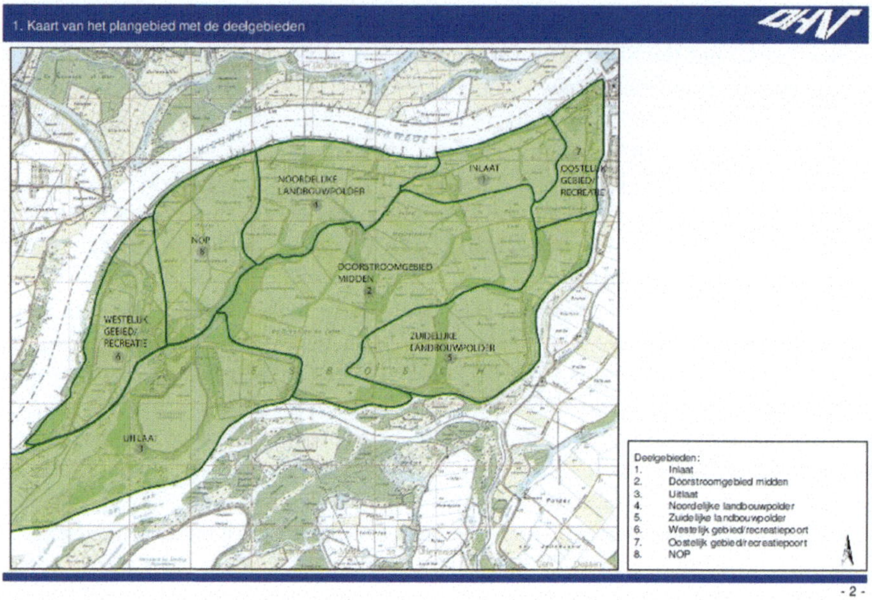

Fig. 5 Map of the plan area, showing its sub-regions: 1. northeast gateway; 2. bypass area; 3. southwest gateway. *Source* Ministry of Transport, Public Works and Water Management, and Room for the River programme office. 2010. Milieueffectrapport. Planstudie ontpoldering Noordwaard. Appendix. p. 31; released under a Creative Commons Attribution-NonCommercial-NoDerivatives 4.0 International License

This recent revival, or reconciliation, with centuries-old means of flood protection not only concerns the Noordwaard but, to a certain extent also envelops the entire RftR program.

The overall RftR strategy of "making room for water" closely captures an adage of one of the oldest written sources on Dutch water management: "not with force, but with sweetness" (Bijker 2007). In the sixteenth century, the dike builder and polder creator, Andries Vierlingh, advised the Dutch to fight the water with a soft hand–that is, to regulate the floodwater–instead of applying brute force (Ministry of Infrastructure and the Environment 2014). Vierlingh understood the opportunities that controlled flooding offered (in agriculture, for example). In the case of RftR, such opportunities were used to help improve spatial quality.

Moreover, the RftR program did not aim to restore uncontrolled river dynamics, but to sustainably protect (in a way that included maintaining local support) the country from the effects of climate change and rising sea levels. Nevertheless, the Netherlands will not be fully climate-proof even after all river-widening projects envisioned are completed. The need to develop additional new programs continues. The success of RftR signals that the future of adaptive flood management will not consist of reconciling with traditional techniques, but rather will call for applying a set of both old methods such as dike improvement and new techniques such as river widening.

Recently, research has warned us that the speed of climate change and sea level rise might exceed our expectations and preparations (Deconto and Pollard 2016). Dutch water state engineers and policymakers admit that the risk of (especially uncontrolled) floods cannot be fully eliminated. Hence, flood risk awareness should be developed among inhabitants of vulnerable areas who have experienced a long period without floods (Borger 2008). The research has shown, moreover, that the consequences of flooding will be devastating and extensive for delta regions like the Netherlands, threatening ever-larger populations and economic assets (Meyer 2016). Therefore, not only does flood protection deserve our attention but damage prevention also must be highlighted. The National Water Plan (of 2009–2015) accordingly introduced this concern as a matter of multi-level safety.

Acknowledging flood risk allows us to think of ways to reduce the possible consequences of flooding. Again, cultural heritage offers inspiration to this effort. The peninsula of Marken, an island until the closure of the Zuiderzee in 1932, still has some of the seventeenth century wooden pole houses that stood above fluctuating water levels (Holland 2018). Another example of what the cultural heritage offers is the twenty-five thousand year-old strategy of living on mounds in flood-prone areas. This approach to building lets the flood water spread out on a larger than usual surface. As a result, the water level stays much lower than it does between dikes; the water is also able to move more steadily and slowly than the occasional explosive flood after a dike breach. Floods themselves also leave behind a cultural heritage, in dike breach ponds or the curved shape of a dike, which can remind younger generations of where and how frequently they occurred.

Acknowledgements The author thanks Dr. Linde Egberts for reviewing his work and supporting the publication of this chapter.

References

Bijker W (2007) American and Dutch coastal engineering: differences in risks conception and differences in technological culture. Soc Stud Sci 37:143–151

Borger GJ (2008) Het tijdelijke comfort van een veiligheidsillusie. In: Van natuurlandschap tot risicomaatschappij: De geografie van de relatie tussen mens en milieu. Amsterdam University Press, Amsterdam, pp 218–25

Bureau Noordwaard (2006) Ontwerpvisie ontpoldering Noordwaard

Burgers T (2014) Nederlands grote rivieren: drie eeuwen strijd tegen overstromingen. Uitgeverij Matrijs, Utrecht

De Langen GJ, Nicolay JAW, Schepers M, Nieuwhof A, Postma D, Bakker M, Varwijk TW, Prummel W, Aalbersberg G, Vos PC (2016) Het terpenlandschap als actueel onderzoeksthema. Tijdschrift voor landschapsgeschiedenis 34:49–58

Deconto Robert M, Pollard David (2016) Contribution of Antarctica to past and future sea-level rise. Nature 531:591–597

Department of Waterways and Public Works (2011) Ontpoldering Noordwaard: Op hoofdlijnen

Disco C (2002) Remaking "nature": the ecological turn in Dutch water management. Sci Technol Human Values 27:206–235

Holland (2018) Best thing to do in Marken, Netherlands. At https://www.holland.com/global/tourism/destinations/provinces/north-holland/marken.htm. Retrieved 26 June 2018

Klijn F, de Bruin D, de Hoog M, Jansen S, Sijmons D (2013) Design quality of room-for-the-river measures in the Netherlands: role and assessment of the quality team (Q-Team). Int J River Basin Manag 11:287–299

Lascaris M, de Kraker A (2013) Dikes and other hydraulic engineering works from the late iron age and Roman times on the coastal area between Dunkirk and the Danish bight. In: Landscapes or seascapes? The history of the coastal environment in the North Sea area reconsidered. Corn Publication Series, Brepols Gent, pp 177–198

Lascaris Michel (2012) Climate change and cultural heritage of dikes. Latvijas Zin ātņu Akadēmijas Vēstis 6:70–77

Marijnissen H (2012) Ontpolderen doe je aan de keukentafel. Trouw February 26

Metz T, van den Heuvel M (2012) Zoet & zout. Water en de Nederlanders. NAi Uitgevers, Rotterdam

Meyer H (2016) Making urbanizing deltas more resilient by design. Int Plan History Soc Proc 3:13–24. Carola Hein (ed) BK Books, Delft

Ministry of Infrastructure and the Environment (2014) Deltaprogramma 2015. Werken aan de delta, De beslissingen om Nederland leefbaar en veilig te houden

Ministry of Transport, Public Works and Water Management, and RftR programme office (2010) Toelichting Rijksinpassingsplan: ontpoldering Noordwaard

Projectbureau Belvedere (2010) Peilwaarden. Omgaan met erfgoed in actuele wateropgaven. Utrecht: Uitgeverij Matrijs

RftR programme office (2006) Planologische Kernbeslissing 4

Te Brake W (2002) Taming the waterwolf: hydraulic engineering and water management in the Netherlands during the middle ages. Technol Cult 43:475–499

Van Dam P (2010) De amfibische cultuur. Een visie op watersnoodrampen. Rede in verkorte vorm uitgesproken (etc.)

Van de Ven GP (2003) Leefbaar Laagland: Geschiedenis van de waterbeheersing in Nederland. Uitgeverij Matrijs, Utrecht

Van der Ham W (2004) Afleiden of opruimen: de strijd om de beste aanpak tegen het rivierbederf, een beschouwing van 300 jaar rivierverbetering in het kader van de spankrachtstudie

Van der Ham W (2003) De Grote Waard: geschiedenis van een Hollands landschap. Uitgeverij 010, Rotterdam

Van der Woud A (2006) Gehoorzame rivieren. In Een nieuwe wereld. Het ontstaan van het moderne Nederland. Uitgeverij Bert Bakker, Amsterdam, pp 197–232

Van Heezik A (2007) Strijd om de rivieren: 200 jaar rivierenbeleid in Nederland. HNT Historische producties, The Hague

Van Rotterdam M (2015) Werelderfgoed van Nederland. Uitgeverij Lias, Hilversum

Van Veen J (1962) Dredge, drain, reclaim! The art of a nation. Martinus Nijhoff, The Hague

Sander van Alphen is a member of the WaterHeritage network. He earned his Master's Degree in Heritage Studies at the Vrije Universiteit Amsterdam. He specialized in historical water management and wrote his premaster thesis on the Dutch water management program, Room for the River, in relation to Dutch heritage.

Suomenlinna Fortress near Helsinki, Wikimedia, released under a Creative Commons Attribution 2.0 Generic licence

Chapter 17
Heritage in European Coastal Landscapes—Four Reasons for Inter-regional Knowledge Exchange

Linde Egberts

Abstract Heritage in coastal landscapes confronts preservationists, spatial planners, policymakers, and politicians with distinctive challenges. Coastal landscapes in all their varieties share common features because humans have interacted in similar ways with their environments on the edge of land and sea, including coastal defence, fishing, shipping, mussel farming, harvesting salt, swimming, boating, and using the beach as a tourist attraction. Coastal communities and societies are historically interconnected with each other and are more like each other than their respective inland societies in language, customs, ways of life, ways of building, and heritage. Moreover, they have a distinctive cultural and spiritual relationship with the sea. Cultural heritage and its management in coastal landscapes can vary greatly from one area to the next, but throughout Europe several key issues and challenges recur. In this chapter, I argue that coastal regions in Europe could manage their heritage resources more efficiently by exchanging expertise and experience. I address the importance of taking each site's regional, spatial, and historical characteristics into account, while not losing sight of their many contemporary economic, social, cultural, and ecological challenges. I address four of these issues: interconnected cultural frontiers; the common challenges of coast-specific heritage; the threats and opportunities of coastal tourism; and the effects of ecological changes on cultural heritage. Each of them is illustrated by an example from one of the European coastal regions. Finally, I consider the roles that coastal heritage plays in the historiography of regions and nations, and how that affects the ways in which the coastal past is remembered, preserved and redeveloped.

Keywords Cultural heritage · Coastal landscapes · Knowledge exchange · Regional governance · Europe

L. Egberts (✉)
Vrije Universteit Amsterdam, Amsterdam, The Netherlands
e-mail: l.r.egberts@vu.nl

C. Hein (ed.), *Adaptive Strategies for Water Heritage*,
https://doi.org/10.1007/978-3-030-00268-8_17

Introduction: Interconnected Cultural Frontiers

Cultural heritage and its management in coastal landscapes can vary greatly from one area to the next, but throughout Europe several key issues and challenges recur. In this chapter, I argue that coastal regions in Europe could manage their heritage resources more efficiently by exchanging expertise and experience, taking each site's regional, spatial, and historical characteristics into account, while not losing sight of their many contemporary economic, social, cultural, and ecological challenges. I address four of these issues: interconnected cultural frontiers; the common challenges of coast-specific heritage; the threats and opportunities of coastal tourism; and the effects of ecological changes on cultural heritage. Each of them is illustrated by an example from one of the European coastal regions. Finally, I consider the roles that coastal heritage plays in the historiography of regions and nations, and how that affects the ways in which the coastal past is remembered, preserved, and redeveloped.

Coastal regions have one important trait in common that has shaped their cultural heritage: they are accessible by boat. Throughout history, travel by boat was less difficult or hazardous than travel across land. Boats often connected places along coastlines and across seas with intensive trade, an exchange of goods that was also an exchange of people, skills, cultural values, ideas, fashions, and, less happily, diseases.

Today, popular culture and the humanities usually depict and imagine the sea from the land, and many coastal communities have changed from a seafaring economy to economies based on tourism or agriculture. When working on heritage management in coastal landscapes, it is important to be aware of this mainly terrestrial view of the coast, which can create a bias in understanding coastal landscapes (Gillis 2012). They consist of both land and sea, and for a long time, the sea has provided easier opportunities for travel, exchange, and cultural interconnectedness. In many cases, these historical connections can tie regions together, though they shift and change continuously.

As travel by land and air has become more convenient than travel by boat, it is easy to forget that port cities were closer to each other by water than places that could only be reached by land. Bergen in Norway was closer to Ipswich than York, which today is only a 340-kilometre trip by road (Pye 2014, p. 48). Jutland would have been quicker to reach from Ipswich than London. And most often the journey by sea was safer than by land. When we look at the land from the sea, our perception of periphery and centre changes. We see seas not as peripheries, but as nodes of culture, innovation, and exchange (Braudel 1949; Broodbank 2013; Horden and Purcell 2000).

A good example of how the sea has historically connected coastal communities is the town of Lekeitio in Basque country (Fig. 1). Although most inhabitants live as farmers, turning their backs to the sea, some have specialized in whale hunting since the early Middle Ages, when they were known to trade whale meat throughout Europe. They probably learned their shipbuilding techniques from Vikings, and hunted for whales as far as Greenland, Canada, and the Faroes (Borja and Collins 2004; Scribano-Ruiz and Azkarate 2015). Basque whale hunters are said to be the

Fig. 1 Coat of arms of the historic whalers' town of Ondarroa, Basque Country. Photograph by author; released under a Creative Commons Attribution-NonCommercial-NoDerivatives 4.0 International Licence

first to undertake whaling as a large, commercial activity, dominating the market until the late sixteenth century. In 1507, the town included the whale in its coat of arms, underlining its economic and political importance (Aguilar 1981). Basque sailors taught the Dutch, Danish, and English how to hunt whales (Du Pasquier 2000, pp. 83–91).

Although the Basques were frontrunners in economically exploiting whaling, this heritage is shared by many other areas on the Atlantic coast, such as Norway, the Netherlands, and France. Moreover, the Dutch, Danish, and English turned around and pushed the Basques out of the market over the course of the seventeenth century (Du Pasquier 2000, pp. 83–91). Whaling heritage can also be found on the Dutch island of Ameland, where giant whale jawbones marked the boundary between domestic and agricultural premises.

Such cultural interconnections can be found between port cities, 'faraway mirrors' of each other, sharing traits rather than exact forms or patterns (Hein 2012, p. 24). Port cities, particularly their waterfronts, have been shaped by port activities that changed dramatically over centuries with technological advances in shipbuilding and harbour construction. The shift from sailing to steam and the move to containerization demanded that cities enlarge ports and deepen waterways. Rotterdam, Antwerp, and Hull, among others, moved port activities out of the city altogether, closer to the sea. Meanwhile, trading companies that operated worldwide developed urban harbour areas for post-industrial functions like tourism and company headquarters, a new phase of similarity and connection between port cities across the globe (Hein 2012, p. 24).

Port cities have long been associated with fears of external influences and diseases from places beyond the seas. Port cities also had (and still have) a not entirely unjustified reputation of extreme and exuberant cultures of hard drinking, prostitution, criminality, and generally loose social morals (Beaven et al. 2016, pp. 1–10). In other words, their reputations as sites of intensive exchange of goods, people, and ideas magnified both the up- and downsides of their connections to other places in the world.

Existing academic research provides a solid basis for grasping the intertwined history of coastal areas, although it focuses on economic and urban aspects. Port cities have received scholarly attention from (urban) historians, who have explored economic interconnectedness—networks, exchange, and flows—mainly between cities in Europe and on the transatlantic coasts. To some extent, they have also attended to the local specificity of many of the places involved (Land 2016). But naval aspects of port cities and cultural relationships to the hinterland have been somewhat overlooked (Beaven et al. 2016, pp. 1–10).

Historically as well as today, naval and mercantile port cities are often nationally prestigious, with military and economic importance, and could thus count on scholarly attention. Yet, the historical interconnectedness via the sea went beyond the cities where the ships docked. In many ways, the hinterlands were also transformed by the opportunities for trade that the ports offered. Inland transport systems developed to bring goods from overseas and export local produce. The demand for supplies for shipping and for raw materials for boat construction influenced the organization of inland trade, agriculture, and industrial production. So even peripheral coastal landscapes often show many traces of interconnectedness with seafaring and areas elsewhere in the world.

This historical interconnectedness of port cities and their surrounding areas—whether the shared characteristics of port cities or the mutual transformations of their hinterlands—means that they have similar and interconnected cultural heritage, and makes it all the more useful for coastal regions to work together to manage that heritage.

Coastal Heritage: Common Heritage, Common Challenges

Due to their comparable functions as places of cultural interaction, trade, fishery, migration, and tourism, coastal regions show many similarities in heritage. What first comes to mind are the lighthouses on many European coasts, but many other heritage structures could be mentioned. Therefore, to a certain extent, the challenges in preservation and development are comparable. Neither trying to be complete, nor disregarding the complexity and differences between coastal regions in Europe, some common characteristics can be highlighted.

Coastal capitals and large port cities like Sydney, Rotterdam, Shanghai, London, Oslo, and Amsterdam have dealt with drastic changes in shipbuilding industries in the past decades, and those changes have registered in the design and locations of their harbours, waterfronts, and wharfs. In many cases, ports have left central districts altogether, leaving historically distinctive sites and their cultural heritage for cities to manage. Commercial docklands did not prove resilient to technological changes in shipbuilding—which are ongoing—and became redundant (Pinder 2003). Similarly, inner-city ports were abandoned when their size and facilities did not keep up with developments in trade, shipping, and shipbuilding. Companies have in most cases relocated shipbuilding and port activities outside the historic centre, as in Rotterdam, Antwerp, and Hull. As a consequence, those cities have lost their direct relationship to their port, and shipping is no longer a visible and tangible aspect of urban life. Meanwhile, companies mechanized loading and other shipment processes, so fewer people worked or lived in harbour areas. (Some harbour cities, such as Kristiansand in Norway, still host harbour activities in their historic locations close to their centres, but struggle with the expansion in capacity that competitive harbours demand.) In response to these changes, many cities turned their large-scale wharfs and historic harbour districts into trendy business and leisure quarters, which profit from the presence of the waterfront and the proximity to a city's historic centre. Iconic examples are legion, mainly in large (former), harbour cities, which often lie far inland.

A first scenario for wharfs that become disused in smaller towns is that they disappear before they become regarded as heritage. In coastal areas that are now peripheral, smaller wharfs are often more isolated than in metropolitan areas, where wharfs are an integral part of large harbour complexes; valuing, preserving, and finding new uses for this historic landscape is a completely different challenge. Moreover, stakeholders, investors, and other social support are much scarcer and more scattered than those working to manage landmarks in harbour metropolises. The wharfs of the European Commission on the Danube in the Romanian town of Sulina, for example, are regarded as heritage on paper, and policymakers consider them to be an integral part of Sulina's town structure. Their policy allows new cultural functions to be housed in the wharfs. But it seems that the current owner has no intention of preserving the complex from ruin, and no one has set out rehabilitate the decaying structures. And they are no longer part of the daily life of the town's citizens. Hopes are set on European collaboration, but it is feared that decay will come sooner than funding.

Some small-scale wharfs have been saved and given new functions and meanings, mainly in cases where initiatives have started from the bottom up by citizens. These often tap into history and count on tourists as their clientele. Re-using the old structures and stories of a wharf to attract paying customers, a town can make both its distinctive heritage and its connection to both land and sea visible. Not far from Ondarroa, in Pasai San Pedro, Spain, the once-shuttered Askorreta shipyard has been re-used for building ships since 2014, under the name Albaola. The yard now builds, exhibits, and demonstrates replicas of the historical boats that once sailed out from this port. Tapping into the famous whale-hunting history of the Basques, Albaola workers are now reconstructing the San Juan, a whaling ship that sank in Red Bay in what is now Canada in 1565. The wharf doubles as a museum and receives some 50.000 visitors per year. Lekuona Architects provided the design for the prizewinning transformation of the disused wharf to a visitor centre and historical boats workshop, which uses regional wood and pallets (Premios Egurtek 2016).

Coastal tourism has changed Europe's coastal landscapes to various degrees and is itself a complex part of their cultural heritage. In the seventeenth century, it was fashionable for the English aristocracy and upper classes to drink seawater and bathe in the cold sea to cure many different ailments (Lenček and Bosker 1998). With the development of the railways, the seaside became a mass-tourism destination, in which people regarded sunbathing on the beach and swimming in the sea as pleasurable pastimes. One could argue that the history of the beach is in many ways illustrative of shifts in cultural values of leisure, gender, health, race, eroticism, and the body in general. Even as people newly engaged with the sea, however, the discovery of the beach as a place of leisure alienated them from the sea as a 'common cultural ground' (Dettingmeijer 1996)—workers and locals having very different experiences than visiting tourists. Typical coastal resort towns developed in a T-shaped morphology: buildings concentrated along an access road running inland from the coast, intersecting with a commercial stretch along the beach, often along a boulevard (Barrett 1958; Pigram 1977), that included facilities such as a grand hotel, an arcade, a railway station, and in some cases also a pier. Coastal landscapes were newly dotted with vacation parks, beach resorts, amusement parks, and all the infrastructure and facilities that modern tourism demanded. As the popularity of destinations in the Mediterranean and further away rose, that of older coastal resorts declined, particularly in England. Scholars study the *tourism area life cycle model* (Butler 1980) to understand the pattern of growth, peak, and decline. In England, ruined sites of early seaside tourism are now being regarded as heritage. Scholars and practitioners are gathering in knowledge networks to integrate this heritage of leisure into the revitalization of run-down coastal towns.[1]

In contrast, in continental Europe, it seems that the modernist architecture of the coastal resort is only sometimes admired, as in post-Tito Adriatic coast (Beyer et al. 2013). This means that an important part of the recent history of coastal landscapes is no longer recognizable. Many coastal regions in Europe are affected by the ebb and flow of the popularity of coastal tourism. They share challenges of how to deal with the ‘ruins of leisure’, especially when large-scale urban revitalization projects are on the table. Regions can exchange strategies and approaches for validating, preserving, re-using, rehabilitating, and redeveloping this type of heritage that is so characteristic of the coast. Even at the level of policymaking, such exchanges could be invaluable: one town might be able to pass along information on how to improve heritage policies; another might convey how to create a careful decision-making process.

A good example of a ruin and resurrection of a leisure site is the Tinside waterfront lido in Plymouth, England. Designed by architect John Wibberley in 1935, it has a semi-circular bath with a diameter of 55 metres that protrudes into the sea (Fig. 2). Car parking is provided on the coastal road; from there, visitors descend to the boldly designed changing rooms and the pool. During the Second World War, its distinctive and recognizable shape is said to have made it a landmark for pilots of German bomber planes. After years of neglect, the pool was closed to the public in 1992. In the years after, a large public campaign was held to save the complex from further decay, and in 1998 it won protection as a Grade II Listed Building; this paved the way for restoring and re-opening it in 2005.

Coastlines have not only been places of friendly contact between travellers and locals, but also of sites of hostilities. Along Europe’s shores, traces of war from various eras still dominate the landscape. In many cases, defence structures and landscapes of trauma cross current-day national and regional borders. Heritage experts see themselves balancing various interpretations of the past—that of heroes and perpetrators, winners and losers, survivors and victims. Moreover, their role is shifting, as the arena in which war memories are shaped and passed on is more multi-vocal and dynamic than ever before. Sharing knowledge with fellow experts can be highly valuable in balancing these conflicting appropriations of the past and in negotiating

[1] Seaside Heritage Subject Specialist Network, http://www.scarboroughmuseumstrust.com/seaside-heritage-network.

Fig. 2 Plymouth's Tinside Lido. Photograph by Lewis Clarke, Wikimedia Commons ShareAlike 2.0 Generic

often fiery discussions on authenticity—better equipping these experts to engage in societal discourse on war heritage. For example, it was fiercely debated whether or not it is ethically appropriate to don Nazi uniforms in live re-enactments of battles from the Second World War. 'Playing a Nazi' is the same as waving away the ideology and horrors of the regime, journalist Joshua Green and historian Andria Orzoff argue in *The Atlantic* (13 October 2010). Others emotionally defend these re-enactment practices for their educational value. Moreover, in managing the heritage of conflict more generally, international cooperation is highly valuable and can lead museums to incorporate multiple perspectives on the past in presentations. Preservation of heritage defence structures that cross-borders would benefit from conversations between local and regional experts, who have worked on their respective parts of its heritage. They can find support from the International Scientific Committee on Fortifications and Military Heritage that assists UNESCO in advising on how the heritage of conflict can be approached and re-used.

A particularly well-known remnant of conflict and current site of narrative controversy is the Atlantic Wall (Fig. 3), an immense fortification system along Europe's continental North Sea and Atlantic coasts. It was built by German forces to halt a potential attack from the West during the Second World War (Rolf and Eckelkamp 1983). They moved buildings to make room for it: in some places, such as Den Helder, IJmuiden, and Hellevoetsluis in the Netherlands, the German occupiers and

Fig. 3 Remains of the Atlantik Wall in Audinghen, France. Photograph by Michel Wal. GNU Free Documentation Licence

their Dutch collaborators demolished parts of historic centres, and in others, such as Petten, they erased entire villages. They also tore down points of orientation such as towers and evacuated citizens on a large scale (Bosma 2006). Today, traces of this extensive structure are still visible along European coastlines (Zaloga and Hook 2009). Each country and region approaches its interpretation, preservation, presentation, and commemoration in its own ways. But historical understanding and inventory of the entire structure are still missing (Beek, N.D.), and the Wall's overarching programme and strategy remain out of sight. It is very difficult for the interested visitor to experience the unity of this historic structure. This large heritage structure exemplifies the interconnectedness of coastal regions. The visitor understanding of the Atlantic Wall remnants would greatly improve if local experts from all its parts would work together in telling stories that bind the local to its vastness.

The Threats and Opportunities of Coastal Tourism

Cultural and natural heritage form an important factor in the attractiveness of regions to tourists. Whereas the imprints of early tourism on coastal landscapes slowly become regarded as heritage, present-day effects of mass tourism are perceived as a threat to heritage (Bourdeau et al. 2015). In coastal areas where mass tourism has matured and stagnated, such as the Costa Brava in Spain, rejuvenation projects are now challenging heritage landscapes (Sardá et al. 2005). In other places, emerging tourism puts pressure on the cultural and natural environment of the coast, with water pollution, waste, and extensive new coastal resorts (hotels, golf courses, entertainment areas, and air strips). Dependence on the car as a mode of transport creates major sustainability issues, as in Southwest England: along with high emissions of CO_2, cars often mean that planners widen roads and site car parks in fragile heritage environments (Howard and Pinder 2003).

On the other hand, it is often assumed that new tourism can also be beneficial, as it can generate employment and new income for local populations, and some coastal regions are pursuing *sustainable tourism*. Sustainable tourism can mean many things: one aspect is matching demand with offers at tourist destinations, so that destinations can be managed in more sustainable ways. And sustainable tourism and preservation have several values in common. They both aim to maintain the integrity and authenticity of places for future generations, increase intercultural understanding and respect, involve stakeholders, protect the environment, and stimulate holistic management that keep the long term in consideration (Brantom 2015).

Heritage tourism is the fastest-growing niche of the tourism industry (Timothy 2011), in line with a growing demand for authentic, unique, and engaged experiences (Egberts and Bosma 2014). Indeed, the tourism industry in coastal regions is in a process of restructuring (Agarwal 2002). Researchers predict the obsolescence of the typical vacations of sun, sand, and surf with a homogenous offer (Lacher et al. 2013). Coastal tourists appreciate local character and heritage, in particular culinary experiences (Lacher et al. 2013). Deliberately integrating heritage and culinary assets in tourist offers can be an important aspect of sustainable tourism (Lacher et al. 2013).

Theoretically, then, sustainable tourism and historical preservation can work well together. But the development of small-scale tourism might not be as economically sustainable for coastal regions as it seems. It can lead to over-supply of accommodation outside the main tourist season, and new businesses often employ low-wage workers from other parts of the world rather than locals. Although seasonal and migrant work is inherently part of coastal histories, it presently does not secure local economies for longer periods of time (Howard and Pinder 2003).

It is partially the task and responsibility of regional policy makers to balance the positive and negative effects of tourism on heritage in coastal landscapes. Exchanging knowledge between regions on building strategies in which local industries and the tourism offer reinforce each other can make a crucial difference.

In order to anticipate and manage the dual impact of tourism on coastal heritage—as a threat as well as an opportunity for development—coastal regions are undertaking action to find a balance between the two. Particularly in the field of World Heritage, exemplary initiatives have been taken in the recent past (Westrik 2015), not least because UNESCO demands detailed management plans and monitoring of sites.

Fig. 4 Suomenlinna Fortress with Helsinki on the horizon. Photograph by Michal Pise. Wikimedia Commons ShareAlike 2.0 Generic

For example, the fortress of Suomenlinna in Finland, listed since 1991, is an irregularly shaped bastion fortification that has a rich and conflicted history that recounts Swedish, Russian, and Finnish governance (Fig. 4). With its 8 million annual visitors, Suomenlinna is one of the most popular tourist sites of Finland, particularly in summer. One of the threats posed by tourism, specifically the number of visitors, is erosion, mainly of the ramparts and fortification. The conservation of the fortress conflicts with its accessibility to tourists, whose presence is resented by local citizens. Therefore, an analysis of strengths, weaknesses, opportunities, and threats was made, leading to a strategy to address problems and strengthen positive qualities. At the core of the strategy is the idea that Suomenlinna is meant to be a living World Heritage Site. Therefore, tourism revenues are now used to benefit the preservation of the site and accessibility during winter is improved to generate income for local entrepreneurs all year round (A Sustainable Tourism Strategy 2015).

Cultural Heritage and Ecology

Caring for cultural heritage in coastal regions does not always go hand in hand with the preservation of nature. In both academic research and policy, there is a dichotomy between the worlds of natural and cultural heritage preservation: world heritage sites, for example, are divided into natural and cultural categories, managed by IUCN and ICOMOS, respectively. This leads to many unwanted side effects for both, as important values remain out of sight in decision-making processes and protection schemes. Some argue that the making the holistic concept *landscape* central to heritage management could contribute to overcoming divides, as it can include natural and cultural values, as well as a strong political and social dimension (Krauss 2015; Egberts 2018).

It is evident to all that coastal heritage is particularly vulnerable to climate change. Climate change concerns heritage specialists of both categories, particularly in coastal areas. Relative sea level rising and changing weather conditions challenge all preservation schemes. In some cases, the effects of erosion are so dramatic that no other strategies can be developed apart from managed decay and retreat, as is the case on England's Jurassic Coast (Howard and Pinder 2003). But sometimes the measures that are taken to prevent flooding can damage historical landscapes (Heimann 2016), and policies that are primarily aimed at preserving biospheres and natural characteristics of coastal landscapes, or stimulating regenerative energy, can also jeopardize heritage values. Experts from various sectors in several countries across Europe perceive the impact of climate change on coastal areas differently; local narratives of identity and heritage influence public perception and thus also political decision-making (Heimann and Mahlkow 2012).

Consider for example the Dutch Wadden Sea Area, part of the trilateral UNESCO World Heritage Site for its geological and natural values. Here the official focus on natural aspects of the site's heritage obscures the importance of its cultural heritage. This results in awkward spatial planning interventions that have negative impacts on the cultural landscape of this coastal region, even when the motivation of these measures are to increase sustainability. One example is the development of a solar park, the largest in the Netherlands, on the Island of Ameland. As the island's municipality seeks to be self-supporting in terms of energy, it has installed 23,000 solar panels and constructed an earthen wall several metres high to hide the panels from view. This wall now crosses the middle of the open landscape between the villages of Hollum and Ballum, obscuring the otherwise open horizon from the sea dike in the south to the dunes in the north. Moreover, the installation detracts from the historical land organization around Ballum, which was in 1916 the first Dutch agricultural area formed by re-parcellation, that is by combining many small properties into larger meadows or fields (Schroor 2000). The solar panel installation therefore jeopardizes the spatial quality of the island, its good beaches, beautiful sea views, and small-scale, even intimate agricultural landscape (Egberts 2016).

Conclusion: Marginalized Pasts in Heritage Preservation?

Can any underlying patterns be discerned in which historic remains of coastal cultures are regarded as heritage? If so, by whom? These issues might not directly address the daily management of heritage, but are essential for understanding the place of coastal heritage in an inter-regional and international perspective.

Heritage refers to those aspects of the past that people give a place in the present. They do so selectively, often in line with what Laurajane Smith calls *Authorised Heritage Discourses* (2006). These discourses legitimize which parts of the past are remembered and which are forgotten. This often happens in a context of national identity politics, and heritage is therefore dominated by national interpretations of the past. Political scientist and historian Benedict Anderson (1982) argued that nation states form their identity by selectively using elements of their community's character. Particularly since the nineteenth century, cultural and historical elements were chosen to reinforce the sense of community of a country's inhabitants: places, stories, and symbols mark common aspects within the group and make claims about how a nation differs from others. Regional identity narratives are often shaped in similar ways, although levels of institutionalization vary from one region to another (Paasi 1986). The process of selection from the past often follows a distinct pattern, in which it is decided what parts of the past are useful for regional and national identity construction and which should be left out (Egberts 2017).

As some of the above examples suggest, coastal heritage is not necessarily part of these nationally authorized heritage discourses (Hundstad 2014). Western European countries like France, Great Britain, Norway, Denmark, Belgium, Sweden, and the Netherlands saw the foundation of museums of national histories in the eighteenth and mostly nineteenth centuries, and of more specialized national maritime and naval museums from the 1910s onwards.[2] Historically, coastal towns and villages often had more cosmopolitan characters than inland areas, making them less suitable as starting points for inventing traditions (Hobsbawm and Ranger 1983) in a national context. Anthropologist Eriksen (2004) describes the fashion sense in a coastal town near Oslo as significantly European and urban during his youth in the 1970s. Hundstad (2014) argues that Norwegian historians and museum professionals noticed a 'coastal culture' in the 1970s, and that their new interest in the histories of sailors, fishermen, and beachcombers was part of a larger movement in research towards democratization and anti-elitism. Similarly, other European countries expressed a new interest in the vernacular past that Sharon MacDonald calls 'the memory phenomenon' (2013). The popularization of coastal livelihood is, for example, illustrated by the famous Dutch TV-series *Sil de strandjutter* ('Sil, the Beachcomber') from 1976, that narrates how a beachcomber finds a shipwrecked girl from Sweden on the shore of the island Terschelling.

Another explanation for the emerging interest in coastal heritage is that things become regarded as heritage once a sense of loss is felt (Hewison 1987). The industrialization and mechanization of fisheries, shipping, navigation, and port activities have ended longstanding patterns of work, housing, and spatial use, directly contributing to a general sense of wanting to hold on to coastal cultures about to be lost for good. This was one of the incentives for the foundation of museums, including the local museum 't Fiskerhuuske in Moddergat, the Netherlands in 1965; the Fisheries and Maritime Museum in Esbjerg, Denmark in 1962; the Scottish Fishery Museum

[2]Spain is an exception, as its *Museo Naval de Madrid* was founded in 1843, highlighting the imperial dominance of the country on the world's seas.

in Ansruther, Scotland in 1976; and the Naval Museum in San Sebastian, Spain in 1991.

These tendencies are interconnected with emerging coastal tourism at the same time, which can be understood as the recreationalization of the coast: the change from use of coast for production (fish, trade) to a conception of the coast for its symbolic values such as heritage, leisure, and nature (Byskov 2007). Coastal regions put the everyday life of fishermen, sailors, lighthouse keepers, and pilots on display. By founding a museum and collecting and displaying objects of an everyday life that is about to be lost, professionals and local communities disconnect them from their former use and meanings, *sacralising* them and using them to stabilize social relationships (Macdonald 2013, p. 148). New museums mark a sense of belonging to the coast (Ballinger 2006). As museums in general express a sense of status and community, it becomes understandable that the musealization of everyday life has more intensively manifested itself in areas like the coasts that once played a central role but then were regarded as marginal in the late capitalist era (Macdonald 2013, p. 160).

Further academic research is needed to bring to light the patterns behind this appropriation of the work of the coastal past in various parts in Europe. The concrete and strategic insights of such an investigation could in turn be highly informative for heritage managers and policymakers in coastal regions.

Acknowledgements This chapter is based on my involvement as an advisory partner in the INTERREG Europe project Hericoast (2016–2020) on behalf of Research institute CLUE+ of the Vrije Universiteit Amsterdam. I thank the partners in the project, particularly Laura Cuevas Ortiz and Kåre Kristiansen, for their valuable feedback on the chapter. I also thank professor Hans Renes for providing comments and discussion on the core ideas of this chapter.

References

(2014). Management PlanFortress of Suomenlinna—Unesco World Heritage Site. Governing Body of Suomenlinna, Helsinki

(2015). A sustainable tourism strategy for Suomenlinna. Governing Body of Suomenlinna, Helsinki

(2016). Premios Egurtek 2016 Edificacion. http://www.portalbec.com/portalbec/comercial/egurtek/2016/premios/EML06_PANEL_PREMIO_ML.pdf. Accessed March 22 2017

Agarwal S (2002) Restructuring seaside tourism: the resort lifecyle. Ann Tourism Res 29(1):25–55

Aguilar A (1981) The Black Right Whale, Eubalaena glacialis, in the Cantabrian Sea. Rep. Int Whal Commn 31:457–459

Ballinger P (2006) Lines in the water, peoples on the map: maritime museums and the representation of cultural boundaries in the Upper Adriatic. Narodna Omjetnost 43(1):15–39

Barrett JA (1958) The seaside resort towns of England and Wales. University of London, London

Beyer E, Hagemann A, Zinganel M (2013) Holidays after the fall: seaside architecture and urbanism in Bulgaria and Croatia. Jovis, Berlin

Beaven B, Bell K, James R (2016) Port Towns and urban cultures: international histories of the waterfront, C. 1700–2000. Palgrave Macmillan, London

Beek M, De Atlantikwall ND (2017) een verdedigingslandschap (erfgoed-essay 3). http://duinenenmensen.nl/de-atlantikwall-een-verdedigingslandschap/. Accessed Feb 22 2017

Borja A, Collins MB (2004) Oceanography and marine environment of the Basque Country, Elsevier Oceanography Series, 70; Elsevier oceanography series, 70. 1st ed., Elsevier, Amsterdam
Bosma K (2006) Schuilstad; bescherming van de bevolking tegen luchtaanvallen. AUP, Amsterdam
Bourdeau L, Gravari-Barbas M, Robinson M (eds) (2015) World heritage, tourism and identity: inscription and co-production. Ashgate, Farnham
Brantom J (2015) World Heritage and Sustainable Tourism: Shared Values? In: Bourdeau L (ed) World heritage, tourism and identity: inscription and co-production. Routledge, London, pp 235–255
Braudel F (1949) La Méditerrané et le monde méditerranéen à l'époque de Philippe II. A Colin, Paris
Broodbank C (2013) The making of the Middle Sea: a history of the Mediterranean from the beginning to the emergence of the classical world. Oxford University Press, Oxford
Burdsey D (2016) Shifting sands? Theories and concepts of contemporary life at the water's edge. In: Race, place and the seaside: postcards from the edge. London: Palgrave Macmillan UK: Palgrave Macmillan, pp 35–77
Butler R (2014) Coastal tourist resorts: history, development and models. Archit City Environ 25:203–227
Butler RW (1980) The concept of a tourism area cycle of evolution: implications for management resources. The Can Geogr 24(1):5–16
Byskov S, Fiskeri- og Søfartsmuseet (2007) Dansk kystkultur under forandring: Kystlandskab, kulturhistorie og naturforvaltning i det 20. århundrede. Esbjerg, Fiskeri- og Søfartsmuseet
Chapman A, Light D (2016) Exploring the tourist destination as a mosaic: the alternative lifecycles of the seaside amusement arcade sector in Britain. Tour Manag 52:254–263. https://doi.org/10.1016/j.tourman.2015.06.020
Davids K (2015) Local and global: Seafaring communities in the North Sea area, c. 1600–2000. Int J Marit History, 27(4):629–646
Dettingmeijer R (1996) The Emergence of the bathing culture marks the end of the North Sea as a common cultural ground. In: Roding J, van Voss LH (eds) The North Sea and Culture 1550–1800, pp 482–91
Du Pasquier JT (2000) Les baleiniers basques. Kronos, 31. S.P.M, Paris
Egberts LR (2016) Voorbij de harde grens: een cultuurhistorisch toekomstperspectief voor het Nederlandse Waddengebied. In: Visies op de Wadden: zes essays ten behoeve van de Beleidsverkenning Toekomstige Rol en Ambitie van het Rijk voor het Waddengebied. Leeuwarden: Waddenacademie, pp 30–47
Egberts LR (2017) Chosen legacies: heritage in regional identity. Routledge, London
Egberts LR (2018) Moving beyond the hard boundary: the nature-culture divide in an agricultural-maritime cultural landscape, the case of the Dutch Wadden region. J Cult Heritage Manag Sustain Dev 9(1):62–73. https://doi.org/10.1108/JCHMSD-12-2016-0067
Egberts LR, Bosma K (2014) Companion to European heritage revivals. Springer, Dordrecht
Eriksen TH (2004) Traditionalism and Neoliberalism: the Norwegian Folk Dress in the 21st century. Properties of culture—culture as property. Pathways to reform in Post-Soviet Siberia In: E Kasten (ed) pp 267–286
Gillis JR (2012) The human shore: seacoasts in history. University of Chicago Press, Chicago
Heimann T (2016) Klimakulturen und Raum: Umgangsweisen mit Klimawandel an europäischen Küsten. Springer VS, Wiesbaden
Heimann T, Mahlkow N (2012, March) The social construction of climate adaptation governance: cultural differences in European coastal areas. Retrieved February 27, 2017, from http://www.adaptgov.com/wp-content/uploads/2012/05/Heimann_Mahlkow_SocialConstruction_CAG.pdf
Hein CM (2012) Port cityspaces: town and harbour development in the global context. In: Hesse FP (ed), Stadtentwicklung zur Moderne: Die Entstehung großstädtischer Hafen- und Bürohausquartiere/Urba Development Towards Modernism: The Birth of the Metropolitan Harbour and Commercial Districts, Hendrik Bäßler Verlag, ICOMOS Hefte des Deutschen Nationalkommittees, LIV, Berlin, pp 24–32

Hewison R (1987) The heritage industry: Britain in a climate of decline. Methuen, London
Hobsbawm E, Ranger T (eds) (1983) The invention of traditions. Cambridge University Press, Cambridge
Horden P, Purcell N (2000) The corrupting sea: a study of Mediterranean history. Blackwell, Oxford UK
Howard P, Pinder D (2003) Cultural heritage and sustainability in the coastal zone: experiences in south west England. J Cult Herit 4(1):57–68
Hundstad D (2014) Kystkultur - et begrepshistorisk perspektiv. (The term Coastal Culture from a conceptual perspective) in Koren ES, Kvalø F (ed) Hundre år over og under vann. Kapitler om maritim historie og arkeologi i anledning Norsk Maritimt Museums hundreårsjubileum. Novus, Oslo
Krauss W (2015) The natural and cultural landscape heritage of Northern Friesland. Int J Herit Stud 11:1:39–52
Lacher RG, Oh C-O, Jodice LW, Norman WC (2013) The role of heritage and cultural elements in coastal tourism destination preferences: a choice modeling-based analysis. J Travel Res 52(4):534
Land I (2016) Doing urban history in the coastal zone. In: Port towns and urban cultures: international histories of the waterfront, c. 1700–2000. Palgrave Macmillan, London, pp 265–281
Lenček L, Bosker G (1998) The beach: the history of paradise on Earth. Viking, New York, NY
Macdonald S (2013) Memorylands: heritage and identity in Europe Today. Routledge, London
Paasi A (1986) The institutionalization of regions: a theoretical framework for understanding the emergence of regions and the constitution of regional identity. Fennia 164(1):105–146
Pigram JJ (1977) Beach resort morphology. Habitat Int 2(5–6):525–541. https://doi.org/10.1016/0197-3975(77)90024-8
Pinder D (2003) Seaport decline and cultural heritage sustainability issues in the UK coastal zone. J Cult Herit 4(1):35–47
Pye M (2014) The edge of the world: how the North Sea made us who we are. Viking, London
Rolf R, Eckelkamp N (1983) Der Atlantikwall: Perlenschnur aus Stahlbeton (Prak, 4; P.R.A.K, 4). Beetsterzwaag: AMA
Sardá R, Mora J, Avila C (2005) Tourism development in the Costa Brava (Girona, Spain)—how integrated coastal zone management may rejuvenate its lifecycle In *Managing european coasts: past, present and future*. Springer, Berlin, Heidelberg, pp 291–314. (Environmental Science)
Schroor M (2000) Ballumer mieden: de eerste ruilverkaveling in Nederland. Noorderbreedte. https://noorderbreedte.nl/2000/12/01/ballumer-mieden-de-eerste-ruilverkaveling-in-nederland/
Scribano-Ruiz S, Azkarate A (2015). Basque Fisheries in Eastern Canada, a special case of cultural encounter in the colonizing of North America. In: Archaeology of culture contact and colonialism in Spanish and Portuguese America. Springer International Publishing: Springer, Cham, pp 239–256
Smith L (2006) The discourse of heritage. In: Uses of heritage. London: Routledge, pp 11–42
Timothy DJ (2011) Cultural heritage and tourism: an introduction (aspects of tourism texts; aspects of tourism texts). Channel View Publications, Bristol
Westrik C (2015) World heritage and sustainable tourism: a multifaceted relationship. In: Albert M-T (ed) Perceptions of sustainability in heritage studies (heritage studies, v. 4; heritage studies). De Gruyter, Berlin/Boston
Zaloga SJ, Hook A (2009) The Atlantic Wall. Osprey, Oxford

Linde Egberts is an assistant professor in heritage studies at Vrije Universteit Amsterdam whose research explores heritage, identity, regions, landscape, and the interplay of tourism and identity. As a member of the CLUE+ research institute, she focuses on heritage in coastal landscapes in Europe.

Part V
Port Cities and Waterfronts

The HafenCity Hamburg, Germany with the Elbphilharmonie, 2018, Carola Hein, released under a Creative Commons Attribution-NonCommercial-NoDerivatives 4.0 International Licence

Chapter 18
The Impact of Planning Reform on Water-Related Heritage Values and on Recalling Collective Maritime Identity of Port Cities: The Case of Rotterdam

Azadeh Arjomand Kermani, Wout van der Toorn Vrijthoff and Arash Salek

Abstract This chapter explores two structural aspects of port-city interaction. First, it studies the evolution of planning policies on post-industrial waterfront spaces in the Netherlands before and after the 2008 financial crisis, focusing on the former shipbuilding company Rotterdamsche Droogdok Maatschappij (RDM). The RDM site in Rotterdam is a significant part of the old port area, and its submarine and shipbuilding legacy has always been present in the heads and hearts of the citizens. Second, the chapter explores how reawakening the nautical culture and marine traditions in Rotterdam can also reanimate the historical links between port and city. It briefly analyses the goals, achievements, and effects of a few heritage projects on the port-city interaction and the maritime identity of this global port-city.

Keywords Water-related heritage · City-port interaction · City Ports Rotterdam · Maritime identity · Waterfront regeneration

Introduction

Bodies of water are an important foundation for economic development, and they facilitate the movement of goods and people around the world; thus, they are a vehicle for globalization (Hein 2016). People have long used water in ways beyond transportation, particularly for leisure and to construct local identities and imagery. Ocean coastlines throughout the world were (re)designed over centuries to create working waterfronts and ports, and cities along them connected water and land.

A. A. Kermani (✉) · W. van der Toorn Vrijthoff
Delft University of Technology, Delft, The Netherlands
e-mail: a.arjomandkermani@tudelft.nl

W. van der Toorn Vrijthoff
e-mail: W.vanderToornVrijthoff@tudelft.nl

A. Salek
Rotterdam Municipality, Rotterdam, The Netherlands
e-mail: a.salek@rotterdam.nl

C. Hein (ed.), *Adaptive Strategies for Water Heritage*,
https://doi.org/10.1007/978-3-030-00268-8_18

In the second half of the twentieth century, ports turned their backs on cities and created geographical barriers between cities and water (Hoyle 2000). They started to develop outside the limits of cities to accommodate larger ships. They abandoned urban waterfronts for deeper water harbours further away from settlements, creating "non-place" areas in the city (Norcliffe et al. 1996, p 126), and they also made ports and their activities disappear from the sight and minds of citizens. Many elements of the established institutional structure of these cities (harbour police forces, insurance companies, tax organizations, etc.) vanished; meanwhile, cities' income from transport- and port-related industries and services radically dropped. In response, urban governments and planners redefined their urban capacities: they no longer identified as port cities, but as urban hubs for various branches of the modern economy like tourism, service, trade and manufacturing. (Ducruet et al. 2009, 112 & 55).

The dissociation of the port from the city has made their respective fates less and less dependent on each other, which has had significant socio-economic and political consequences. The challenging problem of how to deal with the vacated space in the heart of port cities around the world resulted in waterfront redevelopment programs within inner city regeneration plans (Desfor et al. 2010; Hoyle et al. 1994; Marshall 2001; Smith and Ferrari 2012). The regeneration of urban waterfronts became a well-established phenomenon in North America in the 1970s and spread to Europe city ports in the 1980s (Bone et al. 1997; Breen and Rigby 1996; Brown 2009; Schubert 2009; Ward 2006).

This chapter explores two aspects of urban life that are structurally involved in port-city interaction. First, it studies the evolution of planning policies on post-industrial waterfront spaces in the Netherlands before and after the financial crisis (2008), focusing on the former shipbuilding company Rotterdamsche Droogdok Maatschappij (RDM). The RDM site is a significant part of the old port area in Rotterdam, and its submarine and shipbuilding legacy has always been present in heads and hearts of the citizens. Second, this chapter explores how reawakening nautical culture and marine traditions can also reanimate the historical links between port and city. Focusing on a few heritage projects in the city of Rotterdam, this chapter briefly analyses their goals, achievements and effects on port-city interactions and the maritime identity of this global port-city.

The Urban Identity of Port Cities

Each port city has a distinctive urban identity that shapes how the port and the city interact (Lee 1998). Over the course of centuries, economic, social, environmental and cultural actors and networks interacted and produced urban life, institutional constellations and infrastructure (Warsewa 2011; Schubert 1994; Hein 2011, 178). The maritime identity of each port-city also formed in response to different stages of urban development and of seafaring and transport/trade-related urban strategies and policies (Boelens 2009, 62). Local and guest cultures, traditions and habits, all affect the spatial identity of the port cities. Therefore, the waterfronts and harbours represent the collective sense of memories and identity (Pleβka 2014, 144). Along with port hardware and infrastructure, other elements of collective memory—music, movies, texts, ceremonies—represent the culture and identity of each port city.

In fact, maritime culture historically has been established not just in physical objects but also in the collective memory of citizens through maritime traditions and ceremonies of nautical culture. Water-related values are hidden not only in the in their physical attributes but also in the cognitive interpretation of the port cities. Therefore in many situations, language, literature, food, film, photography and music may play a larger role than architecture and urban infrastructure in forming and representing the maritime identity of each port city. (Assmann and Czaplicka 1995, 129). Previously, the representation of the port city in the media conveyed the special sense of time and locality of casual seaman labour, jazz and pop music and migration, along with more negative portrayals of drugs and human trafficking and organized crime (Mah 2014, 10).

As the port detached from the city, the image of the port-city and the familiar stereotypes of harbours and waterfronts of earlier times—the longshoremen with colourful language, the containers and the working cranes—disappeared from urban spaces (Winslow 1998, 14). These changes mobilized networks of citizens, heritage organizations, artists and media to protect the dimming maritime feeling of urban spaces and the nautical legacy of harbour areas (Dündar et al. 2014, prologue). Over the last three decades, local communities have become more aware of the need to protect authenticity and water-related heritage, and public enthusiasm and civic participation in regenerating elements of nautical culture have increased (Feyen et al 2008; Stocker and Kennedy 2009).

Today, the governments of many port cities, in cooperation with several NGOs and private organizations, have set up new institutions to safeguard maritime heritage and protect naval customs. Some famous examples are the Association of European Maritime Heritage, the Maritime Heritage Association of Australia and the Scottish Maritime Heritage Association (Smith et al. 2015, 414). These institutions safeguard tangible maritime heritage in maritime museums and intangible heritage in seaport festivals, cultural excursions and historic sightseeing, paintings and postage stamps (Jinliang 2012, 47; OECD 2014, 136; Alegret et al. 2014).

Port and City Interrelation in Rotterdam

Over the centuries, the port and the city of Rotterdam have experienced a range of interrelationship patterns, from total interdependency up to spatial and strategic detachment (Meyer 1996; Boelens 2009). For most of that time, the port and the city were indistinguishable from each other. Rotterdam's waterfronts and harbours were located in the heart of the city or its nearby peripheries. Rotterdam's docks and harbours were not only port-city infrastructure, but they were also the site of commercial interactions between the city and the rest of the world (de Goey et al.

2004). Major port development here started in the nineteenth century, due to evolving maritime technologies and improvements in cargo handling systems; new small quays adjacent to the city centre reached to the sea. World War II seriously damaged the port and its infrastructure, including some main warehouses and crane facilities. But the port quickly grew to the west in the 1950s and after; the construction of the Botlek and Europoort areas came at the cost of the destruction of some villages and nature reserves. During the 1970s, the port of Europoort and the Maasvlakte was developed 20 km out of Rotterdam by reclaiming land from the North Sea, becoming one of the biggest harbour facilities in the world.

In the early 1980s, port activities including businesses and industries moved tens of kilometres away to new lands rising from the North Sea, dubbed Maasvlakte II. Some prominent parts of the social structure of the port city—seamen, sailors, shipbuilders—disappeared from urban society (Steenhuis-Meurs 2009; Stouten 2010).

Reinvention of the Seaport Identity of Rotterdam

Soon after, the negative consequences of the outward movement of the port came to the attention of municipal authorities. The municipality was responsible for both urban and port development, and the city council had always invested heavily in the expansion of the port; the urban area and port facilities were closely interlinked, as officials believed that the success of the port would make Rotterdam a big city. But with the removal of the port to a site distant from the city, the vacant former port facilities stood as spatial barriers between the city centre and the Maas River, while the river itself literally divided the city into a rich north bank and a poor south bank.

From 1981, the city of Rotterdam was involved in an urban renewal project (*stadsvernieuwing*), in which it redeveloped historic waterfronts and port infrastructure in the urban core, and made some neoliberal changes in the socioeconomic structure of the city (Ministry of Economy 2009). These transformations considerably altered Rotterdam's identity. When the port physically and culturally left the city, maritime institutions waned in importance of the identity of the port city. Today, the port of Rotterdam is facing an image problem, or rather, the lack of any image and like many port-cities across the world, it is looking for ways to evaluate water-related heritage to build a resilient and competitive port-city relationship (Aarts et al. 2012). In recent decades, the municipality of Rotterdam has invested tens of millions of euros to reinvent and memorialize Rotterdam's maritime identity and to recover the mutual relationship between the city and the port to make the faded image of the port vivid once more.

Turn in Rotterdam's Urban Policies

In order to create a common urban identity for city inhabitants, the port and the city had to foster new mutual benefits. Both were willing to invest in innovative solutions for decreasing urban congestion, increasing mobility issues, diversifying the economy and developing housing. The Rotterdam administration courted private investments and slowly changed from supply-led urban planners to development-led facilitators (Wigmans 1998). After the energy crises of the 1970s, Rotterdam shifted from a social housing structure to more liberal housing policies (Mak and Stouten 2014, 1). By the end of the 1980s, Rotterdam's urban policies had taken an entrepreneurial turn. This so-called first wave of waterfront programs saw the abandoned port areas near the city centre as an excellent opportunity to rethink the identity of the city. The municipality prepared a large-scale master plan which recognized the River Maas as "the DNA of the city" (Aarts et al. 2012, p 13).

The regeneration of vacant harbour zones inside the urban core of Rotterdam (Fig. 1) started with the transformation of the *Oude Haven* (Old Port) into new quality housing areas. Soon it expanded to other areas near the city centre, more specifically the *Leuvehaven* (Fig. 2) *Wijnhaven* and *Zalmhaven* neighbourhoods into commercial and office areas. Later, the *Scheepvaartkwartier* and *Parkhaven,* a harbour zone and light industry zone, were turned into a tourist resort and high-quality residential areas. With the *Kop van Zuid project*, Rotterdam developed a new central business district on the south bank of the River Maas. The expansion of the metro line to the other side of the river and the construction of the monumental Erasmus Bridge in 1996 also gave the *Kop van Zuid* project national and international publicity, even though most of the initial office spaces had to be changed to residences due to the lack of market interest (Fig. 3).

In 2002, planners from the Municipal Department of Economy, Port, and Environment (*Economie, Haven en Milieu*) were drafting a promotional report, Port Vision 2020 (*Havenplan 2020*). International study trips to the port cities of Hamburg and London inspired them with a new vision of the port, which included three significant decisions: expanding the port (*Maasvlakte 2*), incorporating the port authority and developing the Stadshavens (*City Ports*) area. The port of Rotterdam wanted to improve its position as the smartest and most sustainable port in Europe; at the same time, city authorities wanted to strengthen the profile of the area by diversifying the economy with new sectors. Space has been created to develop new activities important to both the city and the port (Aarts et al. 2012) in the approximately 1600 ha area in the periphery. Here, inside the city's highway ring, port activities had animated the waterfront in the twentieth century; some port areas remained. The area included in the second phase of the Stadshavens project were *Merwehaven* and *Vierhaven* on the north side of the Maas River; *Waalhaven*, *Eemhaven* and the RDM campus on the south bank; and *Rijnhaven* and *Maashaven* on the eastern part of the plan (Fig. 4).

The Rotterdam Stadshavens Development Company (*Ontwikkelingsmaatschappij Stadshavens Rotterdam*) was established later that year as a new incorporated organization to initiate, facilitate and stimulate the transformation of vacant port industrial

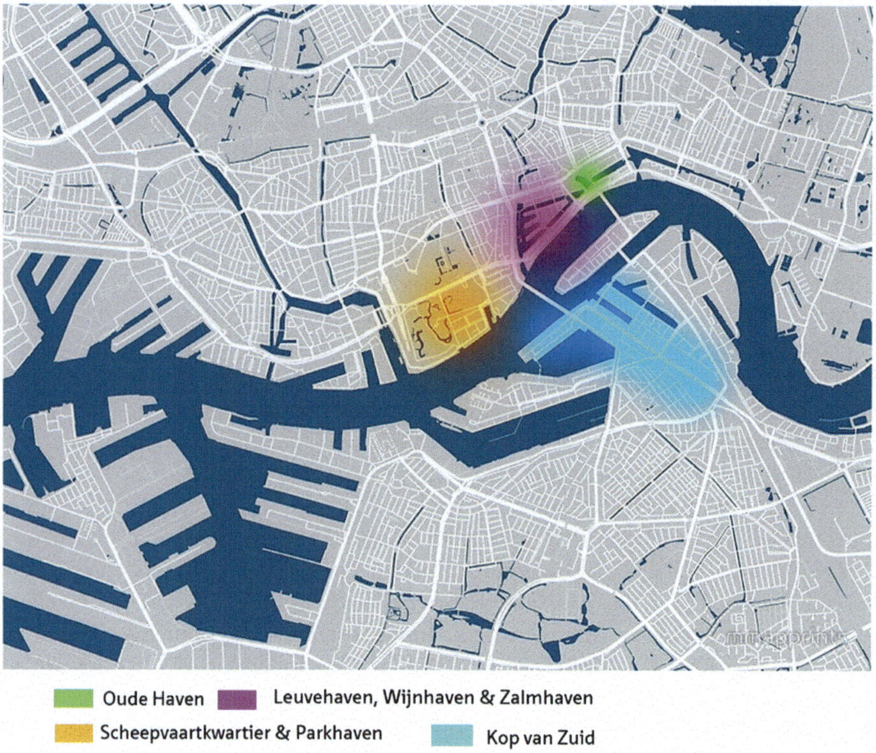

Fig. 1 First wave of waterfront regeneration programmes in Rotterdam, credit Azadeh Arjomand Kermani; released under a Creative Commons Attribution-NonCommercial-NoDerivatives 4.0 International Licence

areas in the west and south-western parts of the city into urban living and working areas by realizing a vision and a plan for their future spatial arrangement (Daamen 2010). In addition, the Company planned an expert seminar for early January 2004 to discuss the added value and potential of the conservation, adjustment, and reuse of the port landscape in the CityPorts area. The CityPorts Company moved its new headquarters to the former office of the Rotterdam Dry-dock Company (*Rotterdamse Droogdok Maatschappij*) in the heart of the CityPorts area.

In May 2005, the CityPorts Company published the Rotterdam CityPorts Development Strategy, a plan for the future spatial claims of the city *and* the port. The document gave an outline of intended changes in the plan areas of Stadshavens, including concrete plans for several locations (OMSR 2005). The strategy focused on the economic renewal of the area with the concept of harbour out/city in, transforming monofunctional areas within the Stadshaven into urban living environments (Structuurvisie-Stadshavens Rotterdam 2011). It valued the water and harbour infrastructure as an attraction for leisure-related activities and urban development, similar to previous waterfront redevelopments in Rotterdam. Moreover, it integrated water-

Fig. 2 Rotterdam Leuvehaven, 2017, credit Arash Salek; released under a Creative Commons Attribution-NonCommercial-NoDerivatives 4.0 International Licence

Fig. 3 View towards Kop van Zuid and the Erasmus bridge, credit Arash Salek; released under a Creative Commons Attribution-NonCommercial-NoDerivatives 4.0 International Licence

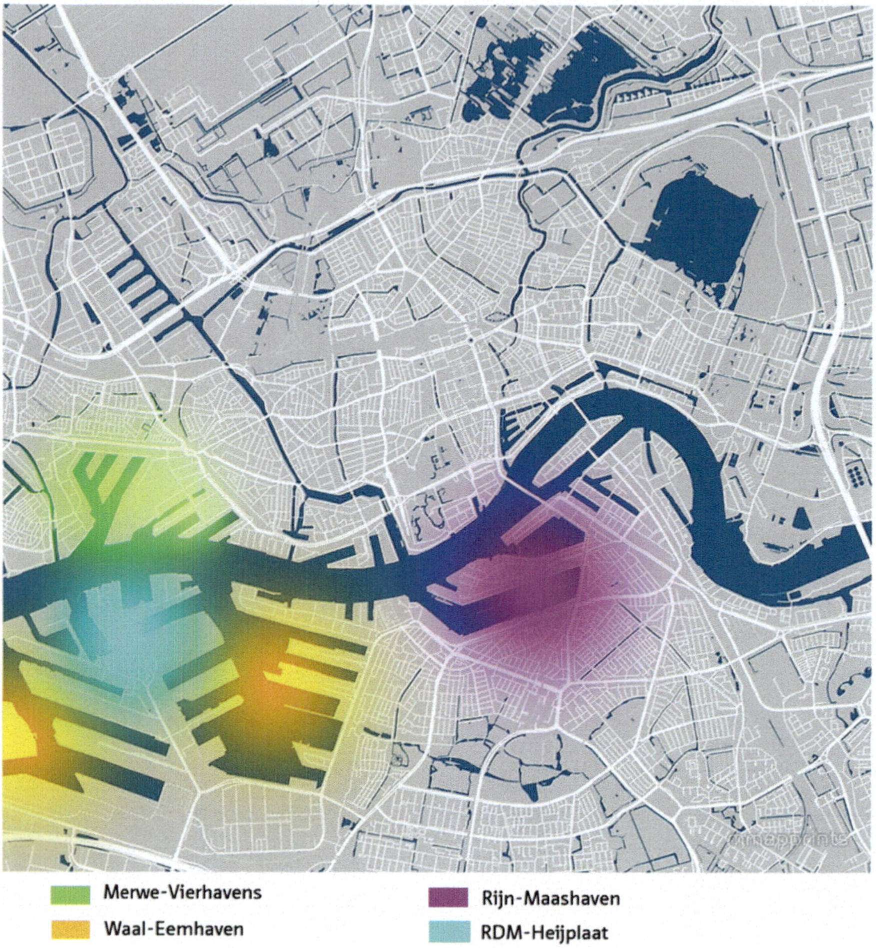

Fig. 4 Second wave of waterfront regeneration program in Rotterdam (Stadshavens area), credit Azadeh Arjomand Kermani; released under a Creative Commons Attribution-NonCommercial-NoDerivatives 4.0 International Licence

related heritage and remnants of port activities into these developments because they are the historic identity of the Rotterdam harbour.

Cruise Ship Tourism as an Instrument of Representing the Maritime Image of a Port City

Yet another approach to revitalizing traditional waterfronts and to re-establish the missing link (Hein and Hillmann 2016) was the reintroduction cruise ships to the centre of Rotterdam. From 1873 to mid-1970, the Wilhelminakade in Rotterdam had been the home of the Holland America Line (HAL). During this period, many steam ships, like the famous "De Rotterdam", "De Nieuw Amsterdam" and "De Statendam", took thousands of passengers from Rotterdam to faraway places. The current Hotel New York was originally the headquarters of the HAL and its passenger terminal. During this period, many liners were operating between the Netherlands and North America and were instrumental transporting more than half a million immigrants from Europe to North America. In the second half of the twentieth century, because of the rapid growth of air transportation, especially the introduction of transatlantic jet air travel, the prominent image of passenger ships and liners disappeared from the Wilhelminakade. In the 1980s, the HAL moved its headquarters from Wilhelminakade to Seattle, and the city lost its dominant position as a global hub of passenger transportation.

This image changed in recent years with the development of cruise travelling (*Financieele dagblad* 3 May 2016). Since 2008, the European market for cruise travelling has grown by 49%. In the late 1990s, the Wilhelmina pier, the historic pier of the Holland America line and its huge passenger terminal building, were owned by the City of Rotterdam. In 2013, the Port and the municipality of Rotterdam invested almost 7 million euros into modernizing the monumental building into Cruise Terminal Rotterdam to bring the water back into the minds and hearts of Rotterdammers and to turn the downtown of Rotterdam into a living and leisure urban area (Rotterdam's City Vision 2030). The renovation was partly motivated by the demands of the cruise industry and tourism organizations, which wanted to double the number of cruise ship arrivals from 28 to 60 per year (Schipper 2010, 65). In May 2016, the Financieel Dagblad [*Financial Daily*] wrote, "The presence of the cruise ships in Rotterdam is good for the port and city imaging; however, for inventing this image, the municipality had to spend lots of money". In his speech at the opening of the renovated cruise terminal, the mayor of Rotterdam, Ahmad Aboutaleb, said that the cruise terminal is a good vehicle for connecting the port and the city and a very beneficial tool for the urban economy. He added that the main reason for investing in a new cruise terminal was to reattach the city to the port and to keep Rotterdammers connected to the port even if it is gone (Trouw 9 July 2015) Today, huge cruise liners enter Rotterdam (Kop van Zuid) from the North Sea to navigate their way down the Maas River through the central business district to the recently renovated International Cruise terminal on the Wilhelmina pier (Fig. 5).

The cruise liners are the remaining image of port activity. They keep the port image in the urban space and memorialize the nostalgic image of port experience. Each week, the once-abandoned waterfronts in the Kop van Zuid area are again full of maritime life and nautical events. While the cruise economy seeks to make best use of the remaining port infrastructure in downtown Rotterdam, the city presents its less discovered tangible and intangible seaport values, local self-image, and cultural identity to the outside world.

Fig. 5 Arrival of the Harmony of the Seas (the second largest passenger ship in the world) into the cruise terminal Rotterdam on 24 May 2016, credit Arash Salek; released under a Creative Commons Attribution-NonCommercial-NoDerivatives 4.0 International Licence

The Redevelopment of RDM Terrain

The Rotterdamsche Droogdok Maatschappij (RDM) shipyard has a long history in the shipbuilding industry. It was founded on the south bank of the Nieuwe Maas in 1902 as a continuation of an older shipyard, Maatschappij De Maas (Crimson 2005). As it gradually flourished, the work force grew and RDM authorities started building housing for the workers; the first development was a 1914 garden village, Heijplaat, which comprised 500 dwellings, 3 churches and other public facilities such as shops, parks and sports fields (Vries 2014). By the Second World War, the area had grown to 40 ha and became one of the largest shipyards in Europe. One ship built there between 1955 and 1959 was the ocean liner *SS-Rotterdam* of the Holland America Line (Crimson 2005). In these prosperous times, RDM wharf employed some 5000 people and therefore added a new housing district: "Het Nieuwe Dorp" (The New Village).

Until the year 2002, the Port of Rotterdam Company is a municipal organization. During the 1980s, the urban management and services of the RDM shipyard and the harbour zones on the south bank of the Maas River were transferred to the port companies. However, many ex-harbour areas in the north bank of Mass River remained part of the city.

After various mergers, the company was taken over by the *Rijn Schelde Verolme (RSV)* Company and the village of Heijplaat was sold to the Woonbron housing corporation in late 1980s. In 1996, shipbuilding ended at the Heijplaat, and submarine maintenance and servicing ended in 1999.

In the early years of the new millennium, the Heijplaat was a partly vacant and relatively barricaded urban district of Rotterdam. At that time, the municipality together with the port-company of Rotterdam and the local housing corporation (Woonbron) realized the imperative need for renovation and redevelopment of this district. In collaboration with the University of Applied Science Rotterdam (which at that time was running short of educational space), the City, Port Company and Woonbron established an organization for redeveloping the Heijplaat. This organization would implement plans and strategies for a sustainable urban district in Heijplaat in combination with an educational campus and a high-tech business complex for start-ups. The renovated head office of the old *RDM* shipyard became well-known as an investment prospect in Rotterdam by the summer of 2005. Its buildings, its raw port industrial surroundings and the neighbouring garden village Heijplaat attracted interest from a variety of visitors. Along with those working on the CityPorts project or related tasks, both the Rotterdam University of Applied Science (Hogeschool Rotterdam) and the Albeda (Technical) College were looking for expansion space as well as connections with businesses. Representatives from Woonbron were looking for opportunities to create high-value living and working spaces. Inspired by notions such as open innovation and informal communities, and by field visits to other renovated historical industrial sites (such as the Zollverein in Essen), city planners formed the RDM Campus as a primary component in the revival of the CityPorts area. The RDM Campus connects research, multi-level education (vocational and higher level), and business—the "golden triangle"—by clustering start-ups, educational institutions, and research firms on the former RDM shipyard. The port authority supported the project: although the campus concept is a new approach and outside of its core business, it addressed the shortage of specialized technical workers and promised to attract other tenants to the area.

Even though the renewal of the Heijplaat was limited to refurbishing existing housing stock, especially rental housing, and new construction targeting higher income groups, Woonbron envisioned the RDM Campus as a hub for renewal, increasing the number of shops and other activity in the village to a critical mass (van Tuijl and Otgaar 2016). In addition, educational institutions and firms at the RDM Campus could also use the village. One example is Concept House Village, a number of houses where temporary users can test all kinds of new living concepts while the Institute of Concept House Construction and management CHIBB, monitors them and their inhabitants in order to analyse their reactions and adapt the technology accordingly.

The RDM development was a joint venture and a public–private partnership with formal partners as key players (educational institutions, Port Authority Rotterdam and Woonbron housing association) playing outside their core business. None of them had experience in large-scale integrated urban development and they all faced new challenges, including persuading others in Rotterdam to support the project and learning integrated area-based marketing (van Tuijl and Otgaar 2016).

As the owner of the buildings at RDM, the Port of Rotterdam Authority developed the site (Hooijer and Muris 2010). A group of architectural historians based in Rotterdam (Crimson) conducted a detailed historical and cultural study of the RDM terrain to clarify the heritage value of the buildings and other historical elements, commissioned by the Stadshavens development company in collaboration with Urban Planning + Housing/Spatial Development/Monuments Rotterdam and Netherlands Department for Conservation.[1] In addition, a detailed historical orientation study was requested for the renovation of each building. Some of the buildings on RDM site were already in the process of becoming a registered national monument. However, the municipality of Rotterdam and the Port of Rotterdam made a gentlemen's agreement to hold the nomination until after the redevelopment and renovation of the site: the developer (the Port) would get more freedom and authority, and it would carefully handle redevelopment with respect to the historical value of the site and the buildings. The Port of Rotterdam started to work on a master plan and business model for the campus.

The educational institutions moved into the renovated central machine hall and the former head office of RDM (now stands for Research, Design & Manufacturing) in 2009. The campus became a place where students and companies collaborated in an open environment and focused on new economic activity and sustainable and innovative solutions in subjects like manufacturing, building and maritime. In the opening ceremony after the development of new office area named as "RDM Innovation Dock", the developers opened the "blue gate", once the main entrance for RDM workers, to local residents; this ritual welcomed them to the site that had been closed to them for so many years. The village of Heijplaat benefited from the renovation, gaining public space, improved access to the waterfront and public transport (a water bus) to the city centre. Residents could eat at the new Dock Café in RDM's former staff canteen (Vries 2014).

The redevelopment of the RDM site was the result of a lucky coincidence: the distinctive site became available and port authorities were looking for a new tenant just as educational establishments had a real need for additional space. Then, the port authority used its traditional funding model to carry out the project: As the landowner and port manager, it actively searched for tenants; after they signed long-term contracts, the port authority renovated the halls according to the wishes of the clients; then, it used the income from the long-term contracts fund further development. In addition, it offered tenants empty plots on the RDM site for new construction. Even though this is a traditional model for the Port Authority, it was a new approach for area development, which normally works with a strict master plan that has been prefinanced with large investments. The Port Authority's approach was more flexible and included less financial risk (van Leeuwen 2013).

[1]Rijksdienst voor de Monumentenzorg.

World Port Days: A Tool for Reinventing Maritime Culture During the Transformation of the Port-City

On the first weekend of September, Rotterdam celebrates its maritime culture and identity during World Port Days, with the harbour as the epicentre of the party. It is Europe's largest port-city festival, attracting thousands of visitors, and it has been happening every year since 1978. Although downtown Rotterdam is no longer a seaport area, it still hosts several maritime and municipal institutions and companies, both local and global.

Looking at the number of visitors and the nautical organizations participating in this event, we can discern its socio-economic benefits (Otgaar et al. 2016, 169). The World Port Days seem to be a prominent podium and an excellent opportunity for bringing public attention to the value of vital interconnection between the port and the city.

During the World Port Days, Rotterdam's maritime identity becomes apparent not just as a folkloric leftover but also as the primary basis for current social and economic development (Van der Berg and Tuij 2016, 218–220). There, the local community and guests gather together and enjoy various activities—ship tours, naval activities, demonstrations, seminars, excursions, exhibitions, music and much more.

Conclusion

Like many port cities across the world, "Rotterdam is looking for ways to reproduce its water-related heritage values to build a responsive, resilient, and competitive port-city relationship". (Stadsvisie Rotterdam 2030 2007). In recent decades, the City of Rotterdam has invested millions of euros to empower the maritime heritage of this famous port city, funding new projects to restore and renovate historic waterfronts, boats, and port infrastructure and to organize related festivals, seminars, and events. The modernization of the Cruise terminal, the annual World Port Days, and the revitalization of Stadshaven—including the redevelopment of the RDM—all shared one goal: reinvention of the links between the port and the city and revitalization of the faded image of port in urban spaces.

Recently, some urban scholars have pointed out the importance of a dynamic maritime identity and water-related heritage values to historic port cities, and have identified the role of a vital port city interconnection in urban renewal and the redevelopment of urban areas because of rapid urban change and socioeconomic makeovers (Warsewa 2011; Dell' Acqua and Wegman 2017, 667).

Each port city needs to consider more wide-ranging approaches for sustainable development, which includes their maritime identity past and present. The redevelopment of the Rotterdam waterfront and the revaluation of its heritage have evolved significantly during the last decade. A Unit Approach, with an integrational perspective and a grand blueprint with governmental financial support, as in the Kop van Zuid project, was replaced by a Chain Approach, a more pluralist view with a public–private partnership within a governance context, as in the RDM Campus project.

But the future trajectory in the Makers Space area breaks with the past and is based on empowering local communities and carrying out the long-term vision on a small scale. The steering will be done by a small unit (Rotterdam CityPorts), a facilitator and a connector between existing and future developers and entrepreneurs. The strategy is shifting towards a flexible, network-based strategy and a more private engagement in development of the coming phases of this project. With its public–private cooperation strategy for the Makers Space, Rotterdam's Stadshavens has created a clear pathway for the redevelopment of the area: promising to be socially inclusive, connect the city to the ports, and create a lively mixed-use area. The new flexible approach in dealing with waterfront redevelopments in Rotterdam considers the use of the places, and the story behind them, more significant than the built environment. At RDM, the planners retained the physical aspects of the buildings and the area to the extent that they can tell their story in the future. They did not aim to make the area attractive or turn it into a museum, but instead highlighted its industrial legacy and its relation to water. The new approach uses local identity and well-known local names and brands to attract more businesses and innovative industries to the area.

This new flexible approach allows for a wide range of users and private investors to invest in the development of this area. However, the risks and opportunities for the private sector are not very clear in this flexible framework. This creates a lot of uncertainty for potential investors and private developers. So, as more new entities and private sectors get involved in the development of the area, adequate control by the monument committee and other heritage bodies within the municipality is becoming necessary. It will also be necessary to make the tools and regulations more clear within this new flexible approach.

References

Aarts M, Daamen T, Huijs M, de Vries W (2012) Port-city development in Rotterdam: a true love story. http://urban-e.aq.upm.es/. Rotterdam, 2013

Alegret J et al (2014) Revisiting the coast: new practices in maritime heritage. Documenta Universitaria, 2012

Assmann J, Czaplicka J (1995) Collective memory and cultural identity. New Ger Crit 65:125–133. https://doi.org/10.2307/488538

Boelens L (2009) The urban connections, an actor relational approach to Urban planning, vol 62. NAI

Bone K, Bone E, Betts MB (1997) The New York waterfront: evolution and building culture of the port and harbor. Monacelli, New York

Breen A, Rigby D (1996) The new waterfront: a worldwide urban success story. Thames and Hudson, London

Brown PH (2009) America's waterfront revival. Port authorities and urban redevelopment. University of Philadelphia Press, Philadelphia

Crimson (2005) Cultuurhistorische verkenning van de RDM

Daamen T (2010) Strategy as force, towards effective strategies for urban development projects: the case of rotterdam city ports. Ph.D. Delft University of Technology
de Goey F et al (2004) Comparative port history of Rotterdam and Antwerp (1880–2000). Aksant
Dell' Acqua G, Wegman F (2017, April 10–12) Transport infrastructure and systems: proceedings of the AIIT international congress on transport infrastructure and systems. Rome, Italy, p. 667
Desfor G, Laidley J, Schubert D, Stevens Q (eds) (2010) Transforming urban waterfronts: fixity and flow. Routledge, London
Ducruet C, Notteboom T, Langen P (2009) Revisiting inter-port relationships under the new economic geography research framework. Ash gate. In: Ports in proximity: competition and coordination among adjacent seaports, Ashgate, pp 11–27
Dündar S et al (2014) New faces of harbour cities. Cambridge Scholars Publishing
Feyen J et al (2008) Water and urban development paradigms: towards an integration of engineering, design and management approaches. CRC Press
Hein C (2011) Port cities: dynamic landscape and global networks. Routledge, New York
Hein C (2016) Port cities and urban waterfronts: how localized planning ignores water as a connector. WIREs Water 3:419–438. https://doi.org/10.1002/wat2.1141
Hein C, Hillmann F (2016) The missing link: redevelopment of the urban waterfront as a function of cruise ship tourism. In: Porfyriou H, Sepe M (eds) Waterfronts revisited. Routledge
Hooijer B, Muris G (2010) RDM campus: an innovative learning and working environment in the port of Rotterdam
Hoyle BS (2000) Global and local change on the port-city waterfront. Geogr Rev 90(3):395–417
Hoyle BS, Pinder D, Husain MS (1994) Revitalising the waterfront: international dimensions of Dockland redevelopment. Wiley, Chichester
Jinliang Q (2012) Protecting China's maritime heritage: current conditions and national policy. J Mar I Cult 2212–6821
Lee R (1998) The socio-economic and demographic characteristics of port cities: a typology for comparative analysis? Urban Hist 25(02):147–172
Mah A (2014) Port cities and global legacies, urban identity, waterfront work, and radicalism. Springer
Mak A, Stouten P (2014) Urban regeneration in Rotterdam; economic and social values. Eur Spat Res Policy 21(1). https://doi.org/10.2478/esrp-2014-0008
Marshall R (ed) (2001) Waterfronts in post-industrial cities. Spon, London
Meyer H (1996) City and port transformation of port cities. London, Barcelona, New York and Rotterdam NAI
Ministry of Economy (2009) Economische visie op lange termijn de mainport Rotterdam
Norcliffe G, Basset K, Hoare T (1996) The emergence of postmodernism on the urban waterfront. J Transp Geogr 4(2):123–134
OECD, Merk O (ed) (2014) The competitiveness of global port-cities: synthesis report. Available at https://www.oecd.org/cfe/regional-policy/Competitiveness-of-Global-Port-Cities-Synthesis-Report.pdf
OMSR (2005) Stadshavens Rotterdam Ontwikkelingsstrategie—concept. OMSR, Rotterdam
Otgaar A et al (2016) Industrial tourism: opportunities for city and enterprise. Routledge, Prologue
Pleßka N (2014) The intelligible metropolis: urban mentality in contemporary London Novels. Transcript Verlag
Schipper (2010) BC AD (English edition). Benthem Crouwel 1979–2009. 010 Publisher
Schubert D (1994) Zwischen Wasser und Land: Neue Entwicklungen in alten Häfen. In: Deutsche bauzeitung, Jg. 128(11): 50–56
Schubert D (2009) Ever-changing waterfronts: urban development and transformation processes in ports and waterfront zones in Singapore, Hong Kong and Shanghai. In: Graf A, Huat CB (eds) Port cities in Asia and Europe. Routledge, London
Smith, HD, de Vivero, JLS, Agardy TS (2015) Handbook of ocean resources and management. Routledge, p 414

Smith H, Ferrari MSG (eds) (2012) Waterfront regeneration. Experiences in city-building. Routledge, London

Stadsvisie Rotterdam 2030, publication of the City of Rotterdam (2007)

Steenhuis-Meurs (2009) Spangen–Rotterdam, cultuurhistorische verkenning. Available at www.rotterdam.nl

Stocker L, Kennedy D (2009) Cultural models of the coast in Australia: toward sustainability. Coast Manag 37(5):387–404

Stouten P (2010) Changing contexts in urban regeneration: 30 years of modernisation in Rotterdam. Techne Press

Structuurvisie Stadshavens Rotterdam, the city of Rotterdam. Sept 2011

Van der Berg L, van Tuij E (2016) Urban festivals as tools for social development, the case of world port days in Rotterdam. In: Nel-lo Oriol, Mele Renata, 2016. Cities in the 21st Century. Routledge

van Leeuwen M (2013) Gebiedsontwikkeling nieuwe stijl standaard voor RDM Heijplaat. vastgoedmarkt

van Tuijl E, Otgaar A (2016) RDM campus. In: Carvalho, L, Van Den Berg L, Galal H, Teunisse P (eds) Delivering sustainable competitiveness: revisiting the organising capacity of cities. Taylor and Francis

Vries IMJ (2014) From shipyard to Brainyard the redevelopment of RDM as an example of a contemporary port-city relationship. In: Alix et al (eds) Port-city governance, pp 107–126

Ward SV (2006) Cities are fun: inventing and spreading the Baltimore model of cultural urbanism. In: Monclús FJ, Guardia M (eds) Culture, urbanism and planning. Ashgate, Aldershot

Warsewa G (2011) Adaption and individuality, the reinvention of the port city. In: Häyrynen S et al (eds) Locality, memory, reconstruction: the cultural challenges and possibilities of former single-industry communities. Cambridge Scholars Publishing

Wigmans G (1998) De facilitaire stad: Rotterdams grondbeleid en postmodernisering. Delft University Press, Delft

Winslow C (1998) Waterfront workers: new perspectives on race and class. University of Illinois Press

Azadeh Arjomand Kermani is a postdoctoral researcher at the Chair of Heritage & Values at Delft University of Technology. Her dissertation explored urban design of the historical built environment, focusing on Iran. Her research interests include the impact of planning policies and governance on the management of both tangible and intangible heritage.

Wout van der Toorn Vrijthoff is emeritus associate professor in real estate management and development at Delft University of Technology. His publications include the book *More Urban Water: Design and Management of Dutch Water Cities*. He is currently working with colleagues from other European universities on heritage planning projects.

Arash Salek is an urban-regional development advisor for the City of Rotterdam. He has also been a policy advisor in heritage and urban history. His background is in architecture, urban history and metropolitan development; his work examines the regeneration of post-harbour urban areas.

The creation of new heritage at Splashdowncentre Grand Turk: commemorating the splashdown of the Friendship 7 capsule off the island's coast in 1962, Supersudaca, released under a Creative Commons Attribution-Share Alike 3.0 Unported license

Chapter 19
From HERITAGE to Feritage: How Economic Path Dependencies in the Caribbean Cruise Destinations Are Distorting the Uses of Heritage Architecture and Urban Form

Supersudaca, Sofia Saavedra Bruno, Martin Delgado and Felix Madrazo

Abstract While the impact of cruise shipping is largely mitigated by the consolidated and diverse economies of port cities, such as Hamburg, Tokyo, and Seattle, it is a key issue in the current transformation of the Caribbean cruise destinations that increasingly depend on tourism. This chapter illustrates how cruise tourism has triggered spatial and sociocultural changes in urban form and architectural heritage in the Caribbean region. It argues that those transformations fall into a path dependency thread, and that we are at a critical juncture whose stakes include the risk that cruise lines might soon just leave heritage sites altogether. The chapter also gives a broader reading of the contemporary modes of cruise tourism exploitation. The "Introduction" describes how previous economic dependencies shaped and conditioned the built heritage (urban form, urban function, and heritage architecture) of Caribbean port cities and how spatial relationships of port, city, and hinterland ultimately followed the spatial logics of colonial exploitation. It describes how this historically established (hence path-dependent) economical patterns are still visible in the current operating modes of cruise tourism in the region. The section "How Historical Political and Socio-economic Dependencies Shaped Both Caribbean Port City Heritage and Current Operating Modes of Cruise Tourism" describes the role of heritage architecture of port cities, in the context of cruise lines' economic interests. The section "Heritage Architecture of Caribbean Cities and Cruise Lines' Economic Interests" looks more specifically at how the cruise lines' original interest in heritage preceded their actual disinterest. If the cruise lines were the first actors to add economic value to Caribbean heritage, the Caribbean cruise experience now sidesteps—if not actually fakes—local culture, cities, and economy.

Supersudaca · S. Saavedra Bruno (✉) · M. Delgado · F. Madrazo
Rotterdam, The Netherlands
e-mail: ssaavedra@supersudaca.org

M. Delgado
e-mail: martindelgado@supersudaca.org

F. Madrazo
e-mail: felixmadrazo@supersudaca.org

C. Hein (ed.), *Adaptive Strategies for Water Heritage*,
https://doi.org/10.1007/978-3-030-00268-8_19

Keywords Path dependency · Cruise tourism · Heritage · Caribbean · Cruise impact · Destination · Destination resilience · Economic lock-in · Expenditures · Economic benefits · Spatial economics · Socio-spatial impact · Waterfronts

Introduction

Since the early sixteenth century, economic considerations have influenced the built heritage of Caribbean port cities. The similarity of the colonial exploitation model and the current operating modes of cruise tourism unveils how historical patterns are repeated in new forms and the essence of power relations—in which the main economic decisions are still in the hands of foreign investors—remains identical. We coined the pun feritage to show simultaneously how contemporary Caribbean cruise destinations are distorting the uses of heritage architecture while resembling spatial and economic practices of colonial times. The title of the paper feritage is a reference to the ongoing deformation of the use of heritage for the purposes of the cruise industry. For example, the constant improvements of replicas are challenging what local tourist boards considered essential and irreplaceable for the tourist purpose of visit: to experience authentic places.

Cruise ship tourism is one of the fastest growing and most stable industries (Rodrigue et al. 2013; Rodrigue and Notteboom 2013), and the landside tourism it generates has transformed urban form, urban function, and heritage architecture around the world (Hein 2013, 2016). Scholarly research on cruise tourism has nevertheless focused on isolated aspects of the cruise industry, notably economics, or on the need for tourist-geared adaptation of the historic built environment and the port facilities; it has not engaged with heritage debates (McCalla 1998; Vaggelas and Pallis 2010; Gui and Russo 2011). In turn, most literature that does explore cruise tourism and heritage is focused on preserving these values in the face of increasing economic pressures (Avrami 2013). The relationship between (cruise) tourism and cultural heritage values of local communities is only starting to be looked at by academics, including Hein (2016) and more briefly Epler Wood (2017). The sociocultural challenges that cruise lines bring to the shores of their destinations and attempts to formulate planning solutions have only recently been explored (Epler Wood 2017). But locals comment strongly and express concern on the impact of cruise shipping on heritage values of local communities, notably in the media. Venice and Barcelona have been at the forefront of recent protests against cruise tourism (Corcoran 2017; Coldwell 2017). Social media has also covered and commented on these protests, but those conversations have yet to be studied.

Dowling (2006) probably offers the most comprehensive overview of academic work on cruise shipping, but the stress in such studies on the industry rather than its destinations is remarkable. A range of significant studies on destination evolution under the influence of tourism (Saarinen 2004; Jaakson 2004) identifies the area in the port visited by tourists as a "tourist bubble" (Jaackson 2004) consisting of a core and a periphery. Nonetheless, a need remains for a more comprehensive investigation of the effects of cruise shipping on historical urban areas (Hein and Hillmann 2013, 2016), and of the transformative effect of the cruise industry on the spatial relations between city, port, and hinterland (Weaver 1993). Through various research formats, our research group of architects and planners, Supersudaca (*Sudaca* is a pejorative term among Spanish people for a Latin American), has investigated the impact of the latest business model of cruise tourism and the spatial relation between city, port, and hinterland in the Caribbean (Saavedra Bruno 2007; Supersudaca 2014); more recently, Supersudaca was asked to advise the government of Turks and Caicos (Saavedra Bruno et al. 2017). In our report, we aimed to unveil the mechanisms behind the changing spatial relationship of the cruise pier with the urban territory as a dynamic relation of interdependence between local and foreign actors; we concluded that new policies are needed to improve and integrate cruise shipping with the local population and their economy. Meanwhile, heritage remains sidelined by strategic positioning of the pier far from the historic center. Yet the role of tourism and heritage within the larger relationships between actors has not yet been investigated. In this chapter, we explore how heritage debates today play out in discussions on cruise ships on the Caribbean Waterfronts.

Over the last twenty years, some of these debates have created a distinctive power balance in which policymakers and planners focus on the economic side of the cruise ship industry and consider urban form and heritage architecture only as supporting elements of the tourist offer instead of seeing heritage as an integral part of the cultural values of the local population. The dominant discourse often adopts a short-term perspective that supports this approach, mainly looking at tourist arrivals and expenditures, leaving aside local actors, their agendas, their interest in urban form and heritage, and their specific identity concerns. But the complex interaction of (cruise) shipping with port, city, and hinterland requires a multifaceted approach that acknowledges long-term development (Hein 2016).

With scholars of historical institutionalism, we argue that the current model of cruise tourism contains a pattern of historical development with trajectories that are inherently difficult to reverse, so-called path dependencies (Hacker 2002; Pierson 2004; Mahoney and Thelen 2010; Sorensen 2015). The decision points during which new institutional configurations are established and new developmental trajectories are launched—usually referred to as "critical junctures"—are crucial to the future direction of each destination (Collier and Collier 1991; Capoccia and Kelemen 2007). In line with Musterd (2012), we propose that city region's attraction to the creative sectors and their potential economic development is influenced by the path of historical developments. Using the concept of path dependence theory and including the built environment as another actor, we analyze the influence of cruise shipping on the development and architectural heritage preservation of port cities in the Caribbean islands.

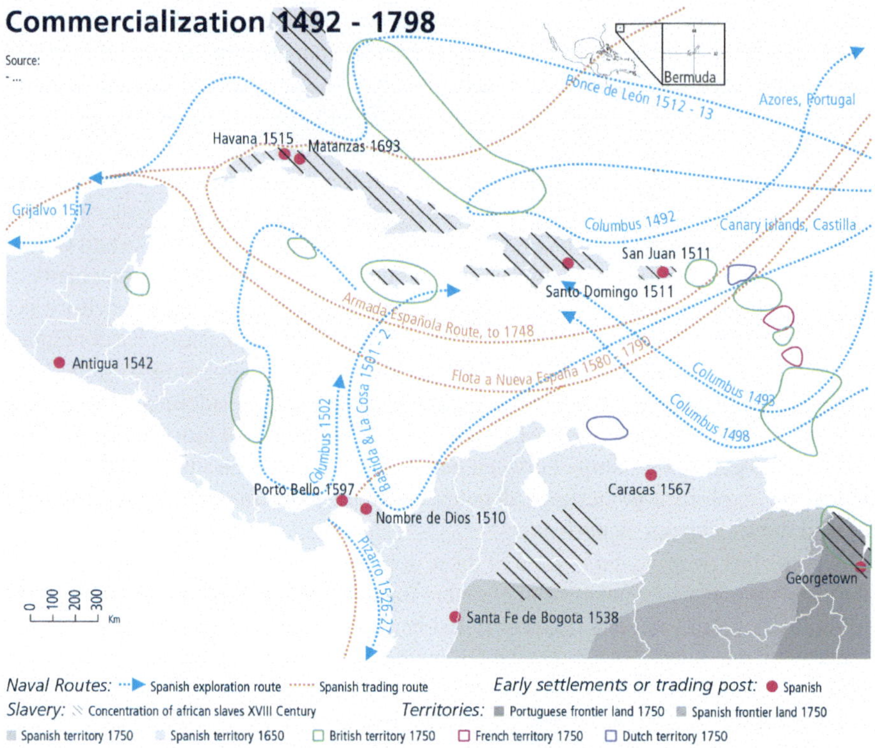

Map 19.1 Commercialization 1492–1798. *Source* Supersudaca, Al Caribe research with auspices of Prince Claus Fund; released under a Creative Commons Attribution-NonCommercial-NoDerivatives 4.0 International License

How Historical Political and Socioeconomic Dependencies Shaped Both Caribbean Port City Heritage and Current Operating Modes of Cruise Tourism

Historical political and socioeconomic dependencies shaped Caribbean port city heritage—both urban form and architectural production—in ways that are still visible in the current operating modes of cruise tourism. This is in part due to the continuation or resurgence of geopolitical structures of the past, but perhaps more interesting is the current relevance of spatial strategies for the cruise industry from that distant past that had as its primary objective the control of flows of capital in the Caribbean. Cruise tourism in Caribbean port cities relies on principles of mercantilism and monopoly control that were normal practice in the region in the sixteenth and seventeenth centuries (see Map 19.1).

What we perceive as the oldest heritage now represents, sometimes in diagrammatic clarity, the economic and cultural policies intended to create an urban system that facilitated trade, security, and stability. Furthermore, the spatial configuration of Caribbean port cities—their grid systems and fortifications—expresses how the market did not allow competition from other places. Culturally, port cities appeared to be neutral spaces but a closer look reveals that they were highly hierarchical, pushing local indigenous populations to the fringes of the system. Emerging models of tourism today echo several of these dynamics.

The mostly European built heritage of the Caribbean islands dates to the beginning of the sixteenth century. This legacy is closely interrelated with the history and interests of the colonial exploitation of the region. The colonization model follows a pattern of discovery and conquest, after which colonizers identified resources to exploit, and, depending on their importance, protect them militarily with city fortifications and (later) force on the high sea. More specifically, once colonizers had discovered a new place, they founded a city: distributing land among conquistadores, building up the infrastructure of extraction, organizing and distributing forced labor, and setting up the logistics of trade to bring the products of exploitation back to the metropole. The model involved the private sector, with strong support and guidance from the state, paralleling today's private–public partnerships to some extent. The newly founded cities might then grow or collapse, depending on the presence and quantity of metal and available labor.

In general, the first phase of Spanish colonization lasted from 1492 until the conquest of Mexico in 1520 and Peru in 1532 (Williamson 1972). The discovery of vast reserves of silver and gold in Mexico and Peru meant drastic change for Caribbean islands and ports, which had to refocus their economies on other activities, such as sugar and tobacco production. As Caribbean port cities became key nodes of logistic trade, bringing precious metals to Europe and importing European products to the colonizers (Lockhart and Schwartz 1983), they accumulated treasures themselves and became more attractive to pirates. This all pushed port cities into a new phase of vulnerability at the end of the sixteenth century. The most drastic change of this phase came in the seventeenth century, after the conformation of the Triple Alliance of 1596 between France, England and the Republic of the Seven United Netherlands. The Treaty of The Hague recognized the Republic of the Seven United Netherlands for the first time, and it implied a common enemy in the Caribbean: Spain. Not long after the French, Dutch, and English too began to claim territory and establish plantations in the region.

From a spatial point of view, the built heritage of several Caribbean port cities corresponds to that of a fortress, most filled with gridded streets. Yet most of these locations did not have fortifications in the first decades of conquest. San Juan de Puerto Rico is a clear example of this, being an open, unwalled city for 130 years and based on an old reference to Plato's disdain for walled cities (Pabón-Charneco 2016). Historically, before the Spanish conquest of America, the port cities of Canary Islands were open structures, that is, unwalled ports promoting a message of free trade in a harbor city. Leonardo Torriani, a sixteenth-century Italian naval engineer, described San Cristobal de La Laguna in Gran Canaria, San Juan's most prominent precedent, as "a city made from peace for peace. No fortresses and no walls" (Pabón-Charneco 2016). The now so-called *Ciudad de La Paz* (*City of Peace*), also known as *Ciudad Maritima* (*Maritime City*) was characterized by the presence of a main square facing the sea (Plaza del Mar), which a grid linked to the main square (plaza mayor) containing the main civic buildings and the church. Later, the Spanish conquistadors used this grid to lay out their new open city, probably for reasons of speed, order, and the availability of rudimentary tools such as cord and ruler (Hardoy 1975; Morris 1994; Lejeune 2005).

The system of colonization moved from a standard strategy of founding cities to one of specializing ports. This had to do more with central planning of the region from Spain rather than with local demands. Ports that dealt with export–import duties to Spain had their duties drastically reduced to single tasks: Veracruz became the ancillary port of Mexico City that controlled the flows of silver from Mexico; Nombre de Dios (also known as Portobelo) in the Panama isthmus controlled the resources (mainly metals) coming from Peru by way of the Pacific Ocean; Cartagena de Indias (now Colombia) served as a stopping point for refueling ships and eventually a hub for trading slaves from Africa. Havana meanwhile was the port where ships coming from Peru and Mexico joined the Spanish naval escort to return to Spain (Williamson 1972). Some ports suffered from this re-configuration: San Juan and Santo Domingo for instance lost some or most of their early importance.

The *Ciudades de La Paz* model was eventually tested by Spain's enemies. British, French, and later Dutch pirates damaged key ports. In 1572, Sir Francis Drake attacked Portobelo; in a turning point in the politics of city defense (Williamson 1972), the Spanish king Philip II responded by commissioning an engineer specialized in fortifications, Battista Antonelli, to improve the security of Portobelo and other key cities, especially of those ports on the main route of import-export monopoly known as the Carrera de Indias: Cartagena, San Juan de Ulua, Havana, and San Juan in Puerto Rico (Williamson 1972). Thus the built heritage in the Caribbean port cities has historically been a product of the Spaniards, who designed cities first to maximize speed of construction and the efficiency of water trade and later for defense and customs.

Today that same heritage is being recycled to maximize cruise tourism, another product of foreign exploitation. Although the theory of path dependency normally refers to a continuous sequence of events, it is worth noting the similarities between these two phases, colonial exploitation and cruise tourism, despite the time that separates them. In colonial times, most of the economies of the Caribbean relied mainly on a single form of exploitation at the regional scale, protected by the monopoly of the market regulated by the Spanish crown through its *Casa de Contratación*, fortress architecture, and naval escort. The relatively recent emergence of tourism in the Caribbean as the main source of the economy also offers one type of product for the region (see Map 19.2). Recent cruise centers in a few Caribbean destinations strikingly recall the fortress strategies of colonial times, and like them are aimed at maximizing control of the economic benefits of the enterprise.

At the same time, it is important to observe the differences between these histories. Tourism differs from mining and sugar industries, with many more economic sectors affecting the business. Cruise tourism is what scholars call a vertical industry, in which giant companies control several sectors of the economy (Supersudaca 2006; Sweenay 2002). That means that we are not talking of monopolistic control

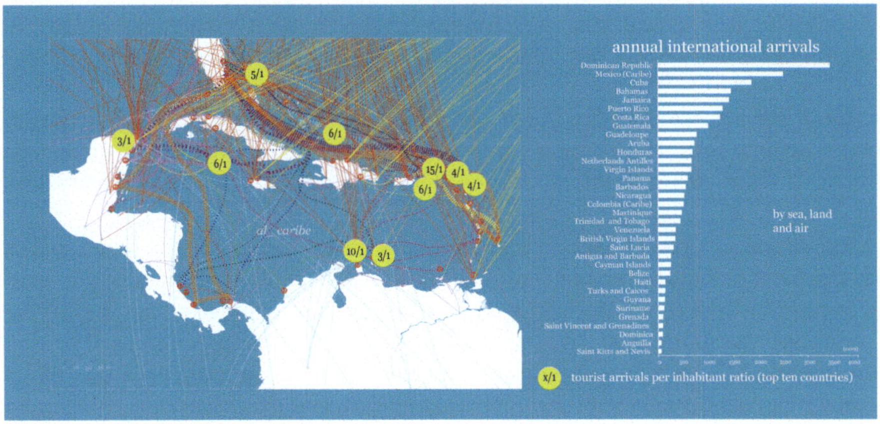

Map 19.2 Tourism dependency Caribbean, flights and cruise itineraries. *Source* Supersudaca, Al Caribe research with auspices of Prince Claus Fund; released under a Creative Commons Attribution-NonCommercial-NoDerivatives 4.0 International License

of one product but of intensive concentration and deformation of the market by few companies. Flows and success or failures of port cities related to cruise tourism are linked to the decisions of foreign-controlled industries, mainly located in Florida. The majority of cruises sail to the Caribbean and serve a primarily North American market, but not a single ship that cruises the Caribbean is Caribbean-owned or for that matter US-owned. Company headquarters are often on US soil, but to avoid taxation most companies sidestep US incorporation and go offshore. Carnival (#1 in terms of market share), which owns Holland America Line (#4), is incorporated in Panama; Royal Caribbean (#2) is "based" in Liberia; and Norwegian Cruise Line (#3) is registered with the Genting Group in Malaysia. It's a matter of economics for the host countries as much as for the cruise companies (Supersudaca 2014). As journalist Elizabeth Becker noted, "during its two-decades-long civil war, Liberia earned at least $20 million every year by acting as the offshore registry for foreign ships" (Becker 2013: 140). Thus historical patterns returned in new forms while the essence of actor power relation remains identical: The main economic decisions are still in hands of foreign investors, whose interest might prevail above local agendas of preservation.

Two foreign-owned cruise lines, each formed by multiple associated brands, monopolize the Caribbean cruise market. Together they hold 70% of the world market share (Sprague-Silgado 2017), and their turnover sometimes triples the GDP of local Caribbean countries. Their power to stabilize economic dynamics in the long run can be termed economic lock-in. As in colonial times, the benefits for the region are clearly not the foreign investors' priority. Although the Caribbean islands are the most active cruise tourist region of the world, the revenues do not correspond to the size of the business. A 2004 report from World Travel and Tourism Council stated that "Given that the Caribbean attracts around 50% of the world cruise market, its contribution to overall tourism earnings for the region is nonetheless relatively insignificant—accounting for between 8 and 10% of international tourism receipts only" (p. 23). Cruise tourists constituted about 42% of all tourists to the Caribbean in 2000, yet the same report stated that they accounted for only 12% of overall tourist expenditures. Nor is the news always good for all destinations. As competition increases so do the problems of growth. Continuous growth of the industry does not automatically guarantee success for all players.

Against the criticism of scarce economic benefits for the region, the report makes the argument that cruise tourism "presents destinations with the opportunity to convert cruise visitors (many of whom admit to being on a familiarization tour of the Caribbean) into future stay over tourists." It calls for "further research as to market perceptions of the two products (cruise and land base tourism), the degree of direct competition and demand substitution between them, and the extent of conversion to stay over visit" (p. 65). But the theory of conversion goes against the current trend of tourists spending less time on shore, which diminishes the chances of the destination to promote itself. Besides cruise lines are controlling shore excursions more and more, as "another source of income for the cruise industry that provide solid revenue for the cruise line in form of sales commission" (Ross 2013, p. 47). The waterfronts and ports catering to mass tourism from the cruise industry are becoming a product controlled by the cruise tourist industry that with its "status as a single sector economy raises the spectre of future regional ruination" (Brouderet al. 2016, p. 03).

Cruise ships are always becoming bigger "floating theme parks" (Wood 2000, p. 358), requiring ports of call to invest more and more money to build, maintain, and modernize big piers. When some Caribbean ports are unable to afford these ballooning expenses, tour operators look elsewhere for destinations that can meet burgeoning demand or they negotiate anchoring fees down to meager sums. In some cases, cruise companies share responsibility and investment for upgrading infrastructure with local partners. In other—more profitable—cases, they develop, manage, and operate ports themselves. Cruise line operators now sometimes work like infrastructure banks, offering loans to governments at local destinations to fund cruise-based infrastructure projects. In 2007, Carnival Corporation PLC and St Maarten signed a $34.5 million agreement for the enlargement of their pier, anticipating that bigger ships will revive the tourist economy in the British Virgin Islands. The loans are calculated based on the head taxes that the governments receive from the flow of cruise tourists. With growing congestion in the Caribbean and stiff competition from emerging economies worldwide, a port unable to upgrade can face abandonment. This threat leaves ports paralyzed: Investing large sums of money to upgrade facilities only risks subsequent obsolescence, but not upgrading means no business at all. (Supersudaca 2014) Once again, as in colonial times, the control of the demand and supply is in foreign hands, but the destinations have no choice: "upgrade their piers or die" (Supersudaca 2014, p. 20).

Heritage Architecture of Caribbean Cities and Cruise Lines' Economic Interests

Since cruise ship tourism depends on heritage, the first actors that were interested in heritage conservation were the cruise industries. Therefore, the preservation strategies for heritage buildings and urban spaces play a major role in marketing the Caribbean islands and are closely related to the attractiveness of the cruise ship industry. In Curacao, for example, the world-famous Dutch canal house—style facades of the waterfront street, the Handelskade, have always attracted cruise tourists, but on the other hand, the Curacao Government learned from a marketing study that their city should appeal to what the tourists have in mind for a "Caribbean" location. They planted palm trees along the public areas near the terminal—but those palm trees are not actually indigenous species of the island, so the government is importing them from Cuba, as payment on an earlier debt. The supposedly Caribbean landscaping of the passage that guides the cruise tourists into the shopping district has led to the "situation that on the same square (Brionplein) two sorts of lamps are used: those paid for by the tourist industry along the path of the cruise tourists and the old and the less kitschy public lamps that remain standing on the square"—all this reflects the absence coordination among the local tourism and planning authorities (Saavedra Bruno 2007, p. 106).

Some of the shops on St Maarten's Front Street literally turned their orientation 180° to face a new walking boulevard for cruise tourists, reorienting urban form; at the same time, when four cruise ships in St Maarten simultaneously unload, their 10,000 passengers instantly cause a traffic jam (Saavedra Bruno 2007, p. 104). In Curacao, the design of public space guides the cruise tourists carefully from the Megapier through a shopping center onto the "swinging old lady" bridge to Punda, the old city center that is now full of luxury duty-free shops. This route literally turns its face away from the main shopping street for locals in Otrabanda, whose "shopkeepers have always expressed that they want to keep orienting themselves to the local client, considering it a more stable factor" (Saavedra Bruno 2007, p. 104).

At first, the interest of the cruise industry in heritage was not only ethical but economically driven, therefore more stable. Yet recent cases point in a different direction, indicating that the cruise industry is comparing the costs and benefits of this model to those of a new model of total control in which they fabricate "heritage" assets elsewhere, preferably far from the city and authentic heritage (see Photo 19.1a, b).

The cruise lines have constructed cruise line-owned shopping destinations in no-man's-lands and leased (or sometimes bought) beaches, creating tourist bubbles (Jaakson 2004) disconnected from actual heritage sites, real cities, and local lives. According to Caroline Cheong, a Ph.D. candidate in planning and geography writing for the World Monument Fund in the Charleston report, "port communities may be more predisposed to commodify their heritage for tourists given the concentration of tourist activity and the revenues generated in the "tourist bubble". She quotes Wood (2000), noting that increased interaction between visitors and local communities

Photo 19.1 Cruise center Costa Maya. *Source* Supersudaca, Al Caribe research with auspices of Prince Claus Fund; released under a Creative Commons Attribution-NonCommercial-NoDerivatives 4.0 International License

furthers "processes of globalization and homogenization," processes that according to Cheong "are sped up within the host community when the number of visitors exceeds that of the local community, a process that Brida (2010) and the UNWTO (2010) note is especially prevalent in the Caribbean" (Cheong 2013, p. 29).

In some isolated islands of the Caribbean, the population spikes in the high season. Cockburn Town, in the Turks Islands, for example, quintuples during peak cruise season, with neither conflict nor negotiation (Supersudaca 2014, p. 22). If the main attraction was first heritage sites, it then became itineraries and ports and later the ship itself in combination with beach and water-related activities, often on leased islands in the middle of nowhere. Yet, according to Cheong, despite cruise lines' separate shopping and beach areas for their passengers, "real heritage sites remain a main attraction for cruise tourists" (Cheong 2013, p. 27). Nonetheless, she notes, "though cruise itineraries and ports of call remain main motivators for cruise travelers (CLIA 2006; Andriotis and Agiomirgianakis 2010)—acknowledging the need to provide satisfactory offshore experiences—the literature indicates that there is a summer-driven shift toward the ship itself acting as the primary attraction" (Cheong 2013, p. 124). Our recent analysis of Caribbean excursions on shore reveals that this tendency goes hand in hand with the tendency that most current excursions ashore in the Caribbean region focus on water and the beach rather than heritage (see Fig. 19.1).

Photo 19.2 Port of Call Costa Maya. *Source* Supersudaca. Al Caribe research with auspices of Prince Claus Fund

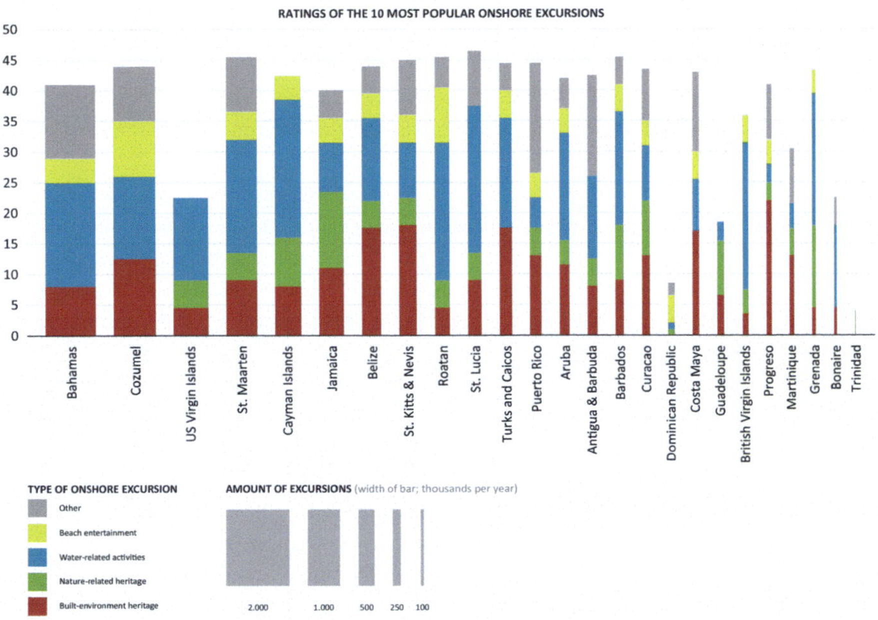

Fig. 19.1 Type of excursion on shore. *Source* Supersudaca recent research 2017; released under a Creative Commons Attribution-NonCommercial-NoDerivatives 4.0 International License

Contemporary Modes of Cruise Tourism in the Caribbean and their Impact on the Heritage of Caribbean Port Cities and their Hinterland

It is hard to shake the impression that cruise ships have become entirely self-sufficient. The exponential increase in ship sizes has turned into a metaphor for the cockiness of the cruise industry: The bigger the boats, the less the companies seem to care about the quality, variety, or authenticity of destinations. If cruising in the Caribbean not long ago meant wandering old colonial cities like San Juan, walking beach promenades in Cozumel, buying goods near the pier, or having a taste of local cuisine of Santo Domingo, the latest cruise development has tended to diminish the importance of destination-specific values.

As cruise companies have succeeded in engineering a diversity of life on board, they call into question the relevance of destinations. The plurality, potential insecurity, and lack of guarantees in real places surely overshadowed any advantage they might offer. In places such as Cozumel, where tourists still are able to reach the local shops, cruise directors warn passengers to avoid the uncertified and unsafe shops of the locals. It even seems that ships could just stop anchoring at local nodes.

One apparently insignificant shift is actually a crucial move changing the role of destinations in the power game of tourist spatial economics. Originally, the pier was the extension of the local economy of a touristic destination. As the extension of the touristic destination, the cruise pier had to lead tourists carefully to the destination charms, seducing then to spend as much time and money as possible in locally owned shops during their short stay. As the cruise industry is now financing, building, and deciding the position of new piers, "the piers have today become extensions of ships" (supersudaca 2014, p. 18).

The dominant new model for handling cruise tourism ashore is to provide a fenced bus terminal and a shopping area attached to the cruise pier, sometimes far from the city or on an unexploited island. This cruise village immediately attached to the cruise pier is providing leisure and (often cruise line owned) shopping wrapped up in duplicates of historical villages, divorced from existing cities and their economies (see Fig. 19.2).

With the new piers increasingly far from historic destinations, the tourist has fewer options to venture into town (and they run the risk of not catching the cruise when it departs). It is easier and easier to stay on the secure grounds of the new ports of call that are under industry control. It is estimated that "at each arrival of the boat to the port, 15% of the passengers never leave the cruise ship" (Lems 2010, p. 51). But this move could not be completed without a revolution on the ship itself. The transformation of the boat has been so massive that many tourists now decide that destinations are less crucial to their experience than in the past.

As ships grew, they "decreased dependence of the ports of call as the ship itself has become the destination" (Wood 2000, p. 358). This has enabled cruise lines to maximize their benefits for the cruise industry. The only remaining role of real places is fulfilling the few tourist wishes unattainable on the boat: authentic experiences of

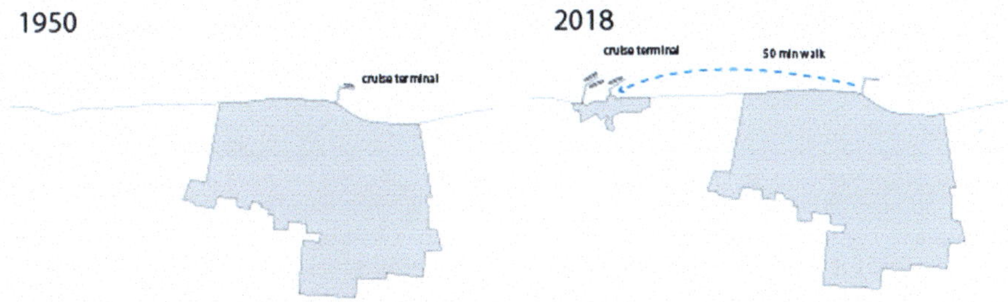

Fig. 19.2 From inner-city pier to tourist bubble. *Source* Supersudaca, Al Caribe research with auspices of Prince Claus Fund; released under a Creative Commons Attribution-NonCommercial-NoDerivatives 4.0 International License

a tropical colonial city, pre-Hispanic archaeological ruins, and an unspoiled beach. Even this remaining niche is now also being contested by the cruise industries that are creating "fantasyscapes on board and on land" (Wood 2000, p. 361). Colonial port destinations where the cruise industry is a big lobby, such as Curacao, control the routing of tourists in a nearly perfectly orchestrated choreography; and to "some extent Caribbean destinations are imitating the cruise ships, introducing theming in port city landscapes (such as in Aruba, whose main street feels very much as a theme park) and creating manmade, artificial attractions, divorced from the geographical environment as in St Maarten" (Wood 2000, p. 363). To meet tourists' demand for exotic architecture, the cruise industry has built a place called "Costa Maya" from scratch and in the middle of nowhere (see Photo 1, Photo 2). It includes a shopping area and restaurants in neo-Mayan style, owned and operated by the industry, while a plaster church tower recalls Spanish colonial times. Fake "stone sculptures and Indian dancers on the shopping plaza recall Mayan culture" (Sofia Saavedra Bruno 2007, p. 106). Not only did Grand Turk promote a replica of the Nasa Friendship 7 capsule, which splashed into the Atlantic in 1962 a few short miles from the island, as one of the island's main excursion attractions on land, but it copied it again when it built the Grand Turk cruise terminal and center. To a significant degree, Wood points out, by extending "the fantasy environment of the ship" the ports also "reproduce in new form the enclave development long characteristic of the region" (Wood 2000, p. 363). Furthering this tendency, the Caribbean region has been a laboratory since the 1970s for the development of all-inclusive resorts, a world parallel to the cities where the locals live.

Ports of call like Curacao and Aruba, where the tourists can walk directly from the terminal into the old city center, became quite exceptional in the Caribbean. When mapping the cruise terminals and their direct surroundings, we found that "most emerging cruise terminals are situated several kilometers away from the closest inner city (see Fig. 19.3)" (Saavedra Bruno 2007, p. 106). But in order to get the local beach environment that tourists demand, one of the last remaining niche for destinations, "cruise lines are also reducing the days in port by buying, or leasing islands or by anchoring at a deserted stretch of beach" (Wood 2000, p. 361). Of the eight major

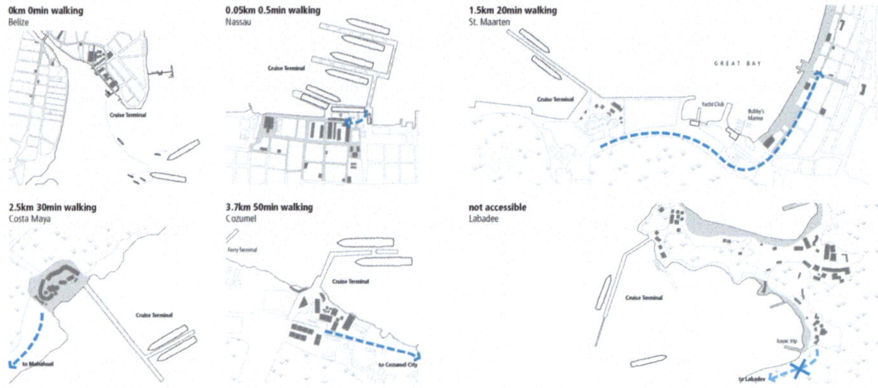

Fig. 19.3 Distance from the historic center. *Source* Supersudaca, Al Caribe research with auspices of Prince Claus Fund; released under a Creative Commons Attribution-NonCommercial-NoDerivatives 4.0 International License

lines that now operate in the Caribbean, 10 own private islands (Supersudaca 2014, p. 18). Royal Caribbean owns Coco Cay in Bahamas, for example, and leases Labadee in Haiti—which they tell tourists is Fantasy Island, in order to not spoil their holidays (Patullo 1996, p. 164)

As cruise companies have developed and refined their business models, they have affected the spatial relation between the city, port, and hinterland by strategically positioning docks in new places outside the old city. Their new constructions and new uses affect the historic spatial development of both urban form and heritage (local identity). But neither public actors—local politicians, tourist agencies, planners, heritage actors—nor citizens themselves have been involved in the ongoing transformation.

Conclusion

Cruise tourism has rewritten the urban form and architectural heritage of the Caribbean region and their functions in the last 20 years with urban form thematization (often disneyfication of the historic inner cities, including waterfronts) and simulations of historic ports. We have analyzed how the most recent cruise business model has affected both urban form and heritage architecture by (1) strategically positioning docking at new places outside of the old city or in the middle of nowhere, (2) reproducing heritage architecture, objects, and landscapes in replicas and simulations, and (3) recodifying heritage to suit the demands of the tourists and to exclude local economies.

The rapid growth of the cruise industry and its concentration in a few companies has established distinctive power relationships between the cruise industry and

Caribbean governments. We understand that this emerging dynamics follow path dependencies (Hacker 2002; Pierson 2004; Mahoney and Thelen 2010; Sorensen 2015), so what now seems like a dynamic process could actually be heading toward a static relationship among the key actors over time, making change increasingly difficult. As Capoccia and Kelemen have argued, "long periods of path dependent institutional stability and reproduction are punctuated occasionally by brief phases of institutional flux—referred to as critical junctures - during which more dramatic change is possible" (Capoccia and Kelemen 2007, p. 341).

We are currently at such a "critical juncture," in which local actors and heritage institutions can both prevent the cruise lines from seizing complete control of heritage areas while simultaneously luring them to stay in heritage sites (and not abandon them altogether). Caribbean governments are increasingly recognizing the role that heritage plays in attracting cruise tourists and the role that cruise tourism could play in preserving heritage and making it valuable in the future. In Havana, for example, the government has increasingly tied the renewal of the waterfronts to cruise tourism. A new port has taken over large-scale transport activities, leaving behind the historic port, which is being redeveloped primarily for cruise tourism. That process had already started, but the recent political opening of Cuba has accelerated it (INTI 2015, p. 24). Similarly, according to the Winning the Future report (Croes 2011) Aruba has recently recognized the power of cruise tourism and has decided to invest part of the revenues from it directly into preserving heritage. The cruise lines are often behind the scenes, still deciding where that money is being invested. The government of Grand Turk, for example, planned to use revenues from cruise tourism to turn an old building in the historic center into Carnival's welcoming cruise center, but the cruise line would not use it without the guarantee that cruise revenues on the other side of the island would be high enough (Saavedra Bruno et al. 2017, p. 45).

Cruise tourism triggers new institutional configurations in Caribbean cruise destinations, including collaboration between heritage and water-related planning institutions to ensure the future of historic port cities and to keep inherited patterns from further distorting uses of heritage architecture and urban form.

References

Andriotis K, Agiomirgianakis G (2010) Cruise visitors' experience in a Mediterranean port of call. Int J Tourism Res 12(4):390–404

Avrami E (ed) (2013) Harboring tourism. Cruise ships in historic port communities. In: Report of an international symposium held in Charleston, South Carolina, USA, 6–8 Feb 2013

Becker E (2013) Destination nowhere: the dark side of the cruise industry. The Saturday Evening Post (November/December 2013). http://www.saturdayeveningpost.com/2014/04/17/health-and-family/travel/the-dark-side-of-the-cruise-ship-industry.html. Last accessed 7 Jan 2018

Brida JG, Aguirre SZ (2008) The impacts of the cruise industry on tourism destinations. In: Castellani V, Sala S (eds) Sustainable tourism as a factor of local development. Tangram Edizioni Scientifiche (Colla), pp. 7–9. Available at SSRN: https://ssrn.com/abstract=1298403

Brida JG, Pulina M (2010) A literature review on the tourism-led-growth hypothesis

Brouder P, Clavé SA, Gill A, Ioannides D (eds) (2016) Tourism destination evolution. Routledge, London
Capoccia G, Kelemen RD (2007) The study of critical junctures: theory, narrative, and counterfactuals in historical institutionalism. World Polit 59(3):341–369
Carolina M, Pau L (2010) Cruise tourists returning to Curaçao for a land-based vacation: a logit model. BNA.WP/10/1
Casellas MR (2009) Soldiers and tourists: the subjective selves of Puerto Rico's modern architecture. In: Moré GL (ed) Archivos de Arquitectura Antillana 034, pp. 50–65
Castillo-Mussot, M, Sprague J, de la Lama García A (2013) Global capitalism and "north-south" unevenness: in light of ranking, statistical correlations, and profits of Forbes world list of top 2000 firms. Perspect Global Develop Technol 12:219–245
Cheong C (2013) Harboring Tourism. Cruise ships in historic port communities. In: Report of an international symposium held in Charleston, South Carolina, USA, p. 29. 6–8 Feb 2013
CIA The World Factbook (2017) Country comparison: GDP (Purchasing Power Parity). https://www.cia.gov/library/publications/the-world-factbook/rankorder/2001rank.html#tk. Accessed 6 Jan 2018
CLIA (2006) cruise market profile: report of findings. Retrieved January 23 (2006): 2008
Coldwell W (2017) First Venice and Barcelona: now anti-tourism marches spread across Europe. The Guardian 10
Collier RB, Collier D (1991) Critical junctures and historical legacies
Collier D, Mahon JE (1993) Conceptual "stretching" revisited: adapting categories in comparative analysis. Am Polit Sci Rev 87(4):845–855
Conzen MRG (1969) Alnwick, Northumberland: a study in town plan analysis. Institute of British Geographers, London, Publication no. 27
Corcoran K (2017) Cruise ships are being banned from sailing through Venice after locals got sick of them dwarfing their city. Business Insider Nederland. https://www.businessinsider.nl/venice-is-banning-cruise-ships-from-sailing-through-the-city-2017-11/?international=true&r=UK. Accessed 6 Jan 2018
Croes R (2011) Measuring and explaining competitiveness in the context of small island destinations. J Travel Res 50(4):431–442
Croes RR, Rivera M, Pizam A, Olson ED, Lee SH, Zhong YY (2011) Winning the future: strategic plan for the development of Tourism
Dowling RK (2006) Looking ahead: the future of cruising. Cruise ship tourism. CAB International, Wallingford, UK
Duval DT (2004) Tourism in the Caribbean, trends, developments and prospects. Routledge, London
Dywer L, Forsyth P (1998) Economic significance of cruise tourism. Ann Tour Res 25(2):393–415
Easterling K (2005) Enduring innocence, global architecture and its political masquerades. MIT Press, Cambridge Massachusets
Epler Wood M (2017) The cruise industry. Empire of the Seas. In: Epler Wood M (ed) Sustainable tourism on a finite planet. Routledge, London, pp 225–260
Gui L, Russo AP (2011) Cruise ports: a strategic nexus between regions and global lines-evidence from the Mediterranean. Maritime Policy Manage 38(2):129–150
Hacker JS (2002) The divided welfare state: the battle over public and private social benefits in the United States. Cambridge University Press, Cambridge
Hall CM (2001) Trends in ocean and coastal tourism: the end of the last frontier? Ocean Coast Manag 44:601–618
Hall CM (2011) Policy learning and policy failure in sustainable tourism gov-ernance: from first- and second-order to third order change? J sustain Tourism 19(4–5):649–671
Hardoy JE (1975) Urbanization in Latin America: approaches and issues. Anchor Books
Hein C (2016) Port cityscapes: conference and research contributions on port cities. Plan Perspect 31(2):313–326
Hein C, Hillmann F (2013) The missing link: redevelopment of the urban waterfront as a function of cruise ship tourism. RETE, Portus, 26 Venice

Hein C, Hillmann F (2016) The missing link. Redevelopment of the urban waterfront as a function of cruise ship tourism. In: Porfyriou H, Sepe M (eds) Waterfronts revisited. European ports in a historic and global perspective. Routledge, London

Hein C, Tourism CS (2016) The missing link. Waterfronts revisited: European ports in a historic and global perspective.

Hjalager AM (2007) Stages in the economic globalization of tourism. Ann Tour Res 34(2):437–457

Hobson J (1993) Analysis of the US cruise line Industry. Tour Manag 14:453–462

International New Towns Institute (INTI) (2015) Project development Plan

Jaakson R (2004) Beyond the tourist bubble? Cruise ship passengers in port. Ann Tourism Res 31(1):44–60

Lejeune J-F (2005) Cruelty and Utopia: cities and landscapes of Latin America. Princeton Architectural Press, Princeton

Lems K (2010) Economic valuation of USVI coral reefs. Is the USVI's tourism industry the seed of its own destruction? IVM Institute for Environmental Studies, p 51

Lockhart J, Schwartz SB (1983) Early Latin America: a history of colonial Spanish America and Brazil. Cambridge University Press, New York

Looijestijn-van der Putten L (2008) Cruise ships and coral reefs. How cruise ship tourists appreciate and at the same time- damage Bermuda's coral reefs. Research project at VU University in Amsterdam. Institute for Environmental Studies

Madrazo F, Delgado M, Saavedra S (2012) Turismos Caribe: super urban Caribbean tourism. (in)forma v6: hiperturismo. Publication University of Puerto Rico, pp 110–157

Mahoney J, Thelen K (2010) A theory of gradual institutional change. Explaining institutional change: ambiguity, agency, and power. Cambridge University Press, New York

McCalla RJ (1998) An investigation into site and situation: cruise ship ports. J Econ Soc Geogr 89(1):44–55

Morris AEJ (1994) History of urban form. Before the industrial revolutions. Pearson Education Ltd., London

Musterd (2012) Inventive city-regions, path dependence and creative knowledge strategies. Urban and regional planning and development series. Ashgate Publishing Ltd., Farnham

Niglio O (2014) Nexus Netw J 16:723. https://doi.org/10.1007/s00004-014-0202-z

Pabón-Charneco A (2016) The architecture of San Juan de Puerto Rico: five centuries of urban and architectural experimentation. Routledge, London

Papathanassis A, Beckman I (2011) Assesing the poverty of the cruise theory hypothesis. Ann Tourism Res 98(1):153–174

Papathanassis A, Lukovic T, Vogel M (eds) (2012) Cruise tourism and society: a socio-economic perspective. Springer Science & Business Media, Berlin

Patullo P (1996) Last resort? Tourism in the Caribbean

Patullo P (2005) Last resorts, the cost of tourism in the Caribbean. New York University Press, New York (first published 1996)

Pierson P (2004) Politics in time: history, institutions, and social analysis. Princeton University Press, Princeton

Rodrigue J-P, Notteboom T (2012) The geography of cruise shipping: itineraries, capacity deployment and ports of call. In: Paper presented at the International Association of MaritimeEconomists (IAME) Conference, Taipei, Taiwan

Rodrigue J-P, Notteboom T (2013) The geography of cruises: itineraries, not destinations. Appl Geogr 38:31–42

Rodrigue J-P, Comtois C, Slack B (2013) Transportation, economy and society. In: The geography of transport systems, 3rd edn. Routledge, New York

Ross AK (2013)Harboring tourism. Cruise ships in historic port communities. In: Report of an international symposium held in Charleston, South Carolina, USA, 6–8 Feb 2013, p 47

Saarinen J (2004) "Destinations in change": the transformation process of tourist destinations. Tourist Stud 4(2):161–179

Saavedra Bruno S (2012) Riviera Maya. (in)forma v6: hiperturismo. Publication University of Puerto Rico, pp 94–95
Saavedra Bruno S/Supersudaca (2007) Caribbean cruisers. Arquine 42:96–107
Saavedra Bruno S, Madrazo F, Delgado M, Roquero P (2017) Fair play. Turks & Caicos feasibility study on the cruise industry
Segre R (2007) Poder y democracia en las obras públicas caribeñas: una per-spectiva histórica. Lecture at the VI FCAA Biennial, Curaçao
Sorensen A (2015) Taking path dependence seriously: an historical institutionalist research agenda in planning history. Plan Perspect 30(1):17–38
Sprague-Silgado Jeb (2017) The Caribbean cruise ship business and the emergence of a transnational capitalist class. J World-Syst Res 23(1):93–125
Supersudaca (2006) Documentos, region del Gran Caribe. Al Caribe. In: Moré GL (ed) Archivos de Arquitectura Antillana 023, pp 47–57
Supersudaca: Delgado M, Koltowska Z, Madrazo F, Saavedra S (2014) Destination whatever: touring the cruise industry of the Caribbean. Harvard Design Mag 39:18–23
Sweenay N (2002) A regional tour operator for the Caribbean? Caribbean Development Bank
UNWTO (18 January, 2010) International tourism on track for a rebound after an exceptionally challenging 2009 UNWTO Press Release, Madrid, Spain. Retrieved April 1, 2010, from http://www.unwto.org/media/news/en/press_det.php?id=5361&idioma=E
Vaggelas GK, Pallis AA (2010) Passenger ports: services provision and their benefits. Maritime Policy Manage 37(1):73–89
Veronneau S, Roy J (2009) Global service supply chains: an empirical study of current practices and challenges of a cruise line corporation. Tour Manag 30:128–139
Weaver DB (1993) Model of urban tourism for small Caribbean islands. Geogr Rev 83(2):134–140
Weaver A (2005) The Mcdonaldization thesis and cruise tourism. Ann Tourism Res 32(2):346–366
Wilkinson PF (2007) Caribbean cruise tourism: Delusion? Illusion? Tourism Geogr Int J Tourism Space Place Environ 1(3):261–282
Wood RE (2000) Caribbean cruise tourism: globalization at sea. Ann Tourism Res 27(2):345-370
World Travel and Tourism Council (2004) The Caribbean. The impact of travel & tourism on jobs and the economy

Sofia Saavedra Bruno is one of the co-founders of Supersudaca, an international network of architects and planners. She is currently pursuing a Ph.D. at KU Leuven. Her publications include the book *Urbanism and Architecture in Chiquitos* (Santillana 2000) and several chapters in the book *Op Grond Van Samenwerking* (Epo 2018), *Kana Nos Kosta* (Louvain Press 2017). Her research focuses on institutional planning analysis, planning, sustainable development of tourist destinations, and social innovation.

Martín Delgado Filippini is Lecturer and Researcher at the Institute of Theory and Urbanism at the University of the Republic, Uruguay. He also is the principal of Re Estudio de Arquitectura in Montevideo. His research focuses on the territorial footprints of contemporary agricultural and tourist systems. He has been a member of Supersudaca, an international network of architects and planners, since 2005.

Felix Madrazo is one of the co-founders of Supersudaca, an international network of architects and planners. He is currently a lecturer at The Why Factory at Delft University of Technology and the principal of IND [Inter.National.Design]. He has published various articles with Supersudaca and is the author of the books *City Shocks* and *Copy Paste*, both by Nai010 publishers.

Lisbon's waterfront seen from the top of the *Padrão dos Descobrimentos*, José M. P. Sánchez, released under a Creative Commons Attribution-NonCommercial-NoDerivatives 4.0 International License

Chapter 20
Using Heritage to Develop Sustainable Port–City Relationships: Lisbon's Shift from Object-Based to Landscape Approaches

José M. Pagés Sánchez and Tom A. Daamen

Abstract Port cities face enormous sustainability challenges. In this chapter, we propose a relational view of these challenges and explore how different models of governance connect the three pillars of sustainable development: economy, environment, and society. We also address the contradictions inherent to new port plans or waterfront projects, zooming in on the case of Lisbon, Portugal to evaluate the role of heritage in the sustainable development of its historic maritime waterfront. We assess the extent to which reusing heritage structures strengthens the Lisbon port-to-city relationship with regard to governance and outcome. Our account shows that the city departed from its earlier object-based approach to adopt UNESCO's approach of Historic Urban Landscapes (HUL). This shift has triggered deeper reflection among key city actors on the connections between city and port in Lisbon, as well as on the role of the waterfront landscape. We argue that its new approach to heritage potentially produces new governance arenas where new port–city coalitions can emerge—coalitions that have the potential to align economic and environmental objectives with the sociocultural motives that underpin the goals of heritage preservation. We conclude by emphasizing both the challenges of public participation and the critical importance of engagement of port authorities. Each is necessary if European port cities are to effectively pursue sustainable relationships.

Keywords Port city · Governance · Historic urban landscape (HUL) · Maritime heritage · Waterfront · Lisbon · Sustainable development

---

J. M. Pagés Sánchez (✉)
Hamburg Hafencity University, Hamburg, Germany
e-mail: jose.sanchez@hcu-hamburg.de; jsanchez@aivp.org

T. A. Daamen
Delft University of Technology, Delft, The Netherlands

C. Hein (ed.), *Adaptive Strategies for Water Heritage*,
https://doi.org/10.1007/978-3-030-00268-8_20

Introduction: Port Cities and the Anthropocene

The world has entered into an era of governance initiatives that aim to reduce the impact human beings have on our planet. Scientists have labeled this new era as the third stage in the Anthropocene, the historical period which began with the Industrial Revolution and during which human behavior has exerted dramatic impact on Earth's crust and ecosystems (Crutzen 2002, 2006). It coincides with the global emergence of the term *sustainability* and, more particularly, that of sustainable development (SD), which the United Nations embraced in 1987 as a leitmotif for a new period of economic growth (as noted in Steffen et al. 2011). The terms Anthropocene and sustainability acknowledge the self-destructive quality of the world's prevailing economic systems. It has become widely understood that unchecked growth will deplete our planet's natural resources and irreversibly damage its environment. The current period is thus characterized by continued efforts to prosper economically—now, in ways that are socially and environmentally sustainable.

If the 2015 Paris Agreement and its aftermath teach us anything, it is that the governance challenges posed by the current stage of the Anthropocene are enormous (UNFCC, 2015; UN, 2015). This is especially true in places where the logic and structure of industries have grown far beyond local authority or state control. Sustainable development asks for governance action in many networks and across many, if not all, scales. Port cities, where industries such as oil and containerized logistics often have had a great impact on a city's economy and on its identity as a place for trade and connectivity on the water's edge (Hein 2011, 2018), experience the emerging transformations especially acutely.

Before the Anthropocene era, relationships between ports and cities were quite symbiotic. Separation started to occur after the early stages of industrialization and continued over much of the twentieth century. Even before the turn of the millennium, however, some scholars started to speculate about a renewal of links between city and port. Although port facilities have largely left their historical waterfront sites behind over the past 50 years, maritime functions and related activities in many of the world's port cities have remained highly urban (Hall and Jacobs 2012). One explanation for this can be found in the relationships valued by port businesses with a wide range of agents: competitors, financiers, insurers, legislators, as well as the schools and universities that supply them with new technologies, insights, and an educated workforce. Recently, scholars have argued that this relational orientation provides a comprehensive account of the diverse developments observed in port cities today (as in Hesse 2017). In contrast to popular spatial-synthetic perspectives, a relational orientation incorporates the fact that ports as part of both competitive global networks and historic landscapes are places with worldly connections, deeply felt cultural meaning, and local symbolism.

In European port cities, relational ties and tensions make governing the future development of a port, its city, and the relationship between them very complex. Responsible agents, like port authorities and city administrations, often have to think and act across multiple scales and sectors. Actors and scholars should not underestimate the challenge of governing the port amid the dynamics of private global logistics, trade, and the politics of public infrastructure provision. Moreover, concerns over the sustainability of current and future port operations render this task ever more demanding.

The unsustainability of current port–city relationships is best explained by economic geographers, who highlight the imbalance between positive and negative externalities that appear with port development (as in, Grossmann 2008). In recent decades, city pollution and congestion due to port operations have been seen to increasingly outweigh the employment and economic added value the port produces in the urban or regional agglomeration to which it belongs (Zhao et al. 2017). Furthermore, in Europe, it has been shown that those economic benefits may be felt far beyond the region, or even the country, which hosts a large seaport (Merk 2013). It can then be more difficult for local politicians to legitimate further port expansion and related public expenditure. As a result, there is a growing awareness among responsible actors that local port-to-city relationships have to be rethought, that, further, new governance approaches are needed if growing stalemates between actors in port–city arrangements are to be avoided (Daamen and Vries 2013; Pagés Sánchez and Daamen forthcoming).

A useful way to study the political tensions and governance dilemmas of contemporary port planning is to explore how different actors try to actively connect the three pillars of sustainable development—economy, environment, and society—and to address the contradictions that inhere to new port plans and projects. We report in this chapter on an on-going research project that focuses on the port city of Lisbon as a case study. We are concerned with the role heritage plays in the development of its historical maritime waterfront and assess the extent to which reusing heritage structures strengthens the port-to-city relationship in Lisbon on issues of governance and outcome. Our account shows that adoption of UNESCO's approach to Historic Urban Landscapes (HUL) has triggered deeper reflection among key city actors on the connections between city and port in Lisbon as well as that of the waterfront landscape. We note that the new approach to heritage management might well produce new governance and, thereby, create opportunities for new port–city coalitions that can align economic and environmental objectives with the sociocultural motives underpinning the goals of heritage preservation.

In order to explain our relational perspective on sustainable port–city development, we elaborate on several historical conceptualizations of these relationships and illustrate them with examples of the evolution Lisbon has undergone. We present the case of Lisbon, identifying two periods that have been marked by a change in heritage management policies: namely, the shift from an object-based to a landscape approach. We conclude by synthesizing the change observed in Lisbon with similar changes that have occurred in other European port cities, in the process, linking heritage management to the governance of sustainable port–city relationships.

Evolving Port-to-City Relationships in Lisbon

Three thousand years ago, where the Tagus River meets the Atlantic, Phoenicians founded Lisbon as a commercial seaport. The Portuguese capital grew in importance, reaching its zenith during the sixteenth century as the capital of a global empire. Many historical depictions show the port of Lisbon in symbiotic relationship to the city, with both its commerce and maritime affairs entrenched in the city's main public spaces. Studies of other ports, which explore the expansion of their associated networks, infrastructure, and the developing relationship between ports and their urban and natural surroundings, often give evidence of similar histories. In fact, geographers often take the spatial evolution of port infrastructure and port–urban waterfront development and redevelopment as a point of departure (Bird 1963; Hayuth 1982; Notteboom and Rodrigue 2005). Hoyle's widely cited six-stage model, for example, is rooted in spatial-historical evolution patterns (2000, p. 405). As is consistent with the case of Lisbon, the first stage of this model runs from ancient times to the nineteenth century. In this stage, cities emerged around port settlements, which were key elements of economic development and urban identity. Relationships between city and port were close and intensive.

The industrial age increased humankind's capacity to change natural landscapes and alter waterways and quays for more and larger ships. This period, dating from the nineteenth to the mid-twentieth century, is indicated, according to Hoyle (2000), by the emergence of break bulk industries in the second stage and early containerization in the third stage. Modern technology and economies-of-scale rationalities triggered changes in the port-to-city relationship, so that ports expanded and separated from the urban core. In this period, the government of Lisbon presented the first port plan (1887)—which included new landfills that distanced the city fabric from the riverfront (Pagés Sánchez 2017)—and implemented it in the following decades.

Hoyle (2000) identifies the fourth and fifth stages in the evolution of port-to-city relationships in the 1960–1990 period. During the fourth phase of the 1960s to the 1980s, container technology revolutionized the maritime sector, pushing the rationalized separation between production and consumption while triggering the emergence of global logistic chains. Lisbon opened its first container terminals during the 1970s and the 1980s close to the city center, and the port expanded along the south side of the Tagus River. Much as in Hoyle's model, new port terminals were located outside historical waterfront areas; in the pursuit of easier access for ever larger ships wider and longer quays on deeper waters were constructed.

Waterfront redevelopment plans, which signify the fifth stage—the 1970s to the 1990s—allowed city authorities to reconnect their urban fabric with the water. The plans also attempted to restructure the city's economy, adding leisure, offices with service functions, more upmarket residential developments, retail, cultural facilities, and public space (Norcliffe et al. 1996; Marshall 2004; Schubert 2008; Schubert 2011). In Lisbon, such redevelopment brought new public attention to the qualities of the waterfront and the city's connection to the river. This change also reinvigorated the port-to-city relationship, alternating moments of opposition to port expansion with proactive collaboration (Rêgo Cabral 2011). During the 1990s, the northeastern section of the waterfront was transformed into a brand-new district for the EXPO98 event, in a manner similar to regeneration projects in other port cities.

The emergence of global logistic chains in the final decades of the twentieth century coincides with a process of corporatization—even privatization—of port authorities throughout continental Europe. As a handful of multinational corporations started to dominate the world's transport network, port authorities quickly became pawns in a game between private shipping and terminal operating firms (Olivier and Slack 2006; Hall 2007). In effect, European port authorities redefined themselves as gatekeepers in a globalized transportation network, seeking to position themselves as strategic partners that controlled vital parts of logistic value chains (particularly in the hinterlands). By the end of the second millennium, many port authorities and related global logistic enterprises no longer concerned themselves with local urban issues—or so it is assumed from the transport economics point of view.

The Lisbon Port Authority became a state-owned limited company in 1998; its changing role and perspective generally follow the larger processes described by Hoyle (2000) and other academic studies. But Hoyle also conceptualizes a sixth stage, the 1980s to the 2000s, expanding his five-stage model (of Hoyle 1989) to include a perceived renewal of port-to-city associations. This observation was later confirmed by scholars who argued that 'ports are more than piers' (Notteboom 2006) and consist of a heterogeneous community of actors that, in many cases, is emphatically anchored in the urban. To what extent such new port-to-city relationships can be observed in contemporary Lisbon is an interesting research question. Decades of spatial, social, and institutional disconnection have created social tensions and local resistance to port presence. In Europe, port authorities have long defended port activities and expansions primarily on the basis of economic indicators such as added value and employment, neglecting sociocultural dimensions and paying little attention to negative environmental impacts (Van Hooydonk 2007). In Lisbon, the port remains physically close to the city; indeed, several terminals are still located on otherwise urbanizing waterfronts. However, looking at this issue from a relational point of view, it does not seem that the port presence is the outcome of joint planning decisions between port and city authorities—which would be more consistent with port–city reconnection efforts observed elsewhere in Europe (Daamen and Vries 2013). Hence, attending to the relationships that shape the implementation of sustainable port and port-related structure and infrastructure, we ask, what governance arrangements drive the spatial development and redevelopment projects that affect the sustainability of Lisbon's port–city associations?

Before we dive into our case study of Lisbon, we will first try to explore the answer to these questions in theoretical terms, taking heritage as a focal point of any port-to-city evolution. The crucial question is: How are port heritage and sustainable development in port cities conceptually related?

Port Heritage and Sustainable Development

Ships, quays, and cranes are some of the most visible parts of the urban environment, shaping international connectivity and a maritime atmosphere. Likewise, port heritage structures can create a sense of pride and belonging, not just for port workers but for all citizens drawn to the waterfront. Although the industrial era has disconnected ports and cities, motives for a rapprochement can be found among ports as well as among urban actors. The global sustainability movement, as adopted and promoted by the United Nations, elevates these motives to sheer necessities. In European port cities, preserving our planet's resources means developing a more balanced, sustainable relationship between the port and the urban.

Although there has been a lively academic debate around the definition of sustainable development (Williams and Millington 2004), many governance initiatives aiming for it still use the definition provided by the famous Brundtland report (World Commission on Environment and Development [WCED] 1987). This definition of sustainable development is based on two precepts. First, development should cover the needs of present-day society without compromising the needs of future generations. Second, development should balance three fundamental pillars in order to be regarded as sustainable: economy, environment, and society. In port cities, we observe that achieving such a balance is particularly daunting given the aforementioned negative externalities associated with most port development plans, the predominantly economic logics driving port evolutions, and the current institutional and sociocultural distance between port and urban actors.

Key port actors, including port authorities, have successfully developed strategies to address tensions between the economic and environmental pillars in the SD Framework (Aregall et al. 2018). However, port authorities have only become concerned with the social pillar relatively recently (Verhoeven 2011). In our conceptualization, this pillar includes the cultural 'soft' values produced by and embedded in the history and identity of a port city, expressed in its intangible aspects, such as traditions, songs, as well as in tangible artifacts, such as heritage structures (consider Van Hooydonk 2007, 2009; Warsewa 2011; Pereira Roders 2013; Mah 2014). For example, Musso and Ghiara (2011) and Schubert (2017) explore the socio-economic and sociocultural interface at ports: they describe *demaritimization* as the gradual loss of economic, social, and cultural significance of ports to their cities—a process which eventually affects the public's acceptance of the port presence inside the urban fabric. In contrast, a strong port–city culture and identity supports innovative waterfront plans that combine port and urban uses. Hence, if port–city communities do not acknowledge the port as a vital element of their DNA, they jeopardize the relationship. Scholars observe a process of *remaritimization:* initiatives and investments that generate new economic and sociocultural links between port and urban development, reinvigorating the local maritime economy and port–city identity (Musso and Ghiara 2011; Schubert 2017).

The European Sea Port Organization (ESPO), the main lobbying organization of European ports, published the European Port Industry Sustainability Report in 2016, a document which confirms that European port authorities are increasingly concerned with the social reconnection between city and port. They often adopt strategies to acquire a Social License to Operate (SLO), a concept explored by Dooms (2014) in which ports maintain or rebuild public support for port activities through social projects such as port festivals or community development programs. International organizations such as ESPO (2010) and the Association Internationale Villes et Ports (AIVP 2015) encourage port authorities to embrace SLO and related concepts, in that way, promoting port identity and culture among local civic and business communities.

In the sections following, we explore to what extent a connection between heritage projects and sustainable development can be observed in the port city of Lisbon. We describe how Lisbon authorities have revitalized some of the city's most prominent maritime heritage structures and evaluate to what extent their recently adopted governance approach stimulates a process that may yield more sustainable results. We studied empirical documents such as plans, policy briefs, media articles, and other research publications, and performed 17 semi-structured interviews with local stakeholders in the period of September 2016 to January 2018. Interviewees included municipal planners and port authority planning experts, presidents of *Juntas de Freguesias*,[1] industry leaders, museum directors, journalists, and leaders of citizen platforms. We asked them about Lisbon's identity as a port city, the city's relationship with the river, and the evolution of the port and its governing organizations. We probed different views on the role of port heritage, discussed the positive and negative externalities of the port, and asked for their view on the future of the port-to-city relationship in Lisbon.

Heritage in Lisbon: An Object-Based Approach, the 1990s to the Early 2000s

The Tagus River is a key element of Lisbon's identity. Numerous artistic representations of Lisbon have historically featured the river, which also feature prominently in the documents and interviews collected for our case study. Interview subjects consider the port to be an important element of the river's imagery, particularly as a reference to its glorious past. Despite this historically significant link between the port and river in Lisbon, most interviewees foresee that heavy port activities in the city will eventually be replaced by green and leisure functions. They consider the modern port to be necessary to the city and region, but prefer that it be located away from the urban waterfront.

The distinction between the historic and modern port in Lisbon is reflected in an object-based approach to port heritage that emerged in the 1990s to the early 2000s. At this time, several port heritage buildings were refurbished to host new cultural

[1] *Freguesias* are the smallest unit of government in Portugal. Its leaders are chosen every four years, parallel to municipal elections. These district bodies, or parishes, are known for their close relationship with local residents.

or leisure programs, including the Royal Factory Cordoaria da Junqueira, the Pedro Alvares Cabral building, and the maritime stations of Alcântara and Rocha Conde d'Óbidos. These projects were executed without explicitly making a connection to Lisbon's maritime history or broader ideas of celebrating its *portuality* (Musso and Ghiara 2011).

The Royal Factory Cordoaria da Junqueira, where ship cords and ropes were once manufactured, is an example of an industrial maritime building reconverted for cultural functions. The building, designed by Reinaldo Manuel during the second half of the eighteenth century, has monumental proportions linked to its former maritime activity: it is four hundred meters long and fifty meters wide (Nabais and Ramos 1987). The Cordoaria was originally on the riverfront, but landfilling activity conducted for port and train expansions in the late nineteenth and early twentieth centuries shifted the coastline one hundred and twenty-five meters away from the building (Ayres dos Santos 2012). The new infrastructure also altered the building's geometry, eliminating two lateral sections of the building. Today, the Cordoaria is managed by the navy and hosts historical archives and spaces for events and exhibitions. In 1996, the building and its immediate surroundings were cataloged as a national monument. In 2008, the municipality came up with ideas to visually reconnect the Cordoaria with the river; nevertheless, new constructions on the riverfront and the infrastructural barrier still obstruct the building's connection to the water (as shown in Fig. 1).

Fig. 1 Cordoaria in 2017. The building is used as a container for events and exhibitions. Before the port and railway landfills were built the south façade of the building (in the picture) was on the river. Author: José M P Sánchez; released under a Creative Commons Attribution-NonCommercial-NoDerivatives 4.0 International License

Located near the main container terminal on the Alcântara waterfront, the Pedro Alvares Cabral building is another example of port heritage refurbished according to an object-based approach. Due to its warehousing function, this large building, designed by João Simões Antunes in 1939, is characterized by very few openings; it features two bas-relief carvings by sculptor Barata Feyo which represent fishing and agriculture (Fig. 2) (Silva 2012). The government closed the facility in 1992, selling it in 2004 to the Fundação Oriente. This foundation engaged architects Carrilho da Graça and Rui Francisco to refurbish the building for the new Museum of the Orient, which opened in 2008.

In the 1930s, more than three hundred fifty thousand people per year arrived or departed from Lisbon by ship (Brito et al. 2007). Lisbon's Port Authority (APL) planned three maritime passenger terminals to handle the intense flow of people. Eventually, only the Alcântara and Rocha Conde d'Óbidos terminals were built, both designed by Pardal Monteiro, a major figure of Portuguese architecture during the twentieth century. They became important examples of the *Português Suave* architectural style. Almada Negreiros painted the interior murals, raising the artistic value of these buildings.

The maritime terminal at Alcântara was inaugurated in 1943; that in Rocha Conde d'Óbidos was built between 1945 and 1948 (Gama and Miranda 1997) (Fig. 3). Not long after this, the airport became Lisbon's main passenger gateway, and the maritime terminals were converted to cruise terminals. In 1985, a new container terminal in Alcântara blocked cruise vessels from the port (Nabais and Ramos 1987). It later became the APL headquarters, which today organizes public visits to see the building

Fig. 2 Pedro Alvares Cabral Building, 2007. Courtesy of José M. P. Sánchez; released under a Creative Commons Attribution-NonCommercial-NoDerivatives 4.0 International License

Fig. 3 Alcântara passenger terminal, 2007. The building hosts today the headquarters of the APL. Author: José M. P. Sánchez; released under a Creative Commons Attribution-NonCommercial-NoDerivatives 4.0 International License

and its murals. The Rocha Conde d'Óbidos station is also owned by APL; it is currently concessioned to Lisbon Cruise Terminals (LCT), which uses it only in exceptional cases.

Along the riverfront are several groups of port warehouses, built during the late decades of the nineteenth century and first of the twentieth. Most lost their original function when port activities changed, but remain the property of the APL, which leases them out as restaurants, clubs, and shops (shown in Fig. 4) (Rêgo Cabral 2011). Some renovations respect the original designs, while others offer a more contemporary interpretation of the warehouse type. Although these are not listed buildings, it is clear that they could be used to contribute to the identity of Lisbon as a port city. Indeed, a governance approach to port heritage that could achieve this has recently emerged. In the next section, we explore this new approach, along with its origins, and present our respondents' perception of its merits in juxtaposition to the object-based practice of the prior period.

Toward a Landscape Approach, the Early 2000s to Today

Since 1988, the relationship between the city, the river, and the port of Lisbon has been discussed publicly several times. In that year, the architectural chamber worked with the municipality, the APL, and the national government to organize a competition for riverfront proposals. The competition brief stipulated that the port would be located inside Lisbon's urban tissue (Brandão 1988). Winning proposals respected

Fig. 4 Former port warehouses refurbished to host bars and restaurants in the docks of Sto. Amaro. Author: José M. P. Sánchez; released under a Creative Commons Attribution-NonCommercial-NoDerivatives 4.0 International License

the port–city identity and would combine public access to the Tagus with working port areas. The APL and municipality have since made several plans in which reconnecting the city with the river is a top priority. These include the municipal master plan (PDM), the strategic plan of 1994, and the failed Plano de Ordernamento da Zona da Ribeirinha (POZOR), presented by the port authority in 1994 and 1995.

During the first decade of the millennium, a new law forced the APL to release land that was no longer suited to port activities to the municipality. This motivated the creation of the Plano Geral de Intervenções da Frente Ribeirinha (PGIFR), a plan which merged several partial plans in an effort to create a coherent vision for Lisbon's 19-km riverfront. Its main goals were to recover the symbolic value of the river, to visually and physically connect the river to the city, to reuse existing heritage structures (possibly changing their use), and to use empty spaces for waterfront regeneration (Câmara Municipal de Lisboa 2008). Following this plan, new public spaces were carefully built around heritage buildings that had already been redeveloped in the city's most central waterfront section. Here, we find the Ribeira das Naus, where the archeological remains of a sixteenth-century shipyard are part of the new green area by the river. Another example is landscaped the Campo das Cebolas, where heritage structures are also part of rearranged public space.

Though very few projects have been realized so far, they represent a broader vision of the central section of the waterfront. They recover the historical connection of the city with the river, which in fact has become part of an overall port and

Fig. 5 Area included in the UNESCO application. The boundary of the property is drawn in red, the buffer zone in blue. In dark red are the walls of Lisbon. *Source* CML (2016); released under a Creative Commons Attribution-NonCommercial-NoDerivatives 4.0 International License

maritime heritage management strategy that has recently taken on more formal shape. In retrospect, the PGIFR can be seen as a first attempt in reconciling some of the interests of the APL, the municipality of Lisbon, and the relationship between them. Today, connecting the city to the river is still one of the municipality's planning priorities, as is expressed in the 2013 municipal master plan and the city's 2016 application to the UNESCO World Heritage List, *Historical Lisbon, Global City* (Câmara Municipal de Lisboa, 2013; 2016). In 2017, UNESCO officially included the city's application on the National Tentative List (UNESCO 2011).

Lisbon's 2016 World Heritage application was not its first. In 2011, Lisbon's municipality put forth a proposal called *Pombaline Lisbon*, which concerned only the historical city center. This application to the World Heritage List was based on a conservation plan for the Baixa (the historical downtown area). The plan was officially approved in 2011, to preserve the physical characteristics of its eighteenth-century reconstruction. After a meeting with UNESCO representatives, the city decided to prepare a new application according to UNESCO's Historic Urban Landscape concept, which triggered a redefinition of the heritage boundary to include the riverfront area and Lisbon's intangible qualities like song and culture that contribute to its identity (as shown in Fig. 5) (Câmara Municipal de Lisboa 2017).

Several municipal departments led the new application, using the HUL concept to determine which properties to include and which arguments to use. Lisbon's maritime identity and port heritage were among the most significant elements of the new UNESCO brief. The application is structured around two historical events in the development of both the city and its port: the discovery era of the fifteenth and sixteenth centuries and the reconstruction of Lisbon in the period after the 1755 earthquake. Other arguments for Lisbon's World Heritage candidacy include the resilience of the urban structure, which was able to overcome the earthquake and other natural disasters; the cultural palimpsest of the city's history of maritime commerce; the identities still visible in its neighborhoods, found in artistic expressions such as the Fado and the azulejos ceramic tiles; and the distinctive light resulting from the reflections in its colorful and rugged urban landscape (Câmara Municipal de Lisboa 2017).

The zone defined for the UNESCO application logically connects actors such as the General Directorate of Cultural Heritage (DGPC), local and national tourism organizations, and associations responsible for preserving artistic expression. The team in charge of the application will consult and coordinate next steps with several organizations responsible for the activities affecting Lisbon's tangible and intangible heritage in the coming years. Although special protection plans already cover up to 70% of the heritage area (Câmara Municipal de Lisboa 2017), the landscape approach is expected to improve current documents, relating existing initiatives to each other—that is, building renovations, public space improvements, and cultural manifestations—that until now were unconnected. Moreover, the approach is seen as an opportunity to establish broader cooperation with additional urban stakeholders, such as the port authority. Whether this cooperation will actually emerge remains an open question.

Conclusion: New Arenas and Next Challenges

In this study, we have placed the present phase in the evolution of port cities within the third stage of the Anthropocene: an era in which humankind has developed the ability to alter the planet's ecosystems and its natural settings in irreversible ways. This ability comes with the responsibility for thinking and acting sustainably, fostering economic growth without depleting resources needed by future generations. We have explained that sustainable development asks for governance arrangements that stimulate key actors to take into account the economic, environmental, and social implications of their actions. In contemporary port cities, where the legacy of the industrial age is often felt heavily, the challenges of sustainable development are paramount. Here, incentives for a rigorous geographical separation of port and city functions are now being balanced by sociocultural and economic forces that seek to renew port-to-city links. Developing a sustainable port-to-city relationship thus implies a governance process in which port and city functions are not regarded as mutually exclusive, but are allowed to prove that they may be elements that reinforce each other.

In Europe, the reuse of maritime heritage can contribute to a renewed and more sustainable port-to-city relationship. We described how in Lisbon, authorities have revitalized some of the city's most prominent maritime heritage structures, evaluating to what extent a recently adopted governance approach to heritage—an approach rooted in UNESCO's Historic Urban Landscape concept—stimulates a process toward (more) sustainable results.

The new approach adopted in Lisbon looks beyond physical buildings and other objects to integrate these with intangible, cultural aspects that define a city and its *portuality*. Port heritage signifies the identity of Lisbon as a port city; and perceiving heritage buildings as part of a larger historic urban landscape helps them to be treated as elements of the sociocultural pillar within the sustainability paradigm. Although Lisbon's application to the World Heritage List was neither a plan nor a vision, it signaled an evolution in how key city actors have learned to understand the urban landscape. It resulted in connecting the old city with the riverfront, creating incentives for the city to renew a dialog with the port and other relevant authorities. The application emphasizes the city-river-port relationship not only as an important historical fact, but also as a framework for the present and the future.

Lisbon is not the only port city in Europe to adopt UNESCO's landscape approach to heritage preservation. In the Italian port city of Naples, for example, this approach has also led to new spaces for dialog between previously separated stakeholders. However, applying the process that moved the city toward a heritage management plan for the historic center excluded experts and elite urban actors participation in these new arenas (De Rosa and Di Palma 2013). Much as in Lisbon, the role of the larger public and that of the port authority—which owns substantial parts of the city's historic waterfront, including heritage objects—still remains unclear. Combining our relational perspective with the premise of sustainable port–city development may be expected to elicit the result that port authorities have sufficient reason to invest and participate in processes with sustainable development themes. However, the agency of these powerful organizations beyond their role in shipping and cargo seems restricted and fraught with both state and European rules and regulations.

Such terms lead us to question what role port authorities should be allowed to play in new sustainable development arenas, such as those focused on heritage management, even as they contribute to the port–cities relationships so desired. Further research in Lisbon and other European port cities is needed to address this governance challenge. Although the HUL approach to heritage management has surely led to more sustainable dialogs in Lisbon, it is important to monitor whether this process also yields more sustainable outcomes.

References

AIVP (2015) Plan the city with the port: guide of good practices. http://www.aivp.org/en/2015/06/30/plan-the-city-with-the-port-guide-of-good-practices-2/. Accessed 18 Dec 2017

Aregall MG, Bergqvist R, Monios J (2018) A global review of the hinterland dimension of green port strategies. Transportation Research Part D: Transport and Environment 59:23–34. https://doi.org/10.1016/j.trd.2017.12.013

Ayres dos Santos S (2012) A flexibilidade na permanência: uma proposta para a reutilização da Cordoaria Nacional. https://www.repository.utl.pt/handle/10400.5/5457. Accessed 15 Dec 2017

Bird J (1963) The major seaports of the United Kingdom. Hutchinson, London

Brandão P (1988) Prefácio. In: Brandão P, Jorge F (eds) Lisboa, a Cidade E O Rio - Concurso de Ideias Para a Renovação Da Zona Ribeirinha de Lisboa. Associação de Arquitectos Portugueses, Lisbon, pp 3–4

Câmara Municipal de Lisboa (2008) Documento de Enquadramento - Plano Geral de Intervenção Para a Frente Ribeirinha. Câmara Municipal de Lisboa - Divisão de Desenvolvimento Urbano, Lisbon. http://www.cm-lisboa.pt/fileadmin/VIVER/Urbanismo/urbanismo/planeamento/prospectivos/ribeirinha/documento_enquadramento.pdf. Accessed 20 Dec 2017

Câmara Municipal de Lisboa (2013) Regulamento do Plano Director Municipal. http://www.cm-lisboa.pt/fileadmin/VIVER/Urbanismo/urbanismo/planeamento/pdm/AF_REGULAMENTO_PDM_Lx.pdf. Accesed 1 June 2019

Câmara Municipal de Lisboa (2016) Formulário Para Submissão À Lista Indicativa - Anexos: Lisboa Histórica, Cidade Global. http://www.cm-lisboa.pt/viver/urbanismo/candidaturas-a-unesco. Accessed 3 Jan 2018

Câmara Municipal de Lisboa (2017) Formulário Para Submissão À Lista Indicativa: Lisboa Histórica, Cidade Global. http://www.cm-lisboa.pt/viver/urbanismo/candidaturas-a-unesco. Accessed 3 Jan 2018

Crutzen PJ (2002) Geology of mankind. Nature 415:23. https://doi.org/10.1038/415023a

Crutzen PJ (2006) The Anthropocene. In Ehlers E, Krafft T (eds) Earth system science in the Anthropocene. Springer, Berlin, Heidelberg, pp 13–18. https://doi.org/10.1007/3-540-26590-2_3

Daamen T, Vries I (2013) Governing the European port-city interface: institutional impacts on spatial projects between city and port. J Transp Geogr 27:4–13. https://doi.org/10.1016/j.jtrangeo.2012.03.013

De Rosa F, Di Palma M (2013) Historic Urban Landscape Approach and Port Cities Regeneration: Naples between Identity and Outlook. Sustainability (Switzerland) 5(10):4268–4287. https://doi.org/10.3390/su5104268

Dooms M (2014) Integrating ‘triple P’ bottom line performance and the license to operate for ports: towards new partnerships between port cluster stakeholders. In: Alix Y, Delsalle B, Comtois C (eds) Port-city governance. ems - Management & Societe, Cormelles-le-Royal, pp 55–76

ESPO (2010) Code of practice on societal integration of port. https://www.espo.be/media/espopublications/ESPOCodeofPracticeonSocietalIntegrationofPorts2010.pdf. Accessed 10 Dec 2018

ESPO (2016) European Port Industry Sustainability Report 2016. http://www.maritime-rdi.eu/media/10488/european-port-industry-sust-rep-2016.pdf. Accessed 7 Jan 2018

Gama E, Miranda I (1997) Lisboa Ribeirinha E as Suas Estações - Síntese Histórica. In Caessa A, Martins MG (eds) Actas Das Sessões Do II Colóquio Temético Lisboa Ribeirinha. Câmara Municipal de Lisboa - Departamento de Património Cultural Divisão de Arquivos, Lisboa, pp 201–224

Grossmann I (2008) Perspectives for Hamburg as a port city in the context of a changing global environment. Geoforum 39(6):2062–2072. https://doi.org/10.1016/j.geoforum.2008.04.011

Hall PV (2007) Seaports, urban sustainability, and paradigm shift. J Urban Technol 14(2):87–101. https://doi.org/10.1080/10630730701531757

Hall PV, Jacobs W (2012) Why are maritime ports (still) urban, and why should policy-makers care? Marit Policy Manag: Flagship J Int Shipp Port Res 39(2):189–206

Hayuth Y (1982) The port-urban interface: an area in transition. Area 14(3):219–224. https://doi.org/10.2307/20001825

Hein C (ed) (2011) Port cities: dynamic landscapes and global networks. Routledge, New York

Hein C (2018) Oil spaces: the global petroleumscape in the Rotterdam/The Hague area. J Urban Hist 00:1–43. https://doi.org/10.1177/0096144217752460

Hesse M (2017) Approaching the relational nature of the port-city interface in Europe: ties and tensions between seaports and the urban. Tijdschrift Voor Economische En Sociale Geografie. https://doi.org/10.1111/tesg.12282

Hoyle B (1989) The port-city interface: trends, problems and examples. Geoforum 20(4):429–435. https://doi.org/10.1016/0016-7185(89)90026-2

Hoyle B (2000) Global and local change on the port-city waterfront. Geogr Rev 90(3):395–417. https://doi.org/10.2307/3250860

Mah A (2014) Port cities and global legacies—urban identity, waterfront work, and radicalism. 1st ed. Palgrave Macmillan, London, UK. https://doi.org/10.1057/9781137283146

Marshall R (2004) Waterfronts in post-industrial cities. Routledge, London. https://doi.org/10.4324/9780203166895

Merk O (2013) The competitiveness of global port-cities: synthesis report. OECD Publishing, Paris. https://www.oecd.org/cfe/regional-policy/Competitiveness-of-Global-Port-Cities-Synthesis-Report.pdf. Accessed 4 Jan 2018

Musso E, Ghiara H (2011) Reshaping the economic landscape of port cities. In: Alemany J, Bruttomesso R (eds) The port city of the XXIst century. New challenges in the relationship between port and city, Rete, Venice, pp 87–101

Nabais AJCM, Ramos PO (1987) 100 Anos Do Porto de Lisboa. Administração do Porto de Lisboa, Lisboa

Norcliffe G, Bassett K, Hoare T (1996) The emergence of postmodernism on the urban waterfront: geographical perspectives on changing relationships. J Transp Geogr 4(2):123–134. https://doi.org/10.1016/0966-6923(96)00005-1

Notteboom TE (2006) Ports are more than piers: Liber Amicorum presented to Prof. Dr. Willy Winkelmans. Antwerp. De Lloyd

Notteboom TE, Rodrigue JP (2005) Port regionalization: towards a new phase in port development. Marit Policy Manag 32(3):297–313. https://doi.org/10.1080/03088830500139885

Olivier D, Slack B (2006) Rethinking the port. Environ Plan A 38(8):1409–1427. https://doi.org/10.1068/a37421

Pagés Sánchez JM (2017) Evolution of Lisbon's port-city relation: from the earthquake of 1755 to the port plan of 1887. PORTUSplus, the Online Journal of RETE, October 20(7)

Pagés Sánchez JM, Daamen T (forthcoming) Sustainable port-city relationships: challenges and opportunities in Europe. Plan Perspect

Pereira Roders A (2013) How can urbanization be sustainable? A reflection on the role of city resources in global sustainable development. Bollettino Del Dipartimento Di Conservazione Dei Beni Architettonici Ed Ambientali 13(1):79–90

Rêgo Cabral N (2011) Governability and sustainability. Redevelopment of the waterfront in Lisbon. In: Alemany J, Bruttomesso, R (eds) The port city of the xxist century. New challenges in the relationship between port and city, Rete, Venice, p 385

Schubert D (2008) Transformation processes on waterfronts in seaport cities-causes and trends between divergence and covergence In: Kokot W, Gandelsman-Trier M, Wildner K, Wonneberger A (eds) Port cities as areas of transition: ethnographic perspectives. Urban Studies. Transcript Verlag

Schubert D (2011) Seaport cities: phases of spatial restructuring. In: Hein C (ed) Port cities: dynamic landscapes and global networks, 1st edn. Routledge, New York, pp 54–69

Schubert D (2017) Ports and urban waterfronts. In: C Hein (ed) The Routledge handbook of planning history. 1st ed. Routledge, New York, pp 336–347

Silva MÂSP (2012) Património industrial em Portugal: inclusão do passado em projectos contemporâneos. http://hdl.handle.net/11067/78. Accessed 28 Feb 2018

Steffen W, Grinevald J, Crutzen P, McNeill J (2011) The Anthropocene: conceptual and historical perspectives. Philos Trans Royal Soc A: Math, Phys Eng Sci 369(1938):842–867. https://doi.org/10.1098/rsta.2010.0327

UNESCO (2011) Convention for the safeguarding of the intangible cultural heritage. In: Intergovernmental committee for the safeguarding of the intangible cultural heritage. Sixth session. https://ich.unesco.org/en/6COM. Accessed 21 Dec 2017

UNFCCC. Conference of the Parties (COP) (2015) Adoption of the Paris Agreement. Proposal by the President. In: Paris climate change conference—Nov 2015, COP 21 21932 (December): 32. FCCC/CP/2015/L.9/Rev.1

United Nations (2015) Transforming our world: the 2030 agenda for sustainable development. https://sustainabledevelopment.un.org/content/documents/7891Transforming%20Our%20World.Pdf, https://doi.org/10.1007/s13398-014-0173-7.2. Accessed 20 Dec 2017

Van Hooydonk E (2007) Soft values of seaports. A strategy for the of public support for seaports. Garant, Antwerp

Van Hooydonk E (2009) Port city identity and urban planning. Portus 18:16–23

Verhoeven P (2011) Social integration of ports: a key task for port authorities. In: Alemany J, Bruttomesso R (eds) The port city of the xxist century. New challenges in the relationship between port and city, Rete, Venice, pp 186–200

Warsewa G (2011) The role of local culture in the transformation of the port–city. Portus plus 2. http://retedigital.com/wp-content/themes/rete/pdfs/portus_plus/2_2011/Temáticas/Laciudadportuariacontemporánea/01_Günter_Warsewa.pdf. Accessed 8 Jan 2018

WCED (1987) Our common future. World Commission on Environment and Development. Oxford University Press

Williams CC, Millington AC (2004) The diverse and contested meanings of sustainable development. Geogr J 170(2):99–104. https://doi.org/10.1111/j.0016-7398.2004.00111.x

Zhao Q, Xu H, Wall RS, Stavropoulos S (2017) Building a bridge between port and city: improving the urban competitiveness of port cities. J Transp Geogr 59:120–133. https://doi.org/10.1016/j.jtrangeo.2017.01.014

José M. P. Sánchez is an architect and Ph.D. candidate in Urban Planning at Hamburg Hafencity University, Germany. His research and publications focus on sustainable port–city relationships, with an emphasis on Lisbon. He is also external strategic adviser in the Association Internationale Ville et Port (AIVP) on policy and research projects.

Tom A. Daamen is an associate professor of Urban Development Management at Delft University of Technology, The Netherlands. He is an expert on governance strategies for the development of waterfront zones and the reshaping of port–city relationships. In addition to his academic research and teaching, he advises port authorities, property developers, and government bodies on urban waterfront change.

Flooding in Port Arthur, Texas on August 31, Wikimedia, SC National Guard—170831-Z-AH923-081 (https://www.flickr.com/photos/scguard/36564448210/), Public Domain

Chapter 21
Toward a Cultural Heritage of Adaptation: A Plea to Embrace the Heritage of a Culture of Risk, Vulnerability and Adaptation

Han Meyer

Abstract We need to tell a new story about urbanizing delta regions. Historically, large-scale 'iconic' hydraulic works and modern industrial ports have been celebrated as showing the power of humans to control and subject nature. The emphasis on this part of cultural heritage tends to bury the remains of engineering and urban development of the previous periods and to erase the ebb and flow of natural processes in the earlier landscape. Instead of emphasizing resistance against nature and victories over nature, we need to embrace mitigation, adaptation, and uncertainty. Many urbanized delta regions, including the Dutch delta, have a rich history of such approaches. We can learn from that history, from the successes as well as from the failures. This chapter is a plea to embrace the dynamic and evolutionary character of delta regions and the cultures of adaptation which have been developed over many centuries. It means that another policy concerning cultural heritage should be stimulated in urbanizing deltas, fostering a heritage with an adaptive approach, not as a complete departure from present ways of doing things, but as a new stage in a centuries-long tradition.

Keywords Heritage of adaptation · Risk · Vulnerability · Uncertainty · Formative power of deltas

Introduction

Mitigation and adaptation are the themes that will dominate the urban agenda of the twenty-first century. *Mitigation* means changing our energy production and consumption, land use, food production, and waste disposal so that climate change and exhaustion of the earth will not lead to disaster. *Adaptation* means adjusting our planning and consumption to the already irreversible effects of climate change, like accelerating sea-level rise. And because we do not know exactly how conditions will change in the future, we must learn to live in *uncertainty*.

H. Meyer (✉)
Delft University of Technology, Delft, The Netherlands
e-mail: v.j.meyer@tudelft.nl

C. Hein (ed.), *Adaptive Strategies for Water Heritage*,
https://doi.org/10.1007/978-3-030-00268-8_21

Mitigation, adaptation, uncertainty: It seems that these themes are new and that they mean a fundamental break with the policies and strategies of the past. That is the dominating story, especially concerning urbanizing deltas. These regions have been developed into important hubs of the fossil fuel economy during the last 150 years. The transshipment, storage, and processing of fossil fuels are required large industrial and port areas and a sophisticated infrastructure of deep sea ports, quay walls, navigation channels, and flood defense systems. Mitigation, including the transition toward new energy systems, will have great impact on both land use and infrastructure in these regions. Next to it, adapting to the effects of climate change, like rising sea level and increasing peak discharges of rivers, requires a fundamental revision of flood defense strategies. Today's uncertain circumstances seem to call for these sharp breaks with the long-term spatial and technical arrangements of delta regions.

But presenting the adaptive approach as a ***new*** approach, breaking with traditions, makes it vulnerable and easy to attack by opponents. Most of the national cultural heritage in the Netherlands (and there are many similar examples in other countries) conveys a message that it is possible to resist, control, and subdue nature: We did it for centuries, with increasing success, so why should we s change to another approach? These monuments are cherished as characteristic reminders of a strong collective culture which has claimed land from the sea step-by-step, making untamed marshlands submit to human will. In particular, the large hydraulic works of the nineteenth and twentieth century are celebrated as important chapters of a story on the supremacy of a nation state, a new technology of steam power, fossil energy, and an industrial economy, tools that definitively control nature (Meyer 2017).

In this chapter, I argue that there is another story about dealing with water throughout the centuries, which is not so different from the 'new' policy of adaptation and dealing with uncertainty. It is a story about awareness of risks, uncertainty, and vulnerability; it is also a story about attempts to understand the dynamics of nature, about awareness of changing conditions and the need to adapt to these changes. If we look carefully at the history of urbanization in urban deltas, we can discover many interesting examples of how people tried to understand the dynamics of these water landscapes and how to make use of them. If we open our eyes to this history, it might be possible to learn from it for the future and to consider the current agenda not as a break, but as a new chapter in a long tradition of continuous adaptation. This attention to another cultural heritage is relevant not only for the Netherlands, but for many other urbanized deltas in the world.

Working with the Formative Power of Delta System

Deltas are complex and dynamic systems, the products of the convergences of rivers and the sea, where sea currents, tides, discharges, wind, sediment deposition, and erosion formed deltaic plains of sandy beaches, dunes, and alluvial lowlands. Vegetation in these lowlands compressed into layers of peat, part of the natural process of land making. This process of land making can be recognized in all deltas in the world, in many different variations. We can distinguish between different types of delta according to the source of their formative sediment: river-dominated deltas, wave-dominated deltas, and tide-dominated deltas (Bradshaw and Weaver 1995).

Regardless of type, deltas are the richest ecosystems of the world. As gradual transitions between land and water and between salt and fresh water, they are the biotopes and nurseries of many species that are crucial for the ecological balance of the world's rivers and oceans. Next to their ecological value in terms of biological productivity and diversity, these ecosystems offer significant economic value in ecosystem services: coastal protection, maintenance of fisheries and wildlife, erosion control, water catchment and purification, carbon sequestration, nutrient cycling, tourism, recreation, education, and research.

Cultures and technologies rose with an awareness of these dynamic processes of the delta. In the Netherlands, the sixteenth-century engineer Andries Vierlingh published a *Tractaet van Dyckagie* ('Treatise on Diking') on the reclamation of land and the construction of dikes, based on learning natural dynamics and making use of them (de Hullu and Verhoeven 1920). Vierlingh explained how original creeks in reclaimed land could be used to drain the land, and how the form, construction, and positioning of dikes might lessen the power of storm surges. His book became an influential manual in reclaiming land and dike construction in the Dutch delta of the sixteenth and seventeenth centuries. Similarly, engineers like Simon Stevin (1548–1620), Jan A. Leeghwater (1575–1650), and Nicolaus Cruquius (1678–1754) researched currents, erosion, and sedimentation (Meyer 2017). They were aware of the dynamic character of these processes, and how they changed conditions for land use and urbanization—indeed, they witnessed many disastrous floods themselves. Buisman counts at least 48 serious floods in the Dutch low lands from the eleventh to the twentieth century, each flood resulting in many victims, land loss, and economic damage (Buisman 1995/2016). Today, we tend to forget all these times that nature hit back, reconquering reclaimed land. Yet in the Southwest delta alone, the sea swallowed more than one hundred villages and small towns over the course of centuries (Stulp 2011). Some traces of these disappeared villages and towns haunt the landscape, like the *Plompe Toren* ('clumsy tower'), the only remnant of the village Koudekerke along the border of the East Scheldt (Fig. 1). Vierlingh warned his readers to be aware of the risks of reclamation and to avoid reckless thinking that nature can be conquered and controlled.

Along the coastline of Holland, as well as along the coasts of the previous Zuiderzee and the Wadden islands, many villages and towns faced unstoppable coastal erosion and an encroaching flood line. These forced them to move further inland. Towns like Egmond and Monster had to rebuild their urban cores in several times across the centuries, after disastrous floods.

Along with floods, ongoing sedimentation was a substantial problem for many cities. A rather famous example is the Flemish city of Bruges, an important seaport until the fourteenth century. After that, the accessibility of the port decreased seriously because of the silting up of the Zwin estuary. The town of Damme, on the same estuary but more seaward, took over the port role from Bruges. When the channel to Damme also silted up, the more seaward located towns of Sluis and St. Anna ter Muiden took

Fig. 1 *Plompe Toren* ('Clumsy Tower') in the Dutch Southwest delta: a remain of a drowned town. Photograph by Han Meyer; released under a Creative Commons Attribution-NonCommercial-NoDerivatives 4.0 International License

over the role of seaports. Finally, when the Zwin silted up completely and Sluis and St. Anna ter Muiden became inaccessible, these towns had to change their economy to agriculture.

The Dutch town Goedereede faced the same fate in the sixteenth and seventeenth century. The famous mapmaker Jacob van Deventer (1505–1575) showed its harbor connected with the sea (Fig. 2). But over the following centuries, the town battled increasing silt by digging a canal to deeper water, as did many other seaports in the Rhine–Meuse–Scheldt delta. Finally, however, the sediment could not be halted and Goedereede lost its position as important seaport (Fig. 3). Today, this small town's remarkable mixture of merchants' houses and farm buildings represents its forced but successful transition from a port economy to an agricultural economy during the seventeenth and eighteenth centuries (Fig. 4).

Not only individual towns and villages, but the whole pattern of urban settlements in the Western part of the Netherlands had to adapt to changing conditions several times. Until the eleventh century, the main string of urban settlements was organized alongside the borders of a branch of the River Rhine now named Oude Rijn (*Old Rhine*). This branch of the river was then the main artery for transport and for the discharge of the drained water from the polders of central Holland. But the mouth of this branch starting silting up in the ninth century, and people had to reroute both transport and drainage. New urban settlements arose along the borders of the Southwest delta and the Zuiderzee: The main body of the Rhine and also the Meuse River now discharged drained water from central Holland into the estuaries of the Southwest delta; in the Northern part of the Netherlands, storm surges had created the Zuiderzee as a new inland sea that people could use for drainage and transport.

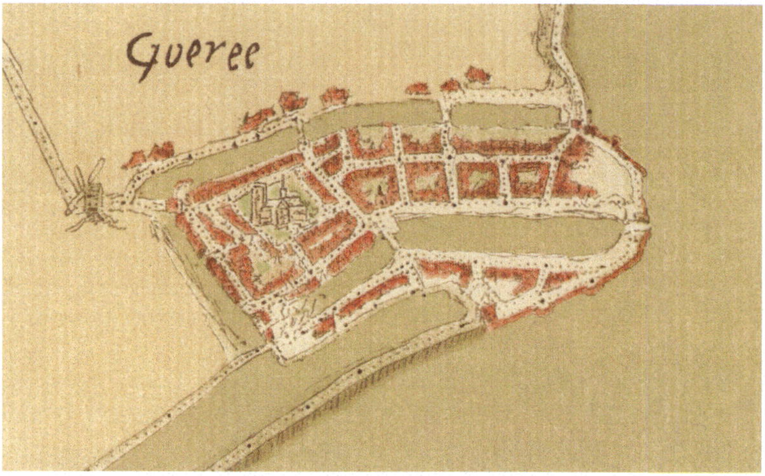

Fig. 2 Town plan of Goedereede, ca. 1550 by Jacob van Deventer; released under a Creative Commons Attribution-NonCommercial-NoDerivatives 4.0 International License

This complete spatial reversal turned central Holland inside out in a period of two centuries (Meyer 2017).

In other deltas, we can find comparable stories of the interrelationship of natural dynamics and the development of human settlements. The American agricultural scientist Franklin Hiram King was fascinated by what he found when he travelled along the coasts of Japan, China, and Korea in the early twentieth century: a strong and centuries-old agricultural culture practicing an economic system which we would call a 'circular economy' today. The communities of the towns and villages used all the waste, including all human feces, to fertilize the land and to stabilize new land created by river sediment (King 1911). Like Vierlingh in the Netherlands, they built on the natural processes of land making by rivers and sea.

In the city of New Orleans in the Mississippi River delta, the 'muddy river' left thick layers of sediment behind on the embankments after each high water event. These layers of mud were considered a public amenity and collected by the Public Works Department to elevate the streets, while private citizens used the mud to elevate their own properties. (Campanella 2006, 2010; Colten 2000). To protect the river mud as a public amenity, the city prohibited building on the waterfront itself. This is the main reason why New Orleans was one of the few American port cities with an open, public waterfront in the nineteenth century (Upton 2008) (Fig. 5).

Thus, awareness of the formative power of delta systems has been important from an ecological point of view, to protect the biodiversity in these areas. It has been especially important to our ability to maintain our cities, ports, and agricultural landscapes in these dynamic territories.

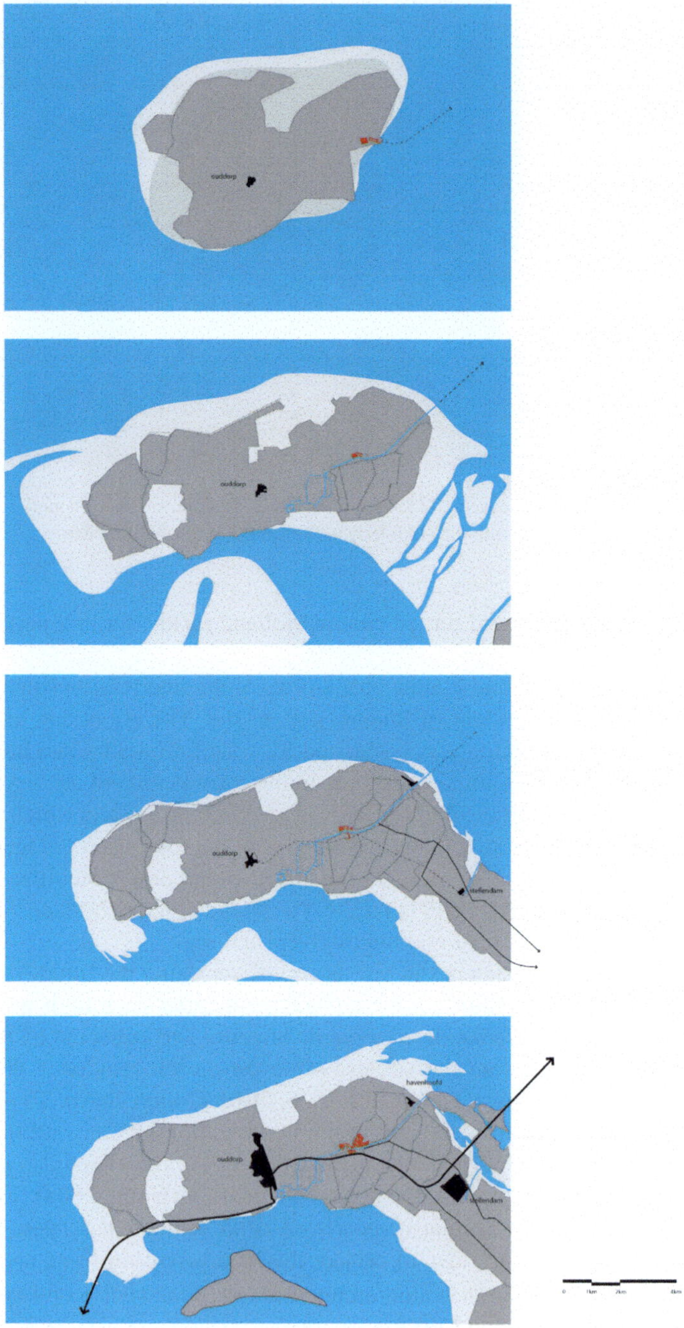

Fig. 3 Island Goeree in the Dutch Southwest delta, in four stages: 1500, 1700, 1900, and 2000. Drawings by Wout Smits, Delft University of Technology; released under a Creative Commons Attribution-NonCommercial-NoDerivatives 4.0 International License

Fig. 4 Town of Goedereede, 2016. The harbor is not accessible anymore. Photograph by Han Meyer; released under a Creative Commons Attribution-NonCommercial-NoDerivatives 4.0 International License

Fig. 5 New Orleans waterfront in 1859, painted by Adrien Persac. *Source* The New Orleans Historic Collection; released under a Creative Commons Attribution-NonCommercial-NoDerivatives 4.0 International License

The Destructive Role of Industrialization and the Colonization of Deltas

In the mid-nineteenth century, steam power and the rise of nation-states created the conditions for large-scale transformations of delta landscapes. New, steam-powered machines were deployed to construct canals, dams, dikes, pumping stations, completely restructuring the complex delta landscapes into artificial and controlled water machines. The management of deltas and river systems became a concern of national states, and the other way around: The management of delta and river systems became a condition for nation-building. This is certainly true for the two nations with the largest expenses in water management, the USA and The Netherlands (O'Neill 2006).

As Mark Twain writes in his memoirs, the Mississippi River basin is 'the body of the nation' (Twain 1883). The basin of the river and its tributaries covers almost 70% of the national territory of the USA; historically, it '*sculpted the landscape, creating the flood plains, canyons, estuaries and deltas that underpin much of America's natural bounty*' (Davis 2011). Moreover, this river system became the nation's most important transport network. After the Civil War of the 1860s, when nation-building became a prime priority for the US federal government, the US Army Corps of Engineers took responsibility for the management of the whole Mississippi River basin (Barry 1997; O'Neill 2006). It focused on improving navigation conditions by constructing levees and narrowing the river channel, creating faster currents which in turn cut deeper channels in the riverbed. The consequences for the delta and for New Orleans, situated in the heart of the delta, were huge. Narrowing the river substantially decreased the capacity of upstream flood plains for storing water, and thereby created higher water levels and a higher risk of flooding in the New Orleans region. The city constructed a high levee on the public waterfront to protect the city from the higher water, but in doing so they robbed the city of its public waterfront and its useful mud. Moreover, the construction of levees closed off the tributaries in the delta from the river, depriving them of a regular supply of freshwater and sediments necessary for the maintenance and resilience of the wetlands.

River sediments had built up the Mississippi River delta over 7000 years to 21,000 km^2, but the new river management erased 4900 km^2 (a full 23% of the land) in just a hundred years (Gramling 2012) (Fig. 6). Because of the role of wetlands as a buffer against storm surges, this loss made New Orleans substantially more vulnerable. At the same time, the population and geography of the city exploded with the exploitation of the oil and gas fields in the Gulf of Mexico. Giant pumping systems drained the marshlands, behind the natural levees of the river, to create dry building conditions for new urban extensions. But the pumping also created irreversible land subsidence, exacerbating the area's vulnerability to flooding (Campanella 2006, 2010; Waggonner et al. 2014).

The story of the Netherlands is even more extreme. Almost the whole country can be considered a delta plain, built up by the Rivers Rhine and Meuse since the last glacial age. From the independence war (1568–1648) to the Napoleonic domination (1795–1813), this territory was governed by the 'Republic of the Seven Provinces,' a

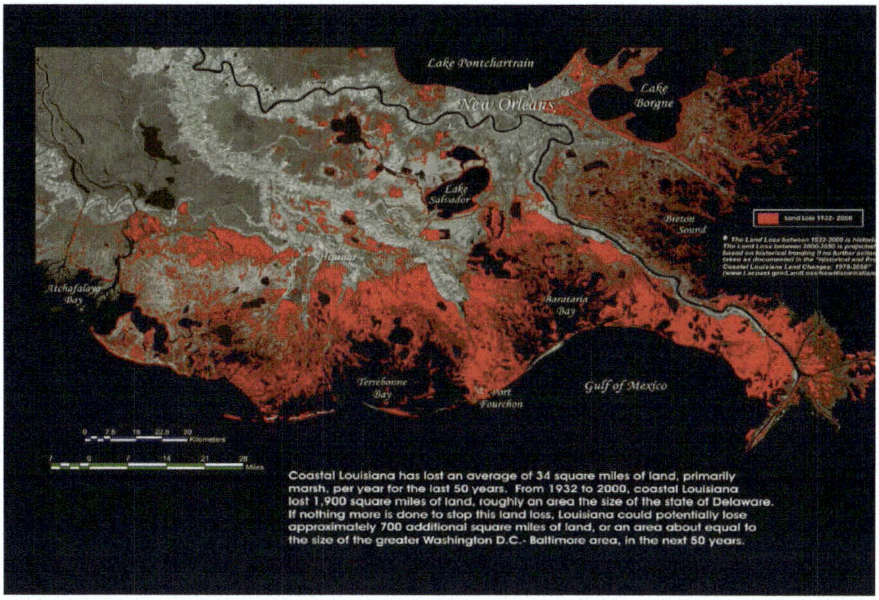

Fig. 6 Mississippi River delta; the red color indicates the expected land loss in the period 1932–2050. *Source* State of Louisiana; released under a Creative Commons Attribution-NonCommercial-NoDerivatives 4.0 International License

loose federation of autonomous city-states and provinces. Many local water boards and polder districts managed water for their local territories. But they rarely coordinated their policies across administrative borders and often fought over jurisdiction with cities and provinces, and the central government of the federative republic was too weak to set overall policy. This situation changed with the introduction of a strong central government during the Napoleonic time and continued and strengthened afterward when the different states of the Netherlands became part of a unified nation-state. The creation of a national water agency (*Rijkswaterstaat)* in 1798 helped transform the Netherlands into an industrial economy in the nineteenth and twentieth century. Like the US Army Corps of Engineers, Rijkswaterstaat made channeling rivers a high priority, and it also dug new canals to the country's major ports, Rotterdam and Amsterdam. The new canals improved the accessibility of the ports and at the same time made room for the overflow from rivers in heavy rains, substantially reducing the flood risk in areas upstream. A crucial intervention in this respect was the construction of the *Nieuwe Waterweg* ('New Waterway') between Rotterdam and North Sea. It solved the problem of the silting up of the *Nieuwe Maas* ('New Meuse'), which was the mouth of both the Rhine and the Meuse. The *Nieuwe Waterweg* was a new, artificial river mouth, making the port of Rotterdam more accessible to ships and accommodating the discharge of the river water of Rhine and Meuse. But it also increased the influence of the sea and the rivers in the city of Rotterdam. The port and adjacent industries in the Rotterdam region grew explosively, as did the urban population in the polders behind the river dikes (Meyer 2017).

In the twentieth century, the large-scale dike construction and reclamation projects of the Zuiderzee works and Delta works created a solid flood defense system, part of larger projects to enormously increase both agricultural land and new national infrastructure. The *Zuiderzee* works (1925—1970) included a new 35-km closure dam (*Afsluitdijk*) and the transformation of the inland sea into 1700 km^2 of new polders, increasing the size of the country by approximately four percent. The new IJsselmeer polders became an international showcase of modern, rationalized agriculture. The Delta works (1960–1990) included a series of dams and storm surge barriers in the Southwest delta, at the confluence of the Rivers Rhine, Meuse, and Scheldt (Fig. 7). The Delta works improved protection against floods in this region, created conditions for industrial development, and improved navigation routes between the ports of Rotterdam and Antwerp. The growth of the port of Rotterdam is linked closely to the construction of the *Nieuwe Waterweg* in the 1870s and to the Delta works. The port area is built on approximately 8.000 ha of artificially raised land and 6000 ha of harbor basin. The height of the land varies 3–6 m above mean sea level. A significant part of the port area (Maasvlakte 1 and 2, appr. 3000 ha) is new land reclaimed from the sea.

The Zuiderzee works and Delta works combined dike construction and land reclamation with the construction of new roads, which tied the different, sometimes very isolated, parts of the delta territory together. The system of hydraulic works not only delivered safety but also economic prosperity and national unification. However, in the 1970s, the disadvantages of this artificial delta machine became the object of public and political debate: ecological degradation, salinization, and substantial land subsidence.

It is a misunderstanding to consider these large-scale hydraulic works as inevitable and necessary projects, as products of a 'natural' Dutch culture. Both projects were already highly controversial from the beginning and the subject of serious scientific, societal, and political debates over many years (Meyer 2017). In the 1950s, the growth of the port along the mouth of the river Nieuwe Maas was considered a risky adventure. Johan van Veen, the secretary of the Delta Committee and 'godfather' of the Delta works, advocated that the country concentrates modern port facilities in the coastal zone of the river mouth. He warned against extending the port upstream on the grounds that it would require deepening the navigation channel in Nieuwe Waterweg and Nieuwe Maas, increasing the effects of tides and storm surges on Rotterdam and of salinization of the hinterland. The province of South Holland argued for concentrating the industrial port south of Rotterdam, partly for the same reasons, partly to maintain natural coastal areas, and partly to avoid dividing the province with a 50-km long strip of industrial port facilities.

The national government and the Port of Rotterdam did not take these arguments seriously and pushed their own preference for completely industrializing the land along the river mouth. But van Veen was right: Dredging the river mouth increased the vulnerability of the Rotterdam area to floods. So the country had to construct one more Delta work: the Maeslant barrier.

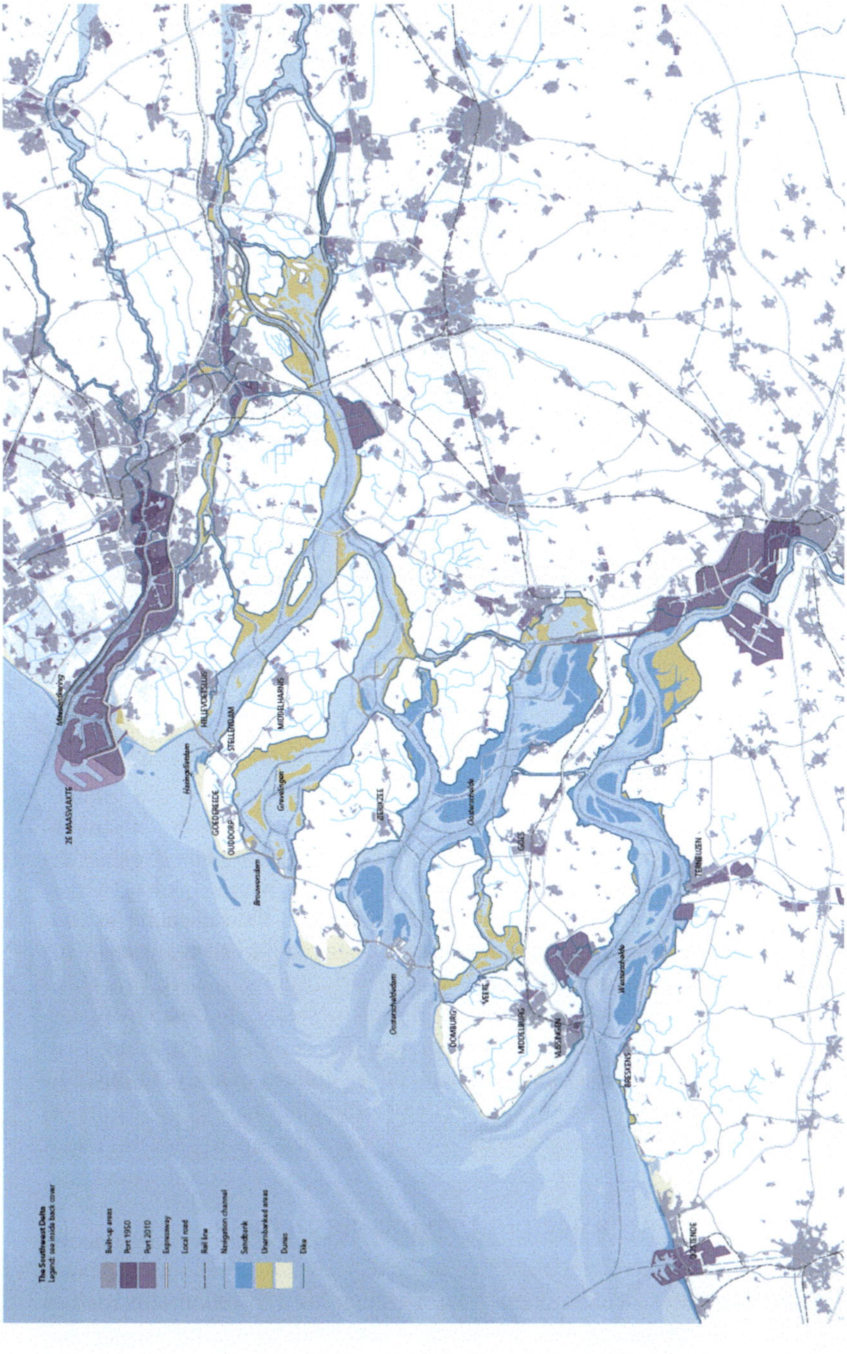

Fig. 7 Southwest delta, the Netherlands, after completion of the Delta works 1960–1990. *Source* Meyer et al. 2015; released under a Creative Commons Attribution-NonCommercial-NoDerivatives 4.0 International License

Despite these systemic problems, the idea of transforming delta landscapes into industrial machines, inspired by the examples of the Netherlands and the Mississippi River delta, was distributed and applied all over the world, with predictably disastrous consequences (Meyer and Nijhuis 2014). During the twentieth century, mass urbanization in Asian, South American, and African deltas put those deltas into crisis. The Intergovernmental Panel on Climate Change (IPCC 2007) identified three major consequences of the depletion of the natural systems of deltas by human intervention. First, dramatic losses of ecosystem services *decreased the formative power of the deltas*. Intense urban and industrial land use, drainage, dredging, reclaiming, and damming have deprived the land–water ecosystems of their capacity to absorb the impact of extreme events and to restore balance after disturbances. Moreover, upstream damming and reservoirs have depleted sediment resources in rivers, deeply eroding delta and coastal landscapes (Mulder et al. 2011; Campanella 2010). In their research on 40 deltas around the world, Ericson et al. show that sediment trapping upstream is the main cause of erosion in 27 of these 40 deltas (Ericson et al. 2006). Second, intense drainage and (industrial use of) groundwater extraction caused *land subsidence*. Urban and agricultural territories in many deltas (including the Nile delta, Rhine–Meuse delta, Mississippi River delta, Jakarta, etc.) have dropped substantially below sea level, making these territories more vulnerable to flooding. This process is still going on in many urbanizing deltas, leading to uncontrollable flood risk (Nicholls et al. 2007). Finally, intense dredging, land subsidence, and sea-level rise combine to diminish the supply of fresh water, *increasing salinization* of the local water. Many urbanizing deltas find themselves in a paradoxical situation: surrounded by water, but lacking *fresh* water for drinking and irrigation (Tessler et al. 2015).

Together, these technologies have vastly increased the vulnerability of deltas to flooding and further salinization, even as external conditions are worsening because of climate change: sea-level rise, increase of peak discharges of rivers, and more extreme precipitation. Recently, the UN-Habitat III Forum agreed in its 'New Urban Agenda' that we must radically change our approach in all urbanized deltas (Meyer and Peters 2016). Ecosystem services should be revitalized; their capacity to contribute to the recovery of delta territories after disturbances should be recovered. This revitalization must occur in other natural land–water transitions as well: beach and dune systems, salt marsh systems, coastal coral reefs systems, and mangrove forest systems. In the long term, the underlying strategy of 'building with nature' will restore conditions in which delta regions can adapt to climate change continuously. The formative power of nature itself is the strategy's foundation.

Toward a Different Approach to Urbanizing Deltas

The increasing awareness of the necessity of a change to a different approach in urbanizing deltas has resulted in new initiatives in several delta regions.

Fig. 8 Sand engine, coast of South Holland, 2011. *Source* Rijkswaterstaat; released under a Creative Commons Attribution-NonCommercial-NoDerivatives 4.0 International License

Public and political debates in the Netherlands on the urgency of drastically changing water management and flood defense led to the agreement of the Dutch government on a report of the Dutch Delta Committee in 2008, entitled '*Working together with wate*r.' Based on this report, the government started a new 'Delta Program' in 2014, which formulated new standards based upon worst-case scenarios of climate change and sea level rise to 2050 (Delta program 2014).

Linked to this policy, the government developed and implemented two programs that sought to 'build with nature': *Coastal Weak Links* (2003–2012) and *Room for the River* (2005–2015). Both programs included new experiments, like the 'sand engine' at the coast of South Holland (Fig. 8) and forty projects in the river area that extended the floodplains of the rivers. Instead of raising dikes and destroying vulnerable ecosystems, the new projects created more room for water and strengthened the ecosystems of land–water transitions. The recovery of natural systems is essential to reversing erosion and land subsidence and stabilizing the delta landscape.

A central question is if and how this new approach of building with nature can be attuned to a continuation of the industrial port in the heart of the Rotterdam delta region. The Port area is one of the most extreme examples of a complete land transformation: At approximately 14,000 ha, it renders the previous landscape invisible. Buried under 3–6 m of sand, the previous landscape was an interesting and dynamic part of the delta, with different forms of sedimentation and erosion; it was considered one of the richest natural environments of the Netherlands. The Rotterdam Port Authority has started to take this hidden history seriously and to investigate how to restore its original dynamics. It seeks to set an international benchmark as the greenest and cleanest port of the world (Port Authority Rotterdam 2014), in part by closely collaborating with both the City of Rotterdam and the World Wildlife Fund. Together they have developed a strategy of 'an open port in a natural delta' (World Wildlife Fund/Port Authority Rotterdam 2016). Part of this strategy is the plan 'The river as tidal park' (Fig. 9), which would achieve flood safety, ecological recovery, and urban recreation space by turning the steep and hard quay walls of the river embankments into gentle slopes of green marshlands. This plan is interesting from the point of view of improving the river's ecosystem and creating a greener environment in the city, but also from the point of view of cultural heritage: It reminds us that there is a longer history of relation between city and river than the remains of the industrial port suggest.

In the long term, restoring the mouth of the river as a tidal estuary would mean that the whole riverbed can silt up in the course of time, forming sandbanks and small islands.

This will be an important contribution to the safety of the Rotterdam area. Instead of a storm surge barrier, natural dynamics can take over the role of protecting the hinterland. The river can discharge its redundant water into the larger estuaries south of Rotterdam. In the long term, too, this 'building with nature' option for the Rotterdam region would diminish the power of storm surges over urbanized areas. It must be acknowledged that the city would lose many of its port-related river quays and would no longer be able to use the river as a deepwater channel. Both aspects would be quite difficult for those who consider the quay walls as important cultural heritage of the port city, and for those who want to maintain the river as a deepwater channel.

But the future of the land use in the port territory is very uncertain because of the unknown effects of energy transition. At this moment, more than 60% of the port territory is used by facilities for the storage, transshipping, or processing of fossil fuels. Changing to a zero-fossil economy will end these uses and will mean fundamentally reconsidering the land use in the port area. It creates the possibility of abandoning the deepwater channels in most of the port area and concentrating the deep sea port in the Maasvlakte ('Meuse Plain') area—as van Veen argued in the 1950s.

Then, it would be easy to implement the 'river as a tidal park' plan, leading to the silting up of the river mouth, creating new islands and wetlands alongside the embankments, and leading to more safety by reducing tidal currents in the river mouth. The result would be a rather radical change of the urban image of Rotterdam, from a hard-core industrial port city to a more gentle, green-blue, and—environmentally friendly region. This is only possible when we are able to consider the industrial period as relatively short episode in Rotterdam's history, not the central characteristic of its past and present identity.

Other port cities are also struggling to form new policies and images. In the Mississippi River delta, the Louisiana Coastal Mitigation Guidebook by the State of Louisiana (2008) proposes approaches for the Mississippi River delta and the Greater New Orleans region that resemble the plans for Rotterdam. Repairing former river branches in the delta by breaking through the levees along the main channel is

Fig. 9 Rotterdam—'The river as a tidal park,' 2016. Drawing by The Urbanisten; released under a Creative Commons Attribution-NonCommercial-NoDerivatives 4.0 International License

Fig. 10 Mississippi River delta, design sketch exploring the possibility of a displacement of the port of New Orleans toward the mouth of the river. Drawing by Hoal et al. 2014b; released under a Creative Commons Attribution-NonCommercial-NoDerivatives 4.0 International License

necessary for the recovery of the wetlands. In the long term, the state also suggests moving the Ports of New Orleans and South Louisiana from their current location upstream from New Orleans, to the mouth of the river downstream from the city. Design studies, showing the possibilities of moving the ports to the deepwater areas outside the delta, have been executed by design teams for the competition 'Changing Courses' (Hoal et al. 2014a) (Fig. 10). They show that sea vessels will not need to pass the City of New Orleans, which means that the maintenance of the deepwater channels in and around the city would no longer be necessary. This would significantly reduce the vulnerability of the city to storm surges.

The European port cities Hamburg and Antwerp are also struggling with their vulnerability to flooding due to frequent dredging of the navigation channels in the estuaries of the rivers Elbe and Scheldt. Studies on reducing this vulnerability all point in one direction: repair of the natural estuaries and an end to dredging (Stokman et al. 2010; Meire and van Dyck 2012). As this will make these estuaries inaccessible to large-size sea vessels, studies recommend that deepwater ports be constructed in deepwater locations, like the *Maasvlakte* in Rotterdam or the Yangshan deep-water port island in the Shanghai region in China. Ending dredging would also make it possible to make the river a tidal park and to repair the historic remains of estuary flood plains and wetlands in the river mouths.

Conclusions: Toward a Rehabilitation of the Heritage of Adaptation

Cultural heritage is neither neutral nor objective; it is a choice, depending on which story we consider important. We choose to maintain a series of buildings, objects, and landscapes, to tell a particular story about the past. The argument of this chapter is that we need a new type of story, one that considers history not as a linear process, but as a sequence of stages, each characterized by specific economic, technical, societal, and climatological conditions. The transition from one stage to another has sometimes resulted in reclaiming new land from the sea; sometimes it resulted in a different economy; and sometimes it resulted in rebuilding cities in a different location. The material heritage of each of these cultures can be made more visible in the official cultural heritage campaigns. The opening of this chapter discussed the example of Goedereede, showing how this town adapted to changing conditions (the silting up of the navigation channels) by changing from a maritime to an agricultural economy. The current town, with its remarkable mixture of merchant's houses and farm buildings, is still a jewel of cultural heritage, showing the capacity of a community to adapt to changing conditions. More attention to examples of adaptation will help us understand the current necessity of an adaptive approach to delta regions and to consider it as a new stage in a long historic process of many centuries. Repairing the dynamic elements of the delta, like wetlands and flood plains, should be an important aspect of a cultural heritage policy even in urbanized and port areas.

Changing our approach does not mean we should forget or deny the importance of large hydraulic works of the industrial age. But we should reframe them in a critical perspective, as parts of a specific period dominated by an industrial economy and its ideology.

Ports and port cities play a crucial role in the change to an adaptive approach in delta regions. Too much emphasis on the close historic relation between industrial port and city and too much emphasis on the short-term interests of stakeholders in the current port territories can blind us to the necessity of radical but necessary choices that will lead to more sustainable and adaptive port cities in the future. Now we find ourselves in another stage of fundamental transition: of the economy, the energy supply, and the climate. Ports, as hubs of transshipment and processing of fossil fuels, have to radically change their role. It is the right moment to link their reconsideration with the move to a more adaptive approach to the natural dynamics of the delta territories. The town of Goedereede can be considered a small-scale forerunner of how to navigate the large-scale transition facing the Rotterdam port area in the next fifty years.

References

Barry JM (1997) Rising tide. The great Mississippi flood of 1927 and how it changed America. Simon & Schuster, New York
Bradshaw M, Weaver R (1995) Foundations of physical geography. Wm.C.Brown Publishers, Boston
Buisman J (1995/2016) Duizend jaar weer, wind en water in de Lage Landen, zes delen. Van Wijnen, Groningen
Campanella R (2006) Geographies of New Orleans. Urban fabrics before the storm. New Orleans
Campanella R (2010) Delta urbanism. American Planning Association, New Orleans, Chicago & Washington
Colten GE (ed) (2000) Transforming new Orleans and its environments: centuries of change. University of Pittsburg Press, Pittsburg PA
Committee Delta (2008) Working together with water. Ministry of Infrastructure, The Hague
Davis M (2011) Nation needs a fresh view of its rivers. https://gulfhypoxia.net/2011/08/23/
de Hullu J, Verhoeven AG (eds) (1920) Tractaet van Dyckagie. Waterbouwkundige adviezen en ervaringen van Andries Vierlingh in de Rijks Geschiedkundige Publicatien. Martinus Nijhoff, Den Haag
Ericson JP, Vörösmarty CJ, Dingman SL, Ward LG, Meybeck M (2006) Effective sea-level rise and deltas: causes of change and human dimension implications. Glob Planet Change 50(1–2):63–82
Gramling C (2012) Rebuilding wetlands by managing the muddy Mississippi. Science 335(6068):520–521
Hoal J, Mins T, Hoeferlin D, de Koning R, Donahue MJ, Etheridge D, Huthoff F, Hunter T, Visser J, Kolker A, Barneveld H, Buenano A, Berlin J, Coburn K, Degenaro N, de Vriend H, Meyer H, Jonkman B, Lyon L, McCorquodale JA, Nepf H, Peters B, Reed C, Taylor Robinson B, Salah-Mars S, Slaughter S, Weiss JP, Ying K (2014a) Studio Misi-Ziibi. Contribution to changing course: navigating the future of the lower Mississippi River Delta. https://www.h3studio.com/competition/
Hoal J, Mins T, Hoeferlin D, de Koning R (2014b) Missi-Ziibi. 3 strategies for the Mississippi river delta, unpublished
Jan Mulder PM, Hommes S, Horstman EM (2011) Implementation of coastal erosion management in the Netherlands. Ocean Coast Manag 54:888–897
King FH (1911) Farmers of forty centuries. Permanent agriculture in China, Korea and Japan. Organic Gardening Press, Emmaus (Penn)
Meire P, van Dyck M (2012) Naar een duurzaam rivierbeheer. Hoe herstellen we de ecosysteemdiensten van rivieren? De Schelde als blauwe draad. UPA, Antwerpen
Meyer Han, Nijhuis Steffen (2014) Urbanizing deltas in transition. Techne Press, Amsterdam
Meyer H, Bregt A, Dammers E, Edelenbos J (eds) (2015) New perspectives on urbanizing deltas. A complex adaptive systems approach to planning and design. MUST Publishers, Amsterdam
Meyer H (2017) The state of the delta. Engineering, urban development and nation building in the Netherlands. VanTilt, Nijmegen
Meyer H, Peters R (2016) A plea for putting the issue of urbanizing deltas on the new urban agenda, report for UN-Habitat III, commissioned by Ministry of Foreign affairs, The Netherlands, Delta Alliance/Delft University of Technology
Nicholls RJ, Wong PP, Burkett VR, Codignotto JO, Hay JE, McClean RF, Ragoonaden S, Woodroffe CD (2007) Coastal systems and low-lying areas. Climate change 2007: impacts, adaptation and vulnerability. In: Parry ML, Canziani OF, Palutikof JP, van der Linden PJ, Hanson CE (eds)

Contribution of working group II to the 4th assessment report of the intergovernmental panel on climate change, Cambridge University Press, Cambridge, UK, pp 315–356

O'Neill KM (2006) Rivers by design. State power and the origins of U.S. flood control. Duke University Press, Durham

Port Authority Rotterdam (2014) Port Compass 2030. http://havenkompas2030.nl/

Stokman A, Köhler B, Rabe S, von Seggern H, Schröder G, Schulz J, Thorhauer T, Zeller S (2010) Tideelbebuch. Landschafts- und stadtplanerische studie zur projekbezogenen umsetzung des tideelbeprojekts. Hamburg Port Authority, Hamburg

Stulp B (2011) Verdronken dorpen in Nederland, deel 5. Falstaff Media, Hoofddorp

Tessler ZD, Vörösmarty CJ, Grossberg M, Gladkova I, Aizenman H, Syvitski JPM, Foufoula-Georgiou E (2015) Profiling risk and sustainability in coastal deltas of the world. Science 349/638:638–643

Twain M (1883) Life on the Mississippi. James R. Osgood & Co., Boston. Republished 2000. Dover Publications Inc., Mineola, NY

Upton D (2008) Another city. Urban life and The New American Republic. Yale University Press, New Haven

Waggonner, Ball Architects (eds) (2013) Greater New Orleans Urban Water Plan. Greater New Orleans Inc., New Orleans. www.livingwithwater.com

Waggonner D, Dolman N, Hoeferlin D, Meyer H, Schengenga P, Thomaesz S, van den Bout J, van der Salm J, van der Zwet C (2014) New Orleans after Katrina. Building America's water city. Built Environ 40(2):281–299

World Wildlife Fund/Port Authority Rotterdam (2016) Een open haven in een natuurlijke delta. https://www.portofrotterdam.com/sites/default/files/open-haven-in-natuurlijke-delta.pdf

Han Meyer is Professor of Urban Design at Delft University of Technology. His work focuses on the fundamentals of urbanism and on Delta Urbanism. His recent books include *Urbanizing Deltas in Transition* (co-edited with Steffen Nijhuis, Amsterdam Techne Press, 2014) and *The State of the Delta* (VanTilt 2017). A forthcoming book is *Urbanism–Fundamentals and Perspectives* (VanTilt 2018).

Index

Page numbers followed by *f* indicate figures.

C. Hein (ed.), *Adaptive Strategies for Water Heritage*,
https://doi.org/10.1007/978-3-030-00268-8

The manufacturer's authorised representative in the EU is Springer Nature Customer Service Centre GmbH, Europaplatz 3, 69115 Heidelberg, Germany. If you have any concerns regarding our products, please contact ProductSafety@springernature.com

Printed and bound by CPI Group (UK) Ltd, Croydon, CR0 4YY
30/09/2026
02225233-0013